T0306140

The Polls Weren't Wrong

Interpreting poll data as a prediction of election outcomes is a practice as old as the field, rooted in a fundamental misunderstanding of what poll data means.

By first understanding how polls work at a fundamental level, this book gives readers the ability to discern flaws in the current methods. Then, through specific political examples from both the United States and the United Kingdom, it is shown how polls famously derided as "wrong" were, in fact, accurate. While polls are not always accurate, the reasons we can and can't (rightly) call them "wrong" are explained in this book.

This book will equip readers with the tools to navigate the mismatch of expectations. It is not intended to replace more technical applications of statistics but is accessible to anyone interested in learning more about how poll data should be understood, compared to how it's currently misunderstood.

Carl Allen is a researcher and analyst of sports and political data. His background includes quantitative analysis for the University of Louisville Baseball, as well as MLB data for Stats Perform. Currently, Carl is the owner of Triple Digit Speed Pitch, a company that partners with schools and youth sports organizations for sports-themed fundraising events in Ohio. Carl's current research interests include inefficiencies in political betting markets and improving polling methods. When he's not analyzing political data, Carl teaches English to adults, and he still finds time for MLB and NFL analysis. Carl earned his BS from the University of Louisville in 2012 (Exercise Physiology, Spanish) and MS in 2013 (Sport Administration). You can find Carl on social media and Substack: @RealCarlAllen.

The Polls Weren't Wrong

Carl Allen

CRC Press
Taylor & Francis Group
Boca Raton London New York

CRC Press is an imprint of the
Taylor & Francis Group, an **informa** business

A CHAPMAN & HALL BOOK

Designed cover image: © Carl Allen

First edition published 2025
by CRC Press
2385 NW Executive Center Drive, Suite 320, Boca Raton FL 33431

and by CRC Press
4 Park Square, Milton Park, Abingdon, Oxon, OX14 4RN

CRC Press is an imprint of Taylor & Francis Group, LLC

© 2025 Carl Allen

ISBN: 9781032486147 (hbk)
ISBN: 9781032483023 (pbk)
ISBN: 9781003389903 (ebk)

DOI: 10.1201/9781003389903

Typeset in Palatino
by KnowledgeWorks Global Ltd.

Contents

Preface

Maybe you've read lots of other books on polls and statistics. Maybe this is the first.

The only assumption I'm making about you as a reader is this: you have some level of interest in politics, polls, election forecasting, and/or statistical literacy. Whether you currently know very much or very little about these topics, you will learn a lot from this book.

The bases of the field of inferential statistics, upon which poll data finds its value, are neither new nor controversial. They are also not too hard to understand for non-experts, even if experts misuse them.

The purpose of this book is to explain everything you need to know about polls without oversimplifying anything. I'll be talking about dogs, lunch, mints, marbles, and other seemingly unrelated topics in my lead up to analysis of political polls. Each of those examples adds one or several vital concepts to understanding how polls work, and how they should be analyzed and understood.

If you find yourself reading this book thinking "just get to the *political* polls already," then you're not ready to analyze political polls. Give the book a chance and your patience won't go unrewarded; there's some good data in here. But what this book does is it gives you the tools to understand *how* to read poll data – and why many perceived experts *continue* to read it wrong.

Just do me a favor and don't skip ahead.

Political polls are, by far, the most difficult class of polls to analyze – for reasons that will be explained – but political polls are not the only type of polls. Many people, even experts, try to analyze that most difficult class of polls without a fundamental understanding of how to analyze more basic ones; it shows in their characterizations and misunderstandings.

Now, this book is not intended to replace advanced statistics classes, nor does it need to. As you'll see, the ability to plug numbers into a complex formula is useless if you don't know what the output means – or use the wrong formula.

If this book had been written 100 years ago, around the time George Gallup started conducting political polls, none of the contents would have been considered controversial or even noteworthy. It'd also have been a lot shorter because it wouldn't be necessary to correct so many misconceptions, falsehoods, and outright lies.

But given that the field *is* permeated with those things, and they have contaminated the minds of experts and non-experts alike, this book goes beyond simply explaining how things "should" be done – it addresses those falsehoods head-on.

I will introduce several new terms and concepts that are both needed and helpful for improving understanding. But neglecting to address how things are "currently done" (and why those new terms and concepts are valuable) would be an incomplete work.

The bluntness of saying someone is "wrong," even while substantiating both why and how, is sometimes perceived as hostile, but in science and academics it is necessary. I would not want to work in, nor would I trust, a field that views its established methods and findings as incontestable or unable to be improved upon. *Saying* someone is wrong is very different from *showing* it.

This leads me to my overall goals, which I hope everyone would agree are positive, if they are met: to improve the field's ability to analyze poll data; to improve the field's ability to explain "what polls tell us"; and to provide a resource that, if authorities in the field and media continue to fail in their duties to properly inform the public, gives the public the ability to inform themselves.

The lessons in this book apply to both US and non-US data. But as you'll see, there are some stunning underlying differences in how US and non-US pollsters tend to report their data, and how US and non-US analysts analyze that data. For that reason – and a few others – they must be analyzed and understood separately.

While this book will conclude with examples and historical data from Brexit, UK, and US elections, it is not simply a discussion of why polls were or weren't wrong, but rather an analysis of how they should have been – and in the future, should be read and judged. Even if the data being analyzed doesn't apply directly to your country (or even political polls at all) doesn't mean there aren't many valuable lessons to be learned from them.

If after understanding what polls *truly* tell us, you conclude those polls were or weren't wrong – that's up to you. But the underlying reasons *why* we can or can't call "the polls" wrong – that's what you will learn if you read this book in its entirety.

The conclusion I drew that is now the title of this book is one I've formed recently, based largely on the understanding and findings I've included in it. You're free to agree or disagree with that conclusion, as are those actively working in the field, all I ask is that you base your agreement or disagreement on valid reasoning.

If you're really interested in diving deep, I'll be giving you some easy, fun, hands-on ways that you can do your own (non-political) polls. I mean, you're more than welcome to ask your classmates, co-workers, or neighbors about their political preferences, but I'm told that's "frowned upon."

If you are a professor or teacher, these hands-on tools will provide some great lessons and generate insightful and valuable discussion.

If you don't really care about the "hands-on" understanding, and you just want to understand polls and data, I've got plenty for you too.

No matter your level of expertise when you picked up this book, whether you agree with my conclusion or not, there's a very good chance you'll be an expert at reading polls by the time you've completed it.

Carl Allen

1

Public Consumption of Data: Some Historical Perspective

Important Points Checklist

- Isolating the variable of interest in research
- Impact of expert opinion on public opinion and other experts
- Use of "error bars" as it relates to the confidence of an estimate
- Challenge of overcoming decades of "contaminated thinking"
- One does not need to be an expert in a field to understand the basics of that field
- "Mediafication"
- It's both healthy and necessary to criticize experts if you can support your criticism with data
- I'm very fun at parties

I think it's fair to say there's a disconnect between the statistics community, who are analysts of data, and the general public, who are consumers of data. In an ideal world, analysts would filter the abundance of data that exists on a subject in a statistically valid and unbiased way and present it to the general public in a way that can be understood, without being oversimplified. In this book, I try to do that with polls.

Many people I talk to about this "idealist" view think it's impossible, but it's not! There are successful examples, historically, of this happening.

For example:

"Smoking causes cancer."

This is one simple, true statement, but it required decades of research to reach that conclusion, studying hundreds of thousands of people, only to be filtered down into one, short sentence.

Now you, I'm assuming not an expert on smoking or cancer, probably believe without question that smoking causes cancer. It does! But could you explain the underlying mechanisms as to why and how? Ever read a

DOI: 10.1201/9781003389903-1

research paper on the topic? What about the fact that we all know someone who knows someone whose great-grandmother smoked her whole life and lived to be 100?

Does the existence of these "exceptions" influence your opinion?

Probably not!

Why not?

Well, thanks to the work of experts in the field, and a basic understanding of statistics, you, an informed consumer of data on this subject, understand that one or even thousands of examples of smokers living to be 100 years old and not getting cancer doesn't change the reality of the fact that "smoking causes cancer."

In fact, even as early as 1990, a whopping 94% of the American public rightly believed smoking caused lung cancer.

> *That 94% of the American public could agree on anything speaks to how effective proper scientific messaging can be.*

Epidemiologists, pathologists, and others in the medical profession – collectively the "experts" on the topic – have taken the decades of data on the subject, most of which is too complex for non-experts to understand, and explained it in a way that makes sense to the general public.

As you can tell by the opinions held by the American public in Figure 1.1, it's not always quick or painless to change public perception. Smoking, for decades, was thought to be "generally safe" and even healthful. This was made especially complicated by the fact that doctors – perceived experts – were used in marketing cigarettes the same way dentists are now used in marketing toothpaste.

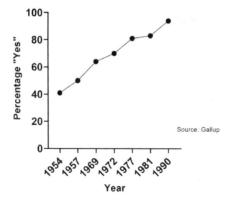

American Public Belief: "Is Smoking A Cause of Lung Cancer?"

FIGURE 1.1
A chart showing the percentage of Americans who believe in the statement "Smoking is a cause of lung cancer." Chart by the author. Data from *Gallup*.

Don't get me started on Big Toothpaste.

While Dr. Isaac Adler may have been the first modern scientist to suggest smoking was dangerous as early as 1912 in his "Primary Malignant Growth of the Lung and Bronchi" study,[1] it didn't receive much attention; at the time, lung cancer was extremely rare. In other words, even though he may have had some data to support his conclusion, the public and even others in the scientific community didn't see the relevance because it didn't impact their lives.

Thoracic surgeon Alton Oschner, one of several early identifiers of the possible "smoking and cancer" link, recalled during his time as a student in 1910 when he went to see the autopsy of a lung cancer patient that he was told he may never see one again because cases were so rare. It was 17 years before he saw another case, and then eight more, which he described as the beginning of an epidemic.[2]

By 1930, while it hadn't yet reached a level of significant concern to the general public, lung cancer was becoming prevalent enough to get the attention of experts in the field. Fritz Lickint of Germany, along with other German pathologists, had identified smoking as a potential or even likely cause of lung cancer. The basis of their findings?

They found that sex differences in lung cancer prevalence were highly correlated with sex differences in smoking rates; in Germany, it was mostly men who smoked, and lung cancer disproportionately killed men. However, in countries where there was a smaller sex difference in smoking rates, there were smaller sex differences in lung cancer rates.[3] While we would rightly not find this data to be *conclusive* proof, it should certainly have been considered powerful evidence that merited more investigation.

Unfortunately, owing to the political climate at the time (Germany, 1930s, not exactly one of collaboration), the world's experts were understandably not particularly interested in those findings, if they were aware of that specific study at all.

Germany wasn't the only place conducting studies with "smoking causes cancer" as one of its conclusions, however. Although many scientists had some data to support this conclusion, none of it was considered authoritative. There were lots of other changes in society that the experts – perceived and actual – contended could be contributing. When someone said "smoking causes cancer," the experts at the time didn't all say, "I don't believe you" but most of them did say, in effect, "I don't think it's that simple."

The Springer Handbook of Special Pathology identified lung cancer as on a concerning rise in its 1930 edition, and that it was most commonly seen in men, but listed many other factors as the likely causes.[4]

With the rise of air pollution via motor vehicles and the development of industries, and the fact that the world had just endured a flu pandemic, plus improving diagnostic techniques made it easier and more reliable to diagnose lung cancer: isn't it possible that these are all contributing to the observed rise in lung cancer?

It would have been highly unreasonable for anyone to conclude that the rise in cigarette smoking was the *only possible explanation* for the increase in lung cancer rates at this time. There was simply not enough quality research to support that conclusion.

Here is a very important place to start for "statistics in general" and will relate directly "how polls actually work."

The question of what variable actually contributes to some outcome (or, if there are multiple variables, which is the *primary* contributor) is a central question to any reputable statistical analysis. At this point, we're looking back at how researchers tackled the variables of "air pollution" and "cigarette smoking" (among others) as it related to lung cancer.

Later, we'll talk about how the "margin of error" and "undecided voters" (among others) can influence the relationship between poll numbers and election results.

This no-longer-debate about smoking and cancer might seem redundant because we have the benefit of a century of hindsight. But put yourselves in a scientist's shoes in the 1930s. You know air pollution is increasing, you know smoking is increasing, and diagnostic techniques are getting better. How can you say for sure which of these variables matter most?

Isolate the Variable of Interest

The gold standard in scientific research is a randomized controlled trial (RCT).[5] The benefit of these kinds of trials is that treatments are randomly assigned, and all other variables that might impact the analysis are controlled for.

In theory, a trial with just a few hundred randomly selected people living in the same environment could provide very powerful data. Randomly assign some people to a "smoking" group; some to a "non-smoking" group; some to a "placebo" group, who receive cigarettes that do not contain tobacco; and follow them for a few decades to see which groups get cancer most frequently. Easy, right?

Of course, in human trials especially, assigning individuals randomly to a "smoking" and "non-smoking" group is ethically problematic, and controlling the environment they live in is practically impossible for a study of that length.

Finding the single factor that has the largest influence on an outcome is hard with human data because you *can't* control all of the variables.

I want you to consider your *current* level of knowledge on the relationship between smoking and cancer. Whether you're a bona fide expert or never studied the issue in your life – consider all the possible factors that could contribute to lung cancer, and estimate for yourself:

1. How much do you think smoking contributes to lung cancer?

2. How certain are you of that conclusion?

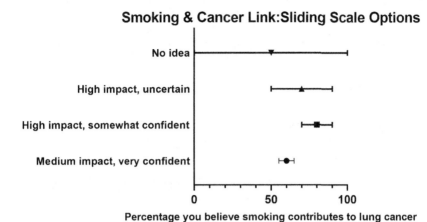

FIGURE 1.2
A "sliding scale" of options that asks the reader to select how much they believe smoking con-
tributes to lung cancer, as a percentage, along with various "error bars" to express uncertainty.
Chart by the author.

All of the examples in Figure 1.2 are samples: you can "slide" any of the
bars to the estimate of your choosing, with the error bars representing your
level of confidence in your answer.

Certainly, I hope, whatever your estimate is, you acknowledge *some amount
of possibility* that your estimate is off. However, given the amount of research
done by experts on the topic – and depending on how much of that research
you're familiar with – the "error bars" you assign to your estimate might be
larger or smaller.

Maybe you believe smoking is a 70% contributor to lung cancer but aren't
particularly confident in that estimate, so you assign very wide error bars to
it, such that if the "true answer" were 50% or 90%, it wouldn't surprise you.

On the other hand, maybe you're quite confident in the relationship
between smoking and cancer, believe it contributes 99% of the cases, and
don't think it could possibly be below 90%.

In either case, if I asserted that the relationship between smoking and lung
cancer were very low, like 2%, you'd rightly be very skeptical. If I said it was
100% – the only factor – you would also be skeptical.

Now, put yourself in the position of an interested citizen in the 1930s: you're
an informed consumer of data. You aren't necessarily going to do your *own*
experiments, but you're knowledgeable enough to read the conclusions of
experts, maybe even a book with the most up-to-date analysis on the subject,
and base your understanding on their research.

How do you think, **in the 1930s**, you'd have estimated how much smoking
contributes to lung cancer?

Importantly, do you think you would've been *more or less confident* in that
estimate in the 1930s than you are today?

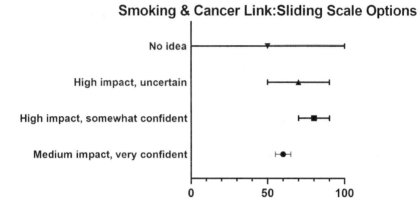

FIGURE 1.3
A "sliding scale" of options that asks the reader to select how much they would have believed smoking contributed to lung cancer, as a percentage, in the 1930s along with various "error bars" to express uncertainty. Chart by the author.

The challenge in the 1930s, when there was very little research on the topic, was that whatever your conclusion was, you were probably highly uncertain of it. You could conclude smoking was a 1% or 99% contributor, or anything in between, and you'd have no ability to say whether or not that conclusion was unreasonable. The "error bars" in your level of confidence would have to be pretty high!

That's the purpose of research. By isolating specific variables on our list of "potential causes," we can refine not only whether a variable contributes 10% or 90% but also the level of confidence we have in that estimate.

Saying that smoking is the primary cause of "5%–95% of lung cancer cases" isn't as useful as "something between 70% and 99%" even if both would agree that 80% is a reasonable estimate.

So in the 1930s, a scientific-minded researcher might want to figure out some way to isolate "smoking" as the only variable that could make a difference in lung cancer prevalence.

In 1932, Adam Syrek, at the University of Krakau in Poland, did just that. He noted that in rural Poland, where cars and industry were extremely rare, lung cancer rates were just as high as elsewhere and noted the same sex-linked rates as Lickint.[6]

That is to say, by 1932, two researchers had independently found convincing evidence of a strong link between smoking and lung cancer and that smoking was *the primary cause* of lung cancer.

In a perfect world, that groundwork would have laid a path to country and eventually worldwide acceptance of this fact, and the worst of the lung cancer pandemic may have been mitigated. At minimum, one would hope that this data would lead to aggressively funding more research, if not immediate action on the existing research. But of course, not long after these studies,

World War II began, and the research coming out of Germany was largely ignored, even by later German scientists.[7]

The first research on the topic that led to governmental and public health action ... outside of Nazi Germany ... didn't come until a 1950 study by Richard Doll and Austin Hill was published in *the British Medical Journal*.[8]

This Doll-Hill study was published two decades after *Springer Handbook of Pathology* identified lung cancer on a dangerous rise.

The major findings of that study concluded that "... the risk of developing carcinoma of the lung is the same in both men and women, apart from the influence of smoking." And that, as found in previous studies, men were substantially more likely to develop lung cancer. This is a powerful finding, even if limited to a sample of approximately 1,000 people, such that – again – one would hope for aggressive funding of more research and/or immediate action.[9]

Finally, in 1953, there was movement within the UK government. The Standing Advisory Committee on Cancer & Radiology (SACCR) consulted with Iain Macleod, Minister of Health. While the committee's opinion on the *extent to* which smoking caused cancer differed, they agreed they needed to make a statement. Macleod, Minister of Health, said the relationship between smoking and lung cancer should be regarded as "established."[10]

So, with the data showing what it did, with multiple studies now having reached the same conclusions, the experts in the United Kingdom went on to immediately warn the general public about the dangers of smoking.

I'm kidding. They did not do that.

While the Minister of Health did say that "It is desirable that young people should be warned of the risks apparently attendant on excessive smoking." He and the SACCR stopped short of an overly powerful statement, hedging with:

> "No immediate dramatic fall in death rates could be expected if smoking ceased" and "there is no firm evidence of the way in which smoking may cause lung cancer or of the extent to which it does so."[11]

Their public conclusion, despite differing internal opinions, was effectively, "smoking might be a factor in lung cancer, more research still needed."

Again, while this "highly uncertain" conclusion wasn't totally indefensible, it wasn't nearly as defensible as 20 years prior. The error bars were getting smaller, thanks to more and better research. It could no longer be denied that, at minimum, smoking was at least *a factor*.

The United States was performing similar research at this time, with small samples of lung cancer patients who "self-reported" their smoking habits.[12]

The samples in these studies were not randomly selected, and the sample sizes weren't very large. Given those methodological concerns, it's fair to say there still wasn't sufficient data to provide *definitive* conclusions about smoking as the primary cause of lung cancer. That's not to say that there was no evidence of some link, nor that it was reasonable to conclude smoking didn't contribute at all, however.

The Doll-Hill Study said:

> Clearly none of these small-scale inquiries can be accepted as conclusive, but they all point in the same direction.[13]

In comes E. Cuyler Hammond, Ph.D.; Daniel Horn, Ph.D.; and the American Cancer Society (ACS).

The Hammond-Horn study was designed to eliminate all of the reasonable questions of validity a study might have: sample size, how subjects were selected, thus isolating the variable of interest: the link between smoking and lung cancer. Beginning in January 1952, Hammond and Horn (and thousands of volunteers with the ACS) enlisted as subjects 188,000 American men aged 50–69 – some of whom smoked and some of whom didn't, from all over the country. Notably, they were choosing from the general population, not exclusively from people who already had lung cancer.

They published their preliminary findings after just 20 months in 1954:

> It was found that men with a history of regular cigarette smoking have a considerably higher death rate ... we are of the opinion that the associations found between regular cigarette smoking and death rates from diseases of the coronary arteries and between regular cigarette smoking and death rates from lung cancer reflect cause and effect relationships.[14]

Two years later, in their final report, they reported what had now come to be unsurprising to the experts in the field. Smokers died at a rate far eclipsing non-smokers, and the most common causes of death among those smokers were coronary artery disease and lung cancer, and of special note:

> an extremely high association between cigarette smoking and death rates ... was found in rural areas as well as large cities.[15]

Given this overwhelming data confirmed with an enormous sample of real Americans they had studied, the experts in the United States went on to immediately warn the general public about the dangers of smoking.

Sorry, they *did not* do that.

By now, there were plenty of actual experts in the field who had been raising concerns for years, and this study marked a major turning point, but the impact of their outreach was complicated by the fact that other *perceived experts* were saying the opposite, that there was not cause for concern.

The public – who might not know how to access the research, or be able to understand it even if they could access it – understandably rely on perceived experts and media to report those findings. Based on that, the public largely believed there was still not a definitive conclusion.

If people the public perceive as experts are saying different things, which experts should you trust? I don't think the public is to blame for not drawing the proper conclusion, but that experts have a responsibility to properly inform the public and should be held to a high standard of accountability.

Changing Public Opinion

In 1954, only about 41% of Americans believed smoking was one of the causes of lung cancer. Note the wording of the question: not that smoking was *the* cause of lung cancer, nor the *primary cause*, but *one of the* causes.[16]

By 1957, that number rose to 50%, largely thanks to the Hammond-Horn study conducted a few years earlier, and the acknowledgement by US Surgeon General Leroy E. Burney said that the evidence pointed to a "causal relationship" between smoking and lung cancer.

That same year, the British Medical Research Council announced a "direct causal connection" between smoking and lung cancer, owing in large part to research done by the aforementioned Doll and Hill, who in continuing their research and improving their methods, opted to collect data from about 35,000 male doctors.[17]

Neither country's leadership did much more than release a public statement at this time.

What They Were Fighting

I should mention what these scientists and governments were up against. It's not as though scientists were publishing data, making statements, and the public was just apathetic. It was an issue that had gained major public interest by the 1950s: only about a quarter of Americans said they "didn't know" if smoking was a cause of cancer. That is, about three in four Americans had a stated belief, yes or no, whether smoking was a cause of cancer.[18]

The whispers of "maybe smoking isn't healthy" didn't start in the 1950s; they had been around for a while. There just hadn't been much scientific research on the topic yet. The tobacco industry made a concerted effort to preempt that scrutiny.

As early as the 1930s, instead of simply marketing their product, many tobacco companies used doctors and cited "medical studies" in their advertising to dismiss concerns about smoking and health.[19]

I can't speak to the expertise of the doctors who did commercials for cigarette companies saying they were safe, but I think it's fair to say just because they were an expert in some related medical topic doesn't necessarily mean they're an expert in this one. It's possible those doctors did wonderful work in their fields, but I don't think it'd be out of line to say, despite their expertise, they were wrong about smoking and cancer.

But to the public, when someone is perceived as an expert on a topic, even if they aren't (or if their expertise is biased or out of date), that doesn't mean the public won't still believe them. The scientists whose studies began in the 1950s, no matter how solid the data, were up against at least 20 years of public perception being molded a certain, specific way.

Not just that, but once the 1950s came, and these reputable studies started making a difference both in the scientific community and the public, the Tobacco Industry didn't simply yield, they fought back.

With the knowledge that the public was, generally, neither qualified nor interested in reading the studies themselves, the first instinct tobacco companies had was to collect statements from scientists who expressed doubt regarding the studies showing the link between smoking and cancer.[20]

But it was a more concentrated and organized approach that gave the tobacco industry the reach and influence that would carry on for decades. The Tobacco Industry Research Committee (TIRC) was established in December 1953.[21]

The TIRC was a coordinated group of the largest tobacco companies, with the explicitly stated goal of spreading the message that, among other things:

> There is no conclusive scientific proof of a link between smoking and cancer.[22]

Instead of simply arguing *against* the studies, which they knew would eventually fail on the merits as well as from a public relations (PR) perspective, the TIRC decided they would fund *their own* studies. This PR campaign disguised as research orchestrated by John W. Hill, President of PR giant Hill & Knowlton (today known as Hill+Knowlton strategies), was initiated by Hill's idea to "create an industry-sponsored research entity."[23]

Even if the research was only loosely related to the topic of concern, the TIRC could truthfully say they were doing research. As such, to the public, the continuing research on the topic lent added credibility to the idea that the science wasn't yet "settled."

To use a concept discussed earlier, and will come up again later when discussing polls, the tobacco companies weren't necessarily trying to influence what "*percentage*" the public estimated as the relationship of smoking to cancer, but in many ways, they were trying to instill *doubt* and influence the size of the public's perceived "error bars."

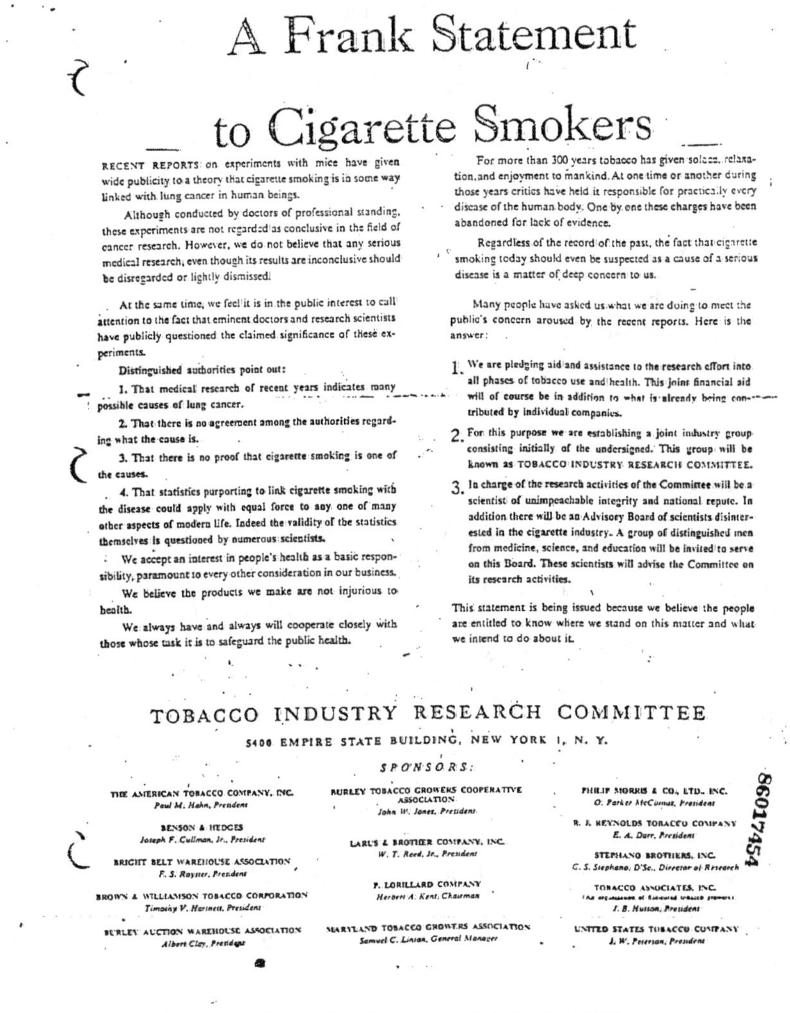

A Frank Statement

to Cigarette Smokers

RECENT REPORTS on experiments with mice have given wide publicity to a theory that cigarette smoking is in some way linked with lung cancer in human beings.

Although conducted by doctors of professional standing, these experiments are not regarded as conclusive in the field of cancer research. However, we do not believe that any serious medical research, even though its results are inconclusive should be disregarded or lightly dismissed.

At the same time, we feel it is in the public interest to call attention to the fact that eminent doctors and research scientists have publicly questioned the claimed significance of these experiments.

Distinguished authorities point out:

1. That medical research of recent years indicates many possible causes of lung cancer.

2. That there is no agreement among the authorities regarding what the cause is.

3. That there is no proof that cigarette smoking is one of the causes.

4. That statistics purporting to link cigarette smoking with the disease could apply with equal force to any one of many other aspects of modern life. Indeed the validity of the statistics themselves is questioned by numerous scientists.

We accept an interest in people's health as a basic responsibility, paramount to every other consideration in our business.

We believe the products we make are not injurious to health.

We always have and always will cooperate closely with those whose task it is to safeguard the public health.

For more than 300 years tobacco has given solace, relaxation, and enjoyment to mankind. At one time or another during those years critics have held it responsible for practically every disease of the human body. One by one these charges have been abandoned for lack of evidence.

Regardless of the record of the past, the fact that cigarette smoking today should even be suspected as a cause of a serious disease is a matter of deep concern to us.

Many people have asked us what we are doing to meet the public's concern aroused by the recent reports. Here is the answer:

1. We are pledging aid and assistance to the research effort into all phases of tobacco use and health. This joint financial aid will of course be in addition to what is already being contributed by individual companies.

2. For this purpose we are establishing a joint industry group consisting initially of the undersigned. This group will be known as TOBACCO INDUSTRY RESEARCH COMMITTEE.

3. In charge of the research activities of the Committee will be a scientist of unimpeachable integrity and national repute. In addition there will be an Advisory Board of scientists disinterested in the cigarette industry. A group of distinguished men from medicine, science, and education will be invited to serve on this Board. These scientists will advise the Committee on its research activities.

This statement is being issued because we believe the people are entitled to know where we stand on this matter and what we intend to do about it.

TOBACCO INDUSTRY RESEARCH COMMITTEE

5400 EMPIRE STATE BUILDING, NEW YORK 1, N. Y.

SPONSORS:

THE AMERICAN TOBACCO COMPANY, INC.
Paul M. Hahn, President

BENSON & HEDGES
Joseph F. Cullman, Jr., President

BRIGHT BELT WAREHOUSE ASSOCIATION
F. S. Royster, President

BROWN & WILLIAMSON TOBACCO CORPORATION
Timothy V. Hartnett, President

BURLEY AUCTION WAREHOUSE ASSOCIATION
Albert Clay, President

BURLEY TOBACCO GROWERS COOPERATIVE ASSOCIATION
John W. Jones, President

LARUS & BROTHER COMPANY, INC.
W. T. Reed, Jr., President

P. LORILLARD COMPANY
Herbert A. Kent, Chairman

MARYLAND TOBACCO GROWERS ASSOCIATION
Samuel C. Linton, General Manager

PHILIP MORRIS & CO., LTD., INC.
O. Parker McComas, President

R. J. REYNOLDS TOBACCO COMPANY
E. A. Darr, President

STEPHANO BROTHERS, INC.
C. S. Stephano, D'Sc., Director of Research

TOBACCO ASSOCIATES, INC.
(An organization of flue-cured tobacco growers)
J. B. Hutson, President

UNITED STATES TOBACCO COMPANY
J. W. Peterson, President

FIGURE 1.4
"A Frank Statement to Cigarette Smokers" published by the Tobacco Industry Research Committee. Archived by the UCSF Library, part of the "Truth Initiative."

Notice how each line is carefully worded not to dismiss findings but to raise doubt.

What's more, while the studies that would eventually receive grants from the TIRC stayed away from the "smoking and cancer" relationship for obvious reasons, they spent $300M over 40 years to fund research into the now pivotal cancer research fields of genetics, virology, pharmacology, and more.[24]

They were spending real money on real research – even though they promoted anti-scientific statements in the media.

Just because they were wrong about the "smoking and cancer" link doesn't mean everything those researchers found was wrong or without value.

It is to say, had they used that money, time, and scientific expertise with a proper understanding of the relationship between smoking and cancer, they could have done **better** research.

The more pressing question during the lung cancer epidemic at that time wasn't to what extent there are genetic or pharmacological links, but a much simpler one: whether smoking was a major cause of lung cancer and if there should be governmental and public health intervention to prevent future deaths.

The American Cancer Society Continues Their Research

In 1959, presumably after years of pulling their hair out and yelling at people seeing that only about 50% of the public believed what they had effectively proven, Hammond and the ACS decided to run another study.

This time, instead of 188,000 subjects, they recruited more than 1 million. Instead of just men as subjects, they recruited men and women. You have to admire this level of scientific dedication, even in the face of lukewarm public acceptance.

With the increasingly worrying rates of lung cancer, and increasingly irrefutable scientific data – regardless of public opinion – showing that cigarettes and tobacco were the primary cause of it, the ACS joined forces with other major health organizations and contacted President Kennedy directly in 1961. Kennedy's Surgeon General, Luther Terry, took the lead – perhaps pressed by his British colleagues.

In 1962, after the release of a widely distributed study from the Royal College of Physicians of London that concluded smoking is a cause of lung cancer and bronchitis and probably contributes to heart disease, Terry formed an advisory committee.[25]

The report from that advisory committee released in January 1964, *Smoking and Health: Report of the Advisory Committee to the Surgeon General*, did no hedging like previous governmental statements. They concluded that even average smokers had a dramatically increased risk of lung cancer, heart disease, emphysema, among other diseases, and that cessation of smoking reduced that risk.[26]

In both the United States and the United Kingdom, whose research was at the forefront on the topic, governmental action was taken rather quickly after this report. The countries banned cigarette ads on TV and placed warning labels on cigarette packets, and these actions had an immediate and measurable impact in the public: by 1972, just 15 years after public acceptance of the link between smoking and cancer reached 50%, that number was 70%. By 1977, it was 80%, and by 1990 that number was over 90%.[27]

It wasn't easy, and it wasn't quick, but when the experts in the field can reliably convey the findings of their research into a simple, true, easy-to-understand statement, the public accepts it. It's not always easy, and it's not always quick, but in the end, better research wins.

The fact that the general public might not be able to explain or understand the exact mechanisms or complexities of the topic doesn't change the fact that they do understand the basic findings of "smoking causes cancer."

Which Brings Me to Polls

Now, lung cancer is a heavy topic, and I don't want you to think I'm insensitive to the difference between these topics. The comparisons I draw between the two are based simply on the disconnect between experts, public opinion, and the influence perceived experts have on the general public (and even other experts) in the face of what happens to be overwhelming data that doesn't support their conclusions.

I find this topic – even if it's a little heavy – to be informative and useful as a starting point for understanding the current state of poll research because of that disconnect between what the data tells and told us and what people believe(d) to be true.

Over time, despite the wide gap in knowledge between the field's experts and the public, the public did come to accept and understand what the experts were able to show, and more importantly, explain.

In case you're curious, the answer to the smoking-and-cancer relationship is considered to be around 90%.[28]

The polling industry itself is not new. George Gallup is credited with running the first statistically valid political survey in 1936 in which his small, random sample outperformed the enormous but not statistically sound sample in the "Literary Digest" poll.[29]

But Gallup would not recognize the polling industry as it exists today – dozens of highly active pollsters months in advance of election time, each with their own techniques, and thousands of "experts" analyzing those polls telling the public what they mean, which polls matter, and how "right" or "wrong" they were.

As it pertains to public opinion relating to the accuracy of political polling, and "expert" discourse regarding poll data, I would say our current state is somewhere between the 1930s and the 1950s "smoking and lung cancer" discussion, in that lots of people are *aware* and *care* about it – and there's a lot of good data to analyze, but the public is being so badly misinformed by experts they don't know what (or who) to believe.

As it happens, unfortunately, the perceived experts on polls all seem to share a similar but poor understanding of what polls tell us. This book offers a very different view that I believe is as different as Gallup and Literary Digest. We are looking at the same data but have reached a different conclusion: one is statistically valid, and one is not.

What About Other Books on Polls?

Fortunately, understanding polls is not a topic of public health. But understanding how public opinion can be molded by misinformation – and how hard it is to overcome the influence of "perceived experts" – will hopefully help you approach the analysis and findings of this book with an open mind.

Unfortunately, in my experience, the existing books and commentary that attempt to tackle this disconnect between analysts and general public as it pertains to polls have what I call a high "barrier of entry" in that they assume (or hope) the reader is comfortable with or interested in varying levels of technical and statistical concepts. To make things worse, many of those articles, books, and even academic research papers make the same fundamental errors regarding how polls work.

Polls are not nearly as complex as medical data. Sure, statistics experts (or, at least, well-read students) have the base of knowledge required to understand polls better than the general public, but just as you don't need to understand carcinogenic and mutagenic properties to understand "smoking causes cancer," you also don't have to be able to calculate the margin of error in your head to know how to read polls.

So with that, a statement I hope gives you some encouragement:

You do not need to be a math or stats wiz to understand polls. This book gives you the tools to be an informed consumer of poll data. While I would strongly encourage anyone who has an interest in more advanced statistics to pursue it, you do not need to be able to build forecasts to understand how polls work.

And a statement that may surprise you but is one of my primary motivations for writing this book:

It seems most who are considered experts in analyzing polls, making forecasts, math and stats wizzes included, don't even understand how polls work.

They analyze polls, they build forecasts, but they don't even understand how polls work? That's right.

Here's what I mean:

The question of "What do polls tell us?" is the most commonly, fundamentally misunderstood topic in the world of political polling and forecasting, and it's not just among the general public, but also the experts (actual or perceived).

Those same experts who the public relies upon to tell them what the polls mean analyze polls in ways that are neither statistically nor logically valid.

To make matters worse, the perceived experts (the "analysts of data") are dabbling in or outright feeding the public ("consumers of data") polls as if they're content. The line between analyst and pundit has become so blurred that it often seems to not exist.

This "**mediafication**" of polling is a double-edged sword: you have more and more people interested in polls and data, click click click, but more and more people are consuming poll data without understanding the basics of how to read them. And it's not the public's fault: the experts they trust to teach them about polls don't understand it either. This is the problem facing the poll industry today.

To use the smoking and cancer example:

Imagine if, instead of a nationwide 188,000-person sample with initial results not released for nearly two years, the historic Hammond-Horn smoking study had released numbers weekly to major media outlets across the country, each with a sample of only hundreds. Inevitably, some of those samples would – just by chance – show little or no difference in the long-term health of smokers versus non-smokers.

Media outlets or radio and television personalities – and in today's world, social media personalities included – with a bias toward a certain conclusion might take one of those snapshot, inconclusive, one-week samples that happens to support their preferred view and make it front page news and talk about it for weeks. Sound like a problem?

If data collected over one week from 500 individuals were released and it showed no difference in cancer risk for smokers versus non-smokers, outlets with a bias toward that conclusion would certainly talk about it, analyze it, and maybe even have an "expert panel" to talk about what this one-week, 500-person sample means for the "smoking and cancer" link. Given the small samples involved, there would inevitably be some of those "one-week" reported results that showed people who don't smoke are at higher risk for cancer!

As an informed consumer of data – with the benefit of hindsight – you might be able to dismiss them as only a small piece of information not nearly enough to draw a conclusion from. But do you do the same with election polls? Does the public do the same?

The public, trusting the experts on the subject, sees that they're discussing data from this one sample seriously, thus lending further credibility not only to the reported data that should be taken with the grainiest of salt, but also to the practice of analyzing one small sample in an intense manner.

Worse yet, if we apply the mediafication theory to this, the outlets need not even have a bias toward a particular conclusion – only a bias for clicks and eyeballs. A headline like:

"New Data Shows No Link Between Smoking and Cancer"

would probably generate more attention than

"Research Ongoing."

This tiny sample overanalysis, combined with mediafication, is precisely what is happening with poll data.

Instead of a proper large-sample conclusion all being reported at one time, as most scientific data is, the public gets lots of smaller sample data which they interpret to be more meaningful than it is: a small sample instead of a big picture.

That's not to say I believe polls should only be reported monthly or annually – but it is to say that it's up to experts and informed consumers of data to understand and explain why individual polls are just one small part of a much bigger picture.

We've reached a point where experts not reading political polls correctly, and improperly presenting them to the public, thus causing the public to interpret polls incorrectly is becoming an urgent problem. The best and most reliable tool we have, by far, for informing predictions and forecasts – polls – is being derided in the media as inaccurate and even pseudoscience, by perceived experts, actual experts, and influential media outlets. It's a problem that I don't think can be overcome quickly, but if we care about statistical literacy, it must be overcome.

It's around this point I should clarify two things, and give you a little insight into my background, and a few philosophies of mine as it pertains to science and statistics.

1. When I first became interested in poll analysis and forecasting, as many people did thanks to Nate Silver and FiveThirtyEight, I assumed the experts knew everything there was to know, a perfectly reasonable assumption I think most people have. I followed his and many others' analysis anxiously but uncritically for multiple election cycles.

   Then, the COVID shutdowns happened, and my usual form of data analysis, sports, was gone. So, looking for an outlet, and political poll data being the only statistics available, I dug into that data to try and make my own forecasts; I did not set out to "improve upon" their work, or even critique it. I was just dreadfully bored and wanted something to do. I'm very fun at parties.

   It did not take me long to arrive at some dramatically different conclusions based on the data I was analyzing. So I read some more. I had to be missing something. I dug back into older posts by Nate Silver, FiveThirtyEight, and other reputable analysts to see how they explained their findings. I had never considered being critical of their work or even questioning it in the slightest because, again, I had just started analyzing poll data and they were the experts. I had no reason to even consider that they could be wrong in their analysis.

   I first assumed that the flaws in the experts' analysis were caused by an inability to break down complex topics in a way the public could understand – so they suffered from oversimplification. Basically, they understood what they were talking about but wanted to convey those complex thoughts in a simpler way to the public, such that it lost a lot of meaning. That happens.

Then I read academic sources, written by experts and for experts; they made the same mistakes. The more I read what they wrote, and the more I interacted with them, it became clear that this "oversimplification" was not the case.

That's not to say they don't do valuable work, it's to say that, with a proper foundation, they could do much better work. While much of what I write will come across as competitive or critical, and some of it is, I believe in the power of collaboration and building. I want to learn, I want you to learn, I want the public to be properly informed, and I expect experts to properly inform the public. When the experts are wrong, they should be corrected.

Creating a new foundation in any field, especially when it's at odds with what is "traditionally accepted," is often met with resistance. Maybe it should be. I don't consider myself an authority on this subject, and I believe every finding I've made – like every finding in any field – should be evaluated and accepted, rejected, or modified according to merit, not authority (perceived or actual).

To my knowledge of what has been written on the subject – both from current analysts and those from the last century – I have come to the conclusion that there has been a stagnation in the poll analysis industry for decades, largely because people (referring here to both general public and experts) are not critical enough of the industry-accepted norms for analysis.

"This is the way we've always done things" is accepted as good enough, even if it's often very imprecise or outright wrong. In doing this, they're making the mistake I made for years: assuming the experts know everything there is to know, **taking what they say at face value without critique**.

Advances in understanding require building on existing foundations and sometimes require questioning the prevailing beliefs of the field's perceived experts. But to build on existing foundation requires a solid foundation, and I believe that foundation is broken. This book offers a new foundation for how polls should be analyzed, and I believe anyone with an interest in polls – especially those actively conducting research on the topic – is capable of understanding it and building on it.

If this new foundation for how polls should be analyzed and evaluated is accepted, while I hope everyone in the public who reads this book will use it in their everyday lives (especially around election season), realistically the onus is on the field's experts to fix their analysis, critiques, and definitions.

That leads me to:

2. The purpose of what you will read is primarily to provide *you* the ability to be an informed consumer of data. This foundation will give you the ability to make informed analyses. While it will necessarily do so in places, the purpose is not just to criticize the people who are considered experts in the polling (or the adjacent-but-separate forecasting) industry. My criticism is academic, not personal.

So with that in mind, I'm requesting that you forget what you know (or think you know) about polling, forecasting, and elections. Leave your biases at the cover. Don't read this book as if I'm an authority telling you what to think about polls, but as an educator trying to teach you how to think about them.

Let's get started.

Notes

1. Adler, I. A. (1980). Classics in oncology: Primary malignant growths of the lung. *CA: A Cancer Journal for Clinicians*, *30*, 295–301. https://doi.org/10.3322/canjclin.30.5.295
2. Ochsner, A. (1973). Corner of history. My first recognition of the relationship of smoking and lung cancer. *Preventive Medicine*, *2*(4), 611–614. https://doi.org/10.1016/0091-7435(73)90059-5
3. Lickint, F. (1929). Tabak und Tabakrauch als ätiologischer Factor des Carcinoms. *Zeitschrift für Krebsforschung*, *30*, 349–365.
4. Hanspeter, W. (2001). A short history of lung cancer. *Toxicological Sciences*, *64*(1), 4–6. https://doi.org/10.1093/toxsci/64.1.4
5. Hariton, E., & Locascio, J. J. (2018). Randomised controlled trials—The gold standard for effectiveness research. *BJOG : An International Journal of Obstetrics and Gynaecology*, *125*(13), 1716–1716. https://doi.org/10.1111/1471-0528.15199
6. Syrek, A. (1932). Zur Häufigkeitszunahme des Lungenkrebses. *Z Krebs-forsch*, *36*, 409–415. https://doi.org/10.1007/BF01627545
7. Proctor, R. N. (2001). Commentary: Schairer and Schöniger's forgotten tobacco epidemiology and the Nazi quest for racial purity. *International Journal of Epidemiology*, *30*(1), 31–34. https://doi.org/10.1093/ije/30.1.31
8. Doll, R., & Hill, A. B. (1950). Smoking and carcinoma of the lung; preliminary report. *British Medical Journal*, *2*(4682), 739–748. https://doi.org/10.1136/bmj.2.4682.739
9. Doll, R., & Hill, A. B. (1950). Smoking and carcinoma of the lung; preliminary report. *British Medical Journal*, *2*(4682), 739–748. https://doi.org/10.1136/bmj.2.4682.739
10. Berridge, V. (2006). The policy response to the smoking and lung cancer connection in the 1950s and 1960s. *Historical Journal*, *49*(4), 1185–1209. https://doi.org/10.1017/S0018246X06005784

11. Berridge, V. (2006). The policy response to the smoking and lung cancer connection in the 1950s and 1960s. *Historical Journal, 49*(4), 1185–1209. https://doi.org/10.1017/S0018246X06005784

12. Mendes, E. (2014). *The study that helped spur the U.S. stop-smoking movement.* American Cancer Society. https://www.cancer.org/research/acs-research-news/the-study-that-helped-spur-the-us-stop-smoking-movement.html

13. Doll, R., & Hill, A. B. (1950). Smoking and carcinoma of the lung; preliminary report. *British Medical Journal, 2*(4682), 739–748. https://doi.org/10.1136/bmj.2.4682.739

14. Mendes, E. (2014). *The study that helped spur the U.S. stop-smoking movement.* American Cancer Society. https://www.cancer.org/research/acs-research-news/the-study-that-helped-spur-the-us-stop-smoking-movement.html

15. Hammond, E. C., & Horn, D. (1984). Smoking and death rates—Report on forty-four months of follow-up of 187,783 men. *JAMA, 251*(21), 2840–2853. https://doi.org/10.1001/jama.1984.03340450056029

16. Blizzard, R. (2021, April 11). *U.S. smoking habits have come a long way, baby.* Gallup.com. https://news.gallup.com/poll/13690/us-smoking-habits-come-long-way-baby.aspx

17. Di Cicco, M. E., Ragazzo, V., & Jacinto, T. (2016). Mortality in relation to smoking: The British Doctors Study. *Breathe (Sheffield, England), 12*(3), 275–276. https://doi.org/10.1183/20734735.013416

18. Blizzard, R. (2021, April 11). *U.S. smoking habits have come a long way, baby.* Gallup.com. https://news.gallup.com/poll/13690/us-smoking-habits-come-long-way-baby.aspx

19. Little, B. (2018). *When cigarette companies used doctors to push smoking.* History.com. https://www.history.com/news/cigarette-ads-doctors-smoking-endorsement

20. Brandt, A. M. (2012). Inventing conflicts of interest: A history of tobacco industry tactics. *American Journal of Public Health, 102*(1), 63–71. https://doi.org/10.2105/AJPH.2011.300292

21. TOPICS INDEX. (1983, December 31). *Council for tobacco research records.* Master Settlement Agreement. https://www.industrydocuments.ucsf.edu/docs/xzcj0007

22. Hill & Knowlton Inc. U.S Exhibit 63,544, Press Release, re: Timothy V. Hartnett named as full-time chairman of Tobacco Industry Research Committee, Hill & Knowlton Inc., Tobacco Industry Research Committee, July 1, 1954. Depositions and Trial Testimony (DATTA); RICO Privilege Downgrades Collection. https://www.industrydocuments.ucsf.edu/docs/jxwb0035

23. Draft Recommendations for Cigarette Manufacturers. (1953, December 22). John W. Hill Papers, State Historical Society of Wisconsin, Archives Division.

24. Tobacco Industry Research Committee, Tobacco Tactics, updated 07 February 2020.

25. Royal College of Physicians. (2021). *Smoking and health (1962).* RCP London. https://www.rcplondon.ac.uk/projects/outputs/smoking-and-health-1962

26. National Institutes of Health. (1964). *The 1964 report on smoking and health. Reports of the surgeon general—Profiles in science.* U.S. National Library of Medicine. https://profiles.nlm.nih.gov/spotlight/nn/feature/smoking

27. Blizzard, R. (2021, April 11). *U.S. smoking habits have come a long way, baby.* Gallup.com. https://news.gallup.com/poll/13690/us-smoking-habits-come-long-way-baby.aspx

28. Ozlü, T., & Bülbül, Y. (2005). Smoking and lung cancer. *Tuberk Toraks, 53*(2), 200–209.

29. George Gallup Biography. (2023). *Encyclopedia of world biography.* https://www.notablebiographies.com/Fi-Gi/Gallup-George.html

2

Polls

Important Points Checklist

- Characteristic, population, and random sample, relating to polls
- What the margin of error is, and why it exists
- Margin of error as it relates to the sample size for very large populations, and also small ones

At this point, you may or may not believe that polls are useful, necessary, accurate, and/or effective, and that's okay. I did say to leave your biases at the cover, after all.

Before we learn about what polls *do*, we have to learn what polls *are*. If you think this section is too easy or simplistic, please stick with it – there are a couple important concepts in these few pages. This section provides the base of understanding required – which many people skip over – when they try to analyze the much more complex political polls that tell us the percentage of people who support each candidate, and they draw avoidably erroneous conclusions as a result.

While people tend to be *most familiar* with political polls, which I suspect is your primary interest in reading this book, I want to start with something a little less intense first. I've found that using non-political examples helps eliminate biases related to understanding what polls do, and what polls are. I want to help you master polls as a concept before applying them to what are the most challenging cases – political polling.

So What *Are* Polls?

Polls are a statistical tool designed to estimate some **characteristic** about a **population** by asking a **random sample** from that population their current position or status.

      DOI: 10.1201/9781003389903-2

Maybe you understand entirely what I just said with a quick skim. On the other hand, if you're anything like me, you probably had to read that definition several times just to digest it.

If only I could take a poll of readers, I'd have a good idea!

In that definition, there are three terms that are important to understand: characteristic, population, and random sample.

Characteristic: the thing you want to know

Population: the *specific* group you want to know about

Random sample: every member of the specific group you want to know something about has an equal chance of being asked

Those first two terms "characteristic" and "population" are the content of the poll, while "random" relates to **how the poll is done**.

The *characteristic* can be anything from a preference to a trait. For example: do you prefer pizza or tacos? Do you own an Android phone? Do you have a dog at home? Are you married?

These are, incidentally, the four most important questions to ask before going on a first date.

As for population, or "specific group you want to know about," you'll note that the term "population" doesn't mean "everyone" in this context, but the specific group a poll is attempting to represent. A population in this context can be of any size, ranging from very large, like "registered voters in the United States," to more specific, like "Kindergarten teachers in the state of Ohio," or very specific, like students in a stats class.

Have I used the term "specific group" enough? It's not by accident. This will be important later.

Polls are created through the **combination** of a characteristic and a population.

"Registered voters in the United States who have a dog at home" is a fine poll!

"Kindergarten teachers in the state of Ohio who use Android phones" also fine!

"The favorite color of a school's stats class" yes, this can also be a poll!

In the simplest terms, polls provide an estimate about some *current characteristic* of a *specific group* by asking a *random sample* from that specific group.

Why Are Polls Done?

When the group you want to know something about is very large – tens of thousands, a few hundred thousand, a million, or millions – it's probably impossible or highly impractical to ask *everyone* in that population. Whether it's registered voters in the United States or Kindergarten teachers in the state of Ohio, it's not very practical (or possible) to contact every single one.

So, instead of trying to ask everyone, we can ask a much more manageable number of people from that population at random and *generalize* those results.

What does it mean to generalize results?

It means that we take the results from our *sample* and apply that to our *population*.

From that poll, you could generate a *very good estimate* of whatever characteristic you asked that group about. What percentage of registered voters in the United States have dogs? What percentage of kindergarten teachers in Ohio use Android phones? You know, the hard-hitting stuff.

If we ask a random sample of 1,000 registered voters in the United States if they have a dog at home, and 500 say yes – 50% – we could say approximately 50% of all registered voters in the United States have a dog at home.

But does that mean the poll says the number of registered voters who have a dog at home should be *exactly 50%*?

Not exactly.

In the process of generalizing a random sample to a population, for the simple fact that we're not asking the entire population, there is a chance the poll's estimate is "off." The result is that you have to pay a statistical price, which is expressed as a range of likely values. What polls truly estimate is that the percentages given by it will match the population, **plus-or-minus some percentage**, with a certain amount of **confidence**. That "plus-or-minus some percentage" is a statistical necessity is known as **the margin of error**, sometimes abbreviated MOE or M.E.

The standard "confidence" level is 95% in poll data. I'll speak briefly about what this means shortly.

If you're a math person and already know this, great. If you're not a math person and don't quite "get it" yet, or the numbers feel overwhelming, don't stress. I will provide some more tangible examples and applications soon.

Naturally, the most obvious use cases for a poll are when your population is very large, and it's not possible or practical to ask everyone and take a true census.

But why limit yourself to only obvious cases?

By performing polls on a smaller population, such as students in a stats class or teachers in a school, we can learn a lot about how polls work and how those numbers apply to the population. A class (or classes) with a total of 100 students would require a random sample of about 86 to get down to a 4% margin of error; at that point, why not just ask 14 more people, complete the census, and have no margin of error?[1]

For science.

By understanding what polls as a tool actually tell us – compared to a knowable and easily conducted census – we can build a better foundation

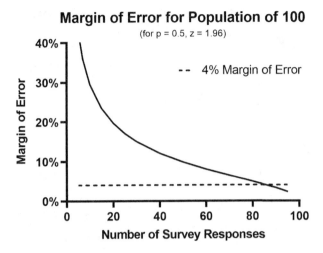

FIGURE 2.1
A chart showing the margin of error for a population of 100 for a given number of responses, and a dotted line to highlight the 4% MOE threshold. Chart by the author.

for what those polls tell us when used on a much larger population, when a census is impractical or impossible.

In order to achieve a 4% margin of error with a confidence level of 95%, comparable to most political polls, Figures 2.1–2.3 show the size of random sample that would be needed for what are considered small or "finite" populations.

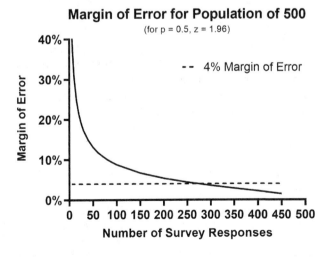

FIGURE 2.2
A chart showing the margin of error for a population of 500 for a given number of responses, and a dotted line to highlight the 4% MOE threshold. Chart by the author.

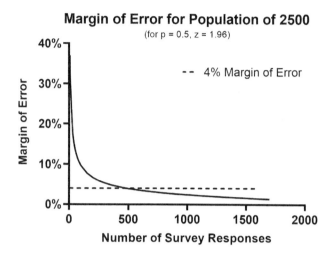

FIGURE 2.3
A chart showing the margin of error for a population of 2,500 for a given number of responses, and a dotted line to highlight the 4% MOE threshold. Chart by the author.

Populations of these sizes are not what we typically think of for "poll data," but there's no reason we can't apply it. Whether you're dealing with a classroom, a school, a city, or a country, the statistical tool known as a poll works the same: if you ask a random sample from a specific population, you can generalize the data to that same population.

The threshold to get below a 4% margin of error for the sample sizes in Figures 2.1, 2.2, and 2.3 is around 86, 274, and 485, respectively.

To this point, things feel rather intuitive: to maintain a reasonably low margin of error, a larger random sample is needed for a larger population.

I'll warn you because I don't want to lose you here, but it's important: once the population becomes very large, things become less intuitive. Much less intuitive. It is so much less intuitive that even "numbers people" find it hard to comprehend. But like all good science, it can be tested.

Andrew Gelman, professor of statistics and political science at Columbia University, wrote an article addressing the unintuitive nature of political polls in *Scientific American* in 2004 aptly entitled:

> How can a poll of only 1,004 Americans represent 260 million people with only a 3 percent margin of error?[2]

It *feels* like far too small of a sample to make a generalization for such a large population, so understandably, people are skeptical of this claim. If you need 485 people to get under a 4% margin of error for a population of just 2,500, how can a random sample of 600 achieve the same for 250 million?

Like I said, it's not intuitive, but it's true. There are plenty of reasons to be skeptical of poll data, but the math underlying the margin of error given by a random sample isn't one of them.

Once the population you're taking a random sample from becomes sufficiently large – whether in the tens of thousands or billions – a sample of the same size provides indistinguishably close margins of error.

A random sample of 600 provides a 3.95% margin of error for a population of 25,000,[3] a 4.00% margin of error for a population of 500,000,[4] and a 4.001% margin of error for a population of one billion![5]

The explanation for how a random sample of this size can give reasonably accurate results for such large populations is purely mathematical.

Admittedly, I'm biased. I think counterintuitive truths discovered in science and statistics are interesting, but other books about poll data digress into how polls work mathematically, and probably lose a lot of readers who aren't interested in that aspect of polling. Now, to reiterate what I said in the previous chapter, if you are interested in these statistics, you would absolutely benefit from understanding them. Go ahead and test it out in Excel or your coding program of choice, if you'd like. But you do not need to memorize or even care about the margin of error formula to be an *informed consumer of poll data*.

My goal is not to make you a stats expert, that's a different book. My goal is to give you the tools necessary to better understand the barrage of *poll data* that comes with every election season – and avoid the mistakes made by almost everyone, even experts in the field.

As you can see in Figures 2.4–2.6, once the population of interest becomes large, the size of random sample needed to achieve the same 4% margin of

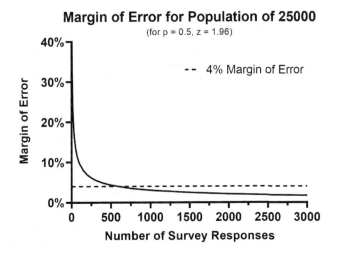

FIGURE 2.4
A chart showing the margin of error for a population of 25,000 for a given number of responses, and a dotted line to highlight the 4% MOE threshold. Chart by the author.

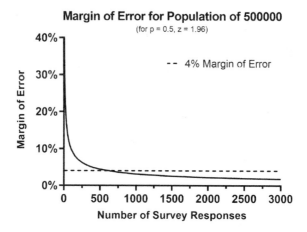

FIGURE 2.5
A chart showing the margin of error for a population of 500,000 for a given number of responses, and a dotted line to highlight the 4% MOE threshold. Chart by the author.

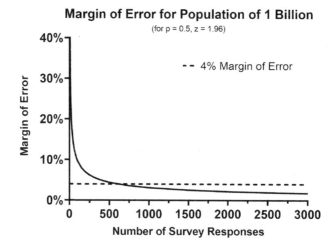

FIGURE 2.6
A chart showing the margin of error for a population of one billion for a given number of responses, and a dotted line to highlight the 4% MOE threshold. Chart by the author.

error effectively does not change. That's a powerful tool, if you use it and interpret it correctly.

The answer to the question "can a sample of just 1,000 give you a reasonably good estimate about a population of millions?" is a resounding "yes."

The reason why I can say that so confidently is because my "bar" for polls is that they provide a *reasonably good estimate*, not that they're infallible or exact.

I suspect you, especially if we talk about a neutral subject like "how many registered voters in your country have dogs at home?" or "how many people in a university stats class prefer blue to red?" you would agree.

Let me explain.

Would a hypothetical poll asking a random sample of 1,000 registered voters in the United States if they have a dog at home, even if it were conducted perfectly by every scientific measure, give a perfect representation, to the decimal point, of the actual number of registered voters across the country who have dogs at home?

What about a poll of 86 random students in a university class of 100? Would it give us the precise percentage of the whole class who prefer red versus blue?

Probably not, nor should we expect it to!

But there's a very strong chance it's closer than your or anyone's best guess.

Again, polls are a *statistical tool* designed to give us *reliable estimates* about some larger population. In cases where the population is very large, if you can identify and ask a random sample of even a tiny fraction of that population, it can provide a reliable estimate. It's not an instrument that claims to be (or tries to be) magic or flawless.

To that end, we should think of polls accordingly: very good estimates, better than guessing.

To illustrate:

Consider the nearest major city to you. What percentage of households in that city do you think have a dog, and what level of confidence would you assign to your guess?

Chances are, without any data, you're making a guess based only on your experience and probably wouldn't be surprised if your guess is off by a lot.

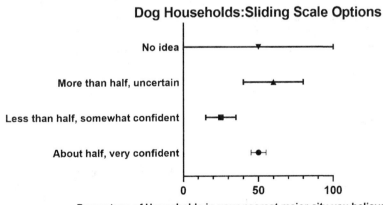

FIGURE 2.7
A sliding scale of options to estimate how many households in your nearest major city have a dog, along with how confident you are in that estimate. Chart by the author.

Now, what if I told you I had conducted a poll of 1,000 random households in that city? Would you use that to inform your prediction, or stick with your gut?

Importantly, which process do you think would be more accurate in the long run, the "best guess" approach, or the "look at a poll" approach?

If you responded, "I'd rather look at a poll to inform my estimate than just make a guess."

I have achieved goal number one. Horray! Polls, imperfect as they are, offer us the ability to make informed estimates about a population – better than a guess without using them.

Chances are you didn't need this abbreviated margin of error lesson to admit "I'd rather have a poll to inform my prediction than to guess without one," but the recognition that some data is better than no data, *even if the data is imperfect*, is a necessary place to start for why polls exist in the first place.

Political Outcomes Have Consequences

Unlike polls regarding if you like red or blue better, or whether you have a dog at home, political polls are about more consequential topics. If my poll regarding households with dogs overestimates or underestimates the actual number by 3%, I don't think anyone is going to be particularly angry, except maybe my boss. Most people wouldn't even find that small discrepancy noteworthy.

But the stakes in political elections are a lot higher than dog ownership percentages, and elections are often decided by just a few percentage points. So, a very small discrepancy between reported poll results and the eventual result that might be forgivable and not even make news on a less consequential topic can balloon into a hotly debated source of skepticism toward the entire polling industry when the polls get the outcome "wrong."

Notes

1. Random sample of 86 is approximately 4% MOE at the 95% confidence level.
2. Gelman, A. (2004, March 15). *How can a poll of only 1,004 Americans represent 260 million people with only a 3 percent margin of error?* Scientific American. https://www.scientificamerican.com/article/howcan-a-poll-of-only-100/
3. $z = 1.96$, $p = 0.5$, $N = 25,000$, $n = 600$. MOE $= 0.98/24.794 \times 100 = 3.95\%$.
4. $z = 1.96$, $p = 0.5$, $N = 500,000$, $n = 600$. MOE $= 0.98/24.51 \times 100 = 4.00\%$.
5. $z = 1.96$, $p = 0.5$, $N = 1,000,000,000$, $n = 600$. MOE $= 0.98/24.495 \times 100 = 4.001\%$.

3

What Makes a Poll "Wrong" Part 1

Important Points Checklist

- Why "closer to result" doesn't always mean "better" or "more accurate"
- The definition of "error" in statistics
- What is meant by "base margin of error" versus other sources of error
- Why poll data is weighted
- The difference between "margin of error" and "total error"

What does it mean for a poll to be "wrong?"

To this point, all I hope we have agreed upon regarding polls is that using a poll to inform an estimate about some population is, more often than not, going to provide a better estimate about that population than a pure guess.

But "better" by what standard?

Maybe you feel that the ability to take a random sample of 600 from a population of millions and have that sample yield a result within 4% of the true value of that population the vast majority of the time is a remarkable example of the power of statistics. Maybe you're unimpressed by it. Maybe you're skeptical of it and believe such a poll is only possible in theory, and not in practice.

Whatever camp you may fall in, your feelings toward the utility of polls as a tool shouldn't impact your understanding of what makes one "right" or "wrong."

My guess is that your opinion of polls is based largely, if not entirely, upon political polls – based on how they're reported in the media and the analysis done by prominent people in the field. As I touched on in the introduction, political polls are the most difficult class of polls, and I will elaborate more about why soon – but we don't need to consider political polls to understand how a poll could be "wrong." It's important that when using terms like "off" or "wrong," we have a shared understanding about what is meant.

DOI: 10.1201/9781003389903-3

The definition of error in polling and statistics is rather uncontroversial and straightforward.

> The error in a poll is the difference between the value obtained, and the true value for the population.[1]

To use the dog-household poll example, if I report that 50% of registered voter households in that city have a dog, and a census reveals that 48% of that specific population has one, that could be called a "2-point" error.

Likewise, if my random sample of 86 students in a stats class says 65% of students prefer blue to red, but when I complete the census by asking all of them, I find 67% do, that's also a 2-point error.

The "direction" of the error, whether the true value is higher or lower than the reported value, is not considered when reporting how much a poll was "off by." The direction of the error matters, but for now it's fair to say each of these polls had a 2-point error or were "off by 2."

This "off by" calculation is the discrepancy between the reported poll result obtained by the random sample and the "true value" for that population. That is, properly defined, what a poll error is.

Don't get ahead of me and start thinking about political polls now. If you're with me so far, you're right where you need to be.

But this statistical error, in this case poll error, isn't a standalone way to judge how "good" a poll is. The poll number not matching the true population number doesn't, in itself, mean a poll is bad – and the number matching doesn't necessarily mean it was good.

Consider if a competing pollster who is far less rigorous than myself had decided to conduct their own "poll" of the nearby major city, knocked on a tiny, nonrandom sample of 25 doors in the same neighborhood – but they happened to report that 12/25 people (48%) in their sample owned a dog. Then, their poll would match the true value of the population, 0% error!

How about if a student in the stats class asked the two people sitting closest to them their favorite color, one says red and the other says blue? Noting their own favorite color is blue, that's 2/3 who prefer blue. Not interested in doing the grueling work that comes with surveying a few dozen more people, that student rounds their findings and reports their poll as 67%, which they later learn matches the true value. 0% error! This poll wasn't "off" at all.

Those pollsters should be rewarded with "A+" grades since they were so accurate, right?

In those instances, I suppose it's better to be lucky than good.

But those methodologies probably wouldn't consistently produce accurate results, if we conducted this poll in a lot of cities or classrooms.

Since they were so accurate (0% error!), the public and even experts tend to give undue consideration in the future to those pollsters on the grounds of "0% error." How they achieved that number – whether through repeatable methods, luck, or perhaps a blend of the two – is rarely given much examination.

While it is very important to understand *how and why* we can say a poll was "off" – it's also important to understand that poll "data" with questionable or poor methodology is probably less reliable in the long run than one with solid methodology. Saying someone, or some poll company, has produced a smaller error in some instances, or several instances – or as is most common, in the most recent election – is not in itself proof they are "better."

As an informed consumer, it's important to understand and be critical not only of a poll company's reported results and history of error – but of its transparency with their methodology.

So far, so good? It's better to use a poll of good quality with a transparent methodology to inform our prediction about a population than it is to guess, and a poll's error is the difference between the reported poll result and true population value.

Let's continue.

The (Base) Margin of Error

As mentioned before, the statistical necessity of the margin of error comes from the fact that whenever you take a sample from a population, there's a chance the sample isn't reflective of the population as a whole. It's the price we pay for the convenience of not conducting a census.

So let's say your nearest major city has 100,000 registered voter households. How many people do you need to survey to get a good estimate, and how could you identify a random sample?

If you have 100,000 households and are able to find the addresses for all of them, you might assign each address a value from 1 to 100,000 and use a random number generator to decide which ones to survey.

To get down to approximately 4% margin of error – comparable to most political polls – you'd need to survey about 600 households.[2]

How will you contact them? Call, mail, and/or knock on their door? Each method has its advantages and disadvantages.

But what if not everyone answers their phone, responds to your mailer, or is home when you knock?

What if you knock, no one is home, but you think you hear a dog barking – does that count?

What should you do if someone doesn't respond, or isn't home? Try again later and keep asking the same household, or choose another of the remaining 99,400 sample at random?

Here, we're introducing another source of error of great interest to political polling: response (and nonresponse) error. This source of error is distinct from the margin of error but can contribute to a poll's **total error**.

If you survey 600 households in the nearest major city, that reported 4% margin of error given by the sample size **only applies if your sample is *truly random*.**

Sampled 600 in the same neighborhood? Not random. The total error of the poll could be much larger.

Made thousands of calls and only included the first 600 who responded to a phone call? Not random. The total error of the poll could be much larger.

In any poll, there can be sources of "bias" and "error" that arise from imperfect sampling methods, intentional or incidental.

To say a survey of 600 households in a city of 100,000 has a margin of error of approximately +/–4% is only true if the sample is random, *and* everyone responds. If the sample is not random, or not everyone randomly selected responds – among other possible factors – even if it's through no fault of the researcher, there is an added source for error not captured in this base margin of error.

Some sources of error could be greater than others.

If you selected a neighborhood near a dog park and knocked on 600 doors, even if 100% responded, your sample is not random. Applying the results of that neighborhood to the city as a whole could be extremely inaccurate – far beyond the "base margin of error" given by the sample size.

On the other hand, if you do identify and attempt to poll a truly random sample, but exclude homes that don't answer the first time you contact them, you may be introducing a **nonresponse error** in which the group of people who don't respond are in some way different from those that do. Any poll that doesn't have a 100% response rate has a risk of nonresponse error.

Sometimes, the underlying source of nonresponse error can be more easily identified than others. If your poll only contacts people by calling them on the phone, you're less likely to get a response from people who don't have a phone, or don't answer it – this much is obvious. But if people who don't answer their phone have some traits in common – whether age, education level, political affiliation, or any number of other factors – the nonresponse error could be substantial **and is very hard to quantify precisely.**

The risk of nonresponse error in the dog-household poll depends largely on how you choose to conduct the poll. The researcher faces some difficult questions: are people with dogs more or less likely to answer their door compared to people without dogs? Are they more or less likely to answer a phone call, text, or online survey? It's hard to say for certain.

As you consider the advantages and disadvantages of various polling methods, and challenges of conducting a poll even after doing the hard work of identifying a random sample, remember: anything that makes the sample in a poll not random could increase the *total error* beyond the base margin of error.

Unlike the statistical margin of error that can be very precisely quantified as plus-or-minus some amount based on the sample size, there are *other* factors, such as nonresponse error, that can increase the total poll error by some, unknown amount.

Fortunately, people who conduct and analyze polls understand the influence factors such as nonresponse can have on their raw poll results. With the near inevitability of some amount of "non-randomness" in their poll samples, the results are **weighted** to produce a result closer to what the sample would have said if it was truly random.

If you're skeptical of how this weighting process works mathematically, that's okay. This is where pollster quality and transparency plays an important role in getting accurate results: pollsters are not only responsible for collecting the raw data but also for how that data is weighted and reported. Weighting is both a necessary and statistically valid practice, though what to weight for and by how much can vary by pollster.

At its simplest, if a poll is conducted in which 45% of respondents were women, but it is known that about 50% of the population are women, the pollster can give slightly more weight to the women's responses to provide a more accurate picture of the population as a whole.

With that being said, given the challenges regarding how to find a random sample; *then* how to contact that random sample; *then* what to do if someone doesn't respond; *then* how and if to weight your responses by age, sex, education, and more – there are too many variables to say with certainty what the *total* error is beyond the *base margin of error.*

But that doesn't mean the total error doesn't matter, that we should just guess at it, or worse, make assumptions that lead us to believe we do know the total error even when we don't.

In a poll, only that base margin of error given by the sample size of the poll is ever reported. This is understandable as it's the only number that can be known with any level of certainty, but hopefully – as an informed consumer of poll data – you understand that other factors can impact how close a poll is to the actual population.

With that, let's sum up what we know about how to calculate poll error so far.

Poll Error = Reported Poll Value – True Population Value.

You got that a long time ago. If the poll reports 50%, but the true value is 48%, that's a 2-point error. If the poll reports 65%, but we do a census and the value is 67%, it is also a 2-point error. Easy stuff.

Let's apply a little algebra here – which is as complex as the math needs to get – and start to quantify the factors that can contribute to poll error.

We've already covered the statistically necessary margin of error; it's the price we have to pay for any poll, a given.

And of course, there's the oft-disparaged nonresponse error we just covered above. Yes, there are more.

But for now, we can say all of the below:

Poll Error = Reported Poll Value – True Population Value.

Poll Error = Statistical Margin of Error + Nonresponse Error + Other Errors.

So, with the power of algebra:

Reported Poll Value – True Population Value = Statistical Margin of Error + Nonresponse Error + Other Errors.

Poll error equals poll error.

Those are a lot of terms, but you know what all of them mean.

I'm not the first to identify or lament the existence of "other" sources of error in polls beyond the statistical margin of error. Quantifying "Total Survey Error" has been done since at least 1979 when the book *"Total Survey Error"* was published, and this term is still used.[3] In fact, as early as the 1940s when using polls as a tool was still a very new practice, W. Edwards Deming published *"On Errors in Surveys"* in which he identified 13 factors (most with multiple parts) which he characterized as factors that can "affect the usefulness of a survey."[4]

While this topic is not new and may be a topic of interest for readers who want to explore deeper, suffice it to say that different researchers include different variables in their definitions of "total survey error," but there's not much debate about the main factors that contribute to it.

It's not *which* factors go into poll error that causes the industry's experts – and by extension how it's reported in the media – to badly mischaracterize how "off" polls are, however. It's their inability to account for them.

We're not to political polls yet (I know I keep saying that, I'm sorry), but I'll give you some insight as to why they're the hardest class of polls: like the relationship between smoking and lung cancer – it has to do with isolating the variable(s) of interest.

Before we could get an accurate measure of the relationship between smoking and lung cancer, researchers had to control for the "other" potential contributors like environment, age, how much someone smoked, and more. Only *then* were they able to provide an answer as to *whether* smoking contributed lung cancer. Much later, with that knowledge as the foundation – and with more data – the researchers could more strictly control for those "other" factors and then were able to account for *how much* each of those factors contributes, including smoking.

If you ask anyone in the relevant fields if poll error *exists*, the answer will be a resounding and consensus "yes." Margin of error, nonresponse error, and others – yep, they exist.

Similar to "smoking is a contributing factor to lung cancer," we are on the same page.

But if you ask *how much* of any poll's error can be attributed to the statistical margin of error, versus nonresponse error, versus improperly weighting a poll, versus still more potential contributors, and so forth, you'll probably get a blank stare. If they're being honest, the answer is "we don't know." This is a very respectable answer in science, but it indicates we need to do more and better research.

Being able to quantify each and every one of these sources of poll error, if only approximately, should be the top priority of anyone who is serious about telling us how "off" polls are or were.

And the problem begins with the fact that there isn't even a consistent definition of what poll error is.

How could that be, didn't we just define poll error earlier? It's simple, reported poll value minus true population value. What did I miss?

Well, experts disagree on what the "reported poll value" is, *and* what the "true population value" is!

If that sounds hard to believe, and I hope it does because it's wild, it's an indisputable reality within the field; I'll explain with examples shortly.

On the topic of what can cause a poll to be "wrong," my first challenge was to **isolate poll error**, the variable of interest. There are a lot of variables that can cause a poll to be wrong – but if you can isolate it down to a single value, that standard could then be applied to all types of polls, including political ones.

Notes

1. Australian Bureau of Statistics. (2023). *Statistical terms and concepts.* https://www.abs.gov.au/statistics/understanding-statistics/statistical-terms-and-concepts
2. $z = 1.96$, $p = 0.5$, N = 100,000, $n = 600$.
   MOE = $0.898/24.569 \times 100 = 3.989\%$.
3. Andersen, R., Kasper, J., & Frankel, M. R. (1979). *Total survey error: Applications to improve health surveys.* Jossey-Bass.
4. Deming, W. E. (1944). On errors in surveys. *American Sociological Review, 9*(4), 359–369. https://doi.org/10.2307/2085979

4

Introducing: Ideal Polls

Important Points Checklist

- An ideal poll and its purpose in understanding poll data
- What 95% confidence is, and what it means for ideal polls
- How dice can be used to illustrate the margin of error and confidence level
- Why including the margin of error (and not just the number given by the poll) is important
- Why ideal polls must be defined and understood before considering "real-world" nonideal ones
- The "theoretical" applications in other fields – and why this field needs them too

If you randomly identify 600 households from your population of interest, and are able to contact all 600 of them to ask them whether they have a dog at home, you can say with a very high level of confidence that whatever percentage of that 600 say "yes" is plus or minus 4% of the actual number of that whole population.

That's pretty cool! To be able to ask just 600 people and get a very good estimate of some characteristic about a population of 100,000 or more is a useful tool.

If 300/600 (50%) say "yes," then we can say the actual number in your city is 50% +/− about 4%, or between 46% and 54%, with 95% confidence.[1]

If 120/600 (20%) say "yes," then we can say the actual number in your city is 20% +/− about 4%, or between 16% and 24%, with 95% confidence.[2]

This hypothetical survey was my first run at defining an **ideal poll:**

1. Defined the characteristic
2. Defined the population
3. Took a random sample from that population

DOI: 10.1201/9781003389903-4

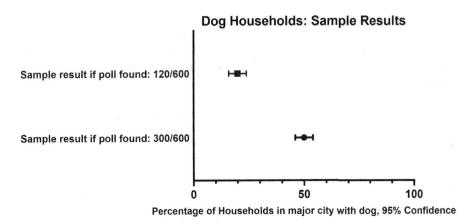

FIGURE 4.1
Two sample poll results plotted with 95% confidence levels to show range of outcomes if we
received a 100% response rate from a random sample. Chart by the author.

4. Achieved a 100% response rate

5. Truthfully reported those results as you observed them, along with
the margin of error and confidence level

Why is that an "ideal poll"?
Because, how could we, as researchers, possibly use this instrument
better?
If you thought, "we could ask more people – take a larger sample size!"
Bonus points for you. You're getting ahead of me, but bonus points
nonetheless.
It's true, we could ask a larger sample or even take a true census and ask
everyone, and that would be better from a pure data perspective, but it would
take a sample size about 10 times larger – 6,000 people for our 100,000-popu-
lation sample – to get our margin of error close to 1%.[3]
Is it worth it?
The word "worth" is the operative term here. Polling takes time and
money. Is it worth a lot more time and money, not to mention the logistical
complications that arise from taking a larger sample, to reduce your reported
46%–54% window to, say, 47%–49%? I suppose that's a matter of opinion.
That's not to say that reducing the margin of error by 1% or 3% doesn't
matter! To the contrary, all else equal, a larger sample size is *always* better.
But how much time and money a researcher can expend on a poll is not
unlimited.
For that reason, as long as the above conditions are met, whether the margin of error is 1%, 10%, or even higher, a poll can (by my definition) still
qualify as ideal.

I use the word "ideal" carefully here: not "perfect" or "flawless" or any other word that implies lack of error. With ideal polls, what I am attempting to define are the factors *within the instrument's natural limitations*. Ideal is the best that can possibly be done **given the tool or instrument's capability**.

A poll being called an "ideal poll" doesn't mean there is *no* margin of error – such a poll is impossible, by definition – the statistical margin of error will always exist unless you conduct a census of 100% of the population of interest. As such, the result reported by an ideal poll can still be "off" from the population.

Importantly, I also *intentionally exclude* weighting methods here. I have no doubt that an expert in this field would sometimes, perhaps often, be able to produce a more accurate estimate of a population with a nonideal poll than many nonexperts could if they conducted an ideal one; weighting methods can be that valuable.

While weighting is a valid and necessary process in polling, and could even be used when a response rate of 100% is achieved, **weighting introduces potential sources of error that are not present in the poll itself**.

The quality and validity of weighting techniques add a variable that cannot be isolated. An "ideal poll" is *not intended to portray the accuracy* of the poll, it is intended to *isolate* the tool's function.

What this means, to summarize as succinctly as possible, is that **an ideal poll's only potential source of error is the statistical margin of error**.

Whether you're right with me at this point or you're hung up on the "ideal poll" construct being, seemingly, only possible in theory – I'll ask you to just roll with it. When your goal is to ultimately figure out "how much" a specific factor contributes to poll error, understanding which factors can (and can't) contribute to it is an important concept to understand.

The "ideal poll" construct allows us to reduce the dozen or more sources of poll error to just one; it isolates the variable of interest.

What Does "95% Confidence" Mean?

While I promised this book wouldn't delve too deeply into statistics, I want to present you with an explanation for which I've seen many lightbulbs come on for people:

A 95% confidence level says that if you did 20 polls, you would expect the *actual population* to be outside the stated margin of error for the poll in only one of them – 19/20 (95%) within the margin of error.

Ninety-nine percent confidence says that if you did 100 polls, you would expect the *actual population* to be outside the stated margin of error for the poll in only one of them – 99/100 (99%) within the margin of error.

The seldom used 99.9% confidence says that if you did 1000 polls, you would expect the *actual population* to be outside the stated margin of error for the poll in only one of them – 999/1,000 (99.9%) within the margin of error.

To most people – me included – grasping the difference between 95% and 99% (or 99% and 99.9%) is a little challenging. They're all pretty strong numbers.

But one out of 20, one out of 100, or one out of 1000 – that's a little easier for us to comprehend the magnitude of difference.

Earlier when I asked you to estimate to what extent smoking contributed to lung cancer, and how many households in your nearest major city have dogs, the size of your "error bars" reflected your level of knowledge on the subject and confidence in your prediction.

The error bars in poll data can be thought of similarly, with a very important distinction:

Instead of us needing to subjectively estimate our level of uncertainty about a topic, with the statistical margin of error, we can do it objectively.

In data at the 95% confidence level, the "error bars" must be wide enough such that the true result is inside of it 19/20 times.

To use an example that I think most people can relate to, consider a standard, six-sided die. The die is numbered 1–6, and given a fair roll, has an equal chance of rolling each number.

But what happens when a *second* die is introduced?

Both dice have an equal chance of landing on each number, but given the possible combinations, sums in the middle of the possible outcomes – 6, 7, and 8 – are far more likely than numbers at the edges – 2, 3, 11, and 12.

Here is a breakdown of the probability of rolling each possible combination with two standard, six-sided dice. For sake of visualization, each probability is expressed as a fraction out of 36.

Sum	2	3	4	5	6	7	8	9	10	11	12
Prob.	1/36	2/36	3/36	4/36	5/36	6/36	5/36	4/36	3/36	2/36	1/36

To most people, these dice facts range from mildly interesting to common knowledge. Like I said, I'm very fun at parties.

But now, let's say someone asked me to *predict* a single roll of these dice. Some sums are more common than others, but none of them are "likely." The most common outcome, 7, occurs only 6/36 times (about 16.7%). While most people would rightly classify such a probability as "unlikely," it is still the most common outcome; I don't think anyone would say my characterization of "rolling a 7 is unlikely" is wrong because 7s are sometimes rolled.

But let's put it in terms of poll statistics.

If someone asked me to express the likely outcome of the roll of two dice and allowed for a 95% confidence interval, I could say:

The roll of the dice will result in a 7, +/– 4.

Here, you understand that there are a couple unlikely but possible outcomes (2 and 12) excluded from my range. When polls are reported by most media outlets and even some pollsters, the 95% confidence part is not reported, but taken as "understood."

Worse yet, the margin of error is also sometimes buried or not reported, such that my dice-roll "poll" would be reported simply as "7." Yikes.

No, the data I provided didn't say "7," it said 7 plus-or-minus 4.

Big difference, right?

As an informed consumer of poll data, you should understand that both the margin of error and confidence interval apply to all polls.

That is to say, about 95% of the time, the roll of two dice will result in 7 +/– 4, or a sum from 3 to 11.

For people who aren't mathematically inclined, and even for many who are, the margin of error and confidence intervals are abstract mathematical concepts, whereas the range of outcomes given a roll of two dice is much more intuitive.

But the margin of error given by the sample size and confidence interval *is not* a hypothetical or abstract concept; it is a testable and measurable one. This example is intended to give a tangible representation not just of what the margin of error is, but what it *means*.

Whether these fractions make perfect sense to you or not, though I'll reiterate these mathematical abilities are valuable to have or acquire, it's not necessary to be able to calculate the possible combination of dice roll outcomes yourself to understand what the range of possible outcomes is.

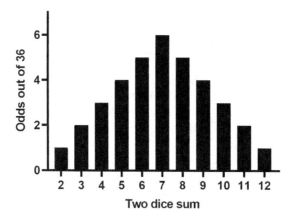

Two (six-sided) dice sum probability, out of 36

FIGURE 4.2

The probability of rolling a given sum with two standard, six-sided dice is shown; expressed in probability out of 36. Chart by the author.

You understand that while 2s and 12s are rare, they're not impossible. When I report the result of the next roll will be 7 plus-or-minus 4, I'm not saying 2s and 12s *never* happen. In fact, given enough chances, 2s and 12s are *inevitable*. The same can be said for poll results and the margin of error.

The industry standard for poll data – and most research data in general – is the 95% confidence interval. To provide data that is within the margin of error 19/20 times (95%) is pretty good! The time, money, and logistical challenges required to achieve a substantially higher level of confidence for the same margin of error – unlike with dice – are usually prohibitive with poll data.

Now let's apply this knowledge to – for now – our hypothetical dog-household polls. Given a 95% confidence interval:

If I conduct the same poll in 20 different cities, conducted with the same underlying methodology, even meeting all the standards of an ideal poll, there will be – more likely than not – a poll "outside the margin of error" *at least once.*

Likewise, if I conduct this same poll 20 times in the *same city*, with all of the same qualifications as above, a result "outside the margin of error" is likely to happen at least once.

There is no way to predict which poll will fall outside the margin of error in advance. A poll that produces an estimate outside the margin of error doesn't mean polls are no longer useful, nor does it even mean the poll was bad. The reality is that an *occasional* poll outside the margin of error is a normal statistical occurrence. Even polls conducted ideally in every way will sometimes produce results outside the margin of error. This is one reason why having *a lot* of good, independent poll data is valuable.

Looking at or citing individual polls that proved to be very far off (i.e. "outside the margin of error") from the true population as evidence that polls can be wrong is a fundamental misunderstanding of how we should understand poll data. Any expert who has ever done this is, and in my experience most have, directly contributes to the misinformation the public consumes.

Can polls be wrong? Absolutely. Can they be unrepresentative, or use non-random samples that are poorly weighted, causing them to report an inaccurate result? Yes. Can the statistically necessary margin of error and accompanying confidence interval cause a very good, even ideal, poll to produce results that don't match the specific population for that poll? For sure. But does any poll, even an ideal poll, that produces results outside the margin of error mean, with no exception, that the *poll was flawed*?

No.

In a city where 50,000/100,000 households have dogs, I could take a random sample of 600, receive a 100% response rate, and most of the time I'll get a result around 50%. But sometimes, for no reason other than random chance, I could get a result that says the true number is somewhere around 44%, or 57%.

And sometimes with the roll of two dice, you can get 2 or 12. It's the same concept.

Consider, if an ideal poll with a known and provable margin of error can sometimes produce a result outside the margin of error, then real-world, nonideal polls will probably produce results outside the margin of error at least that often.

We shouldn't be disheartened or surprised when polls occasionally produce results outside the margin of error. We should *expect*, some small percentage of the time, that this would happen. It should not be front-page news, it's barely notable. I don't question whether "the dice" are accurate every time I roll a 2 or 12 in Monopoly – it happens. Yet, in their work, experts feel compelled to cite the fact that a pollster "rolled a 2" as evidence of their wrongness, with *no consideration given* to why or how, and those experts need to do better. The fact that someone rolled a 2 is not reason enough to criticize the accuracy of their "dice."

To reiterate, that is not me saying any specific poll *couldn't have been* wrong, it's to say the statistical equivalent of "their most recent roll was a 2" *isn't even close* to sufficient evidence to reach the conclusion that the "dice" are or were inaccurate; the current standards used by the field's experts see no problem with this. I believe those who participate in that level of reasoning[4] should take responsibility for the misinformation they produce and do better in the future.

Note: The odds of rolling a 2 or 12 with two dice is 1/18, while the odds of an ideal poll compared to the true population producing results outside the margin of error is 1/20. While these numbers aren't the same, I believe this is a useful enough analogy to forgive the 0.55% difference.

Regarding the Ideal Poll Standard

Having spoken to qualified experts, and many other stats and politics veterans, I suspect you might be thinking as they did: this is all fine, "but a 100% response rate is *impossible in political polls,* so this ideal poll thing isn't useful."

And this is why the book doesn't jump into political polling immediately, and I insisted that readers – even very well-read ones – not skip ahead.

The fact that our primary exposure to polls is in the form of political polls *contaminates* our ability to understand what polls tell us, and through an *avoidable domino effect,* leads to a fundamental misunderstanding of what *political polls* tell us.

In any scientific field, understanding both the intended function and limitations of your instrument – in this case, a poll – seems like it should be an obvious prerequisite to analyzing how accurate that instrument is.

So, as a scientist, researcher, or interested reader, that requires understanding how that instrument operates in *ideal* circumstances – otherwise, you will not be properly equipped to correct or **calibrate for nonideal ones**.

There is currently no definition (or even a framework)
for what an ideal poll is.

As a result, it's not just students, learners, and the public who are left struggling to understand what polls are supposed to do, experts in this field are working from different, incorrect, and even blatantly contradictory assumptions of what polls "should" tell us.

Even if the ideal poll concept were purely theoretical (it's not) that still wouldn't mean it's not useful. To the contrary, it is the *fundamental* lesson anyone who wants to analyze polls must understand: polling 101. You *cannot* accurately analyze a nonideal poll without a functional understanding of what an ideal one does – yet that is somehow the standard practice for this field. In other fields, learning about how things operate in "ideal" circumstances is already established.

For example, there is not a chemistry course in the world that could call itself complete without teaching the ideal gas law. Maybe you remember PV=nRT? Indeed, in every university I could find, it is part of their "Chemistry 101" curriculum.

The ideal gas law has some major shortcomings, however:

> In reality, there are no ideal gases ... ideal gases are strictly a theoretical conception.[5]

Well, does that mean studying and understanding them isn't helpful or useful? Try that one on your chemistry teacher.

Of course that doesn't mean it's not useful. It is more than useful; it is absolutely necessary. Experts and learners alike know that if you want to study or understand gases, understanding *ideal gases* is not just some mildly useful teaching device: it is a *prerequisite* to being able to understand how nonideal or "real world" gases work.

I have never encountered a student (never mind an expert) who has questioned why learning about "ideal gases" is necessary, *even if* it's "just" theoretical. Chemistry, in this instance, has a few hundred years of "status quo" working in its favor. Unfortunately, there are about a hundred years of contradictory status quos going against my work; hence, the reasonable but misplaced belief that an ideal poll standard isn't necessary or useful.

The fact that I have to provide reasoning to support my assertion "understanding what a poll does is a requirement to measure its accuracy" only exists because the "poll accuracy" calculations currently done in this field are built on bad science that no one has thought to (or been allowed to) correct.

The **first step** to understanding how polls work and fixing the flawed and contradictory assumptions made by experts is to establish what an ideal poll is. I outlined my initial proposal in numbered form at the beginning of this chapter; here, it is more traditionally defined, with two major additions:

In simplest terms – but without oversimplifying – **if you poll a specific population, the sample is truly random, you receive a 100% response rate, those polled understood the question – and their responses were recorded and reported accurately, the only potential source for error, with regards to that specific population, is the statistical margin of error itself. That is an ideal poll.**

You'll note, I'm using that term "specific population" again. If a poll's random sample comes from a population even slightly different from the one it is being generalized to, it might yield erroneous results.

The goal of a poll is to make a generalization about a larger population, but the **same population** as the one sampled for the poll. That's important to remember.

The two major "additions" were assisted by my research into other fields' "ideals," plus some real polls I conducted myself, and an understanding that being *complete* is more important than it is to not be too theoretical. So, is it possible to know with philosophical or "moral" certainty that:

1. Everyone you asked understood the question, and

2. Responses were recorded, counted, and reported accurately?

No. But those sources of error, while *usually* negligible or zero, are assumptions that must be made for *any poll* involving humans – ideal or otherwise. They are constants. Even if this concept were "strictly theoretical," **that doesn't mean it isn't the fundamental basis upon which the math that underlies poll data is built**.

One could even participate in a meta debate about what it means for something to be truly "random" as many people kind enough to provide me feedback have. This debate is irrelevant, missing the forest for the trees.

Identifying a slightly flawed randomization procedure, a couple of people who did not understand the question, or a clerical error in how the numbers were recorded or reported, is another way of saying, "but this poll wasn't ideal."

Good! Identifying the assumptions made by a method and why they don't apply to a specific case, some specific cases, or any specific case, would signal an enormous leap forward for the field. In fact, it's this exact process of identifying assumptions that don't always (or ever) hold true that allows me to say the existing poll error calculations are wrong – getting slightly ahead of myself.

But again, there currently exists no standard for what an ideal poll is. My proposed definition is nothing more than an attempt to quantify what a poll is, and what a poll does. If someone can do better – and I'm not too prideful to admit that's a good possibility – they should.

> The calculation of "poll accuracy" *depends* on understanding
> what an ideal one would measure.

It happens that the ideal gas law makes assumptions that are known to be untrue,[6] but it nonetheless remains useful across a wide range of applications, both as a starting point for learners and actively in the field. The ideal poll standard is extremely useful too – and it doesn't make assumptions nearly as bold. Fortunately, poll math is much simpler than thermodynamics.

And even if a poll couldn't be ideal, that doesn't mean they can't be very, very close. "How nonideal is it?" is a question you can only ask if you have an ideal standard to compare it to. If my sample wasn't random, how much error did the nonrandomness contribute, compared to the ideal? If I didn't achieve a 100% response rate, how much error did nonresponse contribute, compared to the ideal?

In the absence of an ideal poll standard, the "status quo" has substituted contradictory unscientific assumptions, and even when presented with evidence those assumptions are/were flawed, maintaining the assumption they couldn't have been wrong.

The ideal poll standard should be built upon or modified as necessary but ultimately understood by learners and researchers alike. It is a fundamental requirement for understanding how polls work. Here is the version I propose.

An Ideal Poll

1. Defines the characteristic
2. Defines the population
3. Takes a random sample from that population
4. Achieves a 100% response rate
5. Truthfully reports those results, as they are observed, along with margin of error and confidence level
6. Everyone asked understands the question
7. Responses are recorded, counted, and reported accurately

Following and achieving all these steps is not a guarantee of a poll's utility, or its accuracy. It is an analysis of a poll's function and limitations. In a vacuum, that doesn't sound useful; in practice, it provides a standard to which any nonideal poll can be compared.

While the ideal poll standard requires making a few assumptions, it does something essential for calculating poll accuracy: it **isolates the many potential sources of error down to the margin of error itself**. That is a standard that can be easily tested and built upon, and I will do so in the coming chapters.

But remember, even ideal polls **can and will** occasionally produce results outside the margin of error. There's one more important concept to present to better understand poll data before starting to put things together.

Notes

1. Given a population proportion of 50%, the MOE would be about 4%.
2. Given a population proportion of 20%, the MOE would be about 3.2%.
   These respective margins of error (4.0% and 3.2%) are different enough to note, but not too different to justify simplifying for illustrative purposes. I believe using different MOEs here risks overcomplicating things for readers who aren't interested in this level of precision – and this level of precision is not required to have a functional understanding of polls.
3. $z = 1.96$, $p = 0.5$, $N = 100,000$, $n = 6,000$
   MOE $= 0.98/79.893 \times 100 = 1.227\%$.
4. Morris, G. E. (2022, July 28). How pollsters got the 2016 election so wrong, and what they learned from their mistakes. *Literary Hub*. https://lithub.com/how-pollsters-got-the-2016-election-so-wrong-and-what-they-learned-from-their-mistakes/
   "The outlier ABC/*Washington Post* poll that missed Joe Biden's margin by 17 percentage points in Wisconsin was called out as one of the biggest misses in the history of polling. It nearly gave the *Literary Digest* a run for its money."
   This is just one example of thousands because it is the most recent example I've come across and problematically compares what is by all accounts a scientific poll (ABC/Washington Post) to a notoriously unscientific one (Literary Digest) solely on the basis of how closely it predicted a result.
5. Tenny, K. M., & Cooper, J. S. (2022). *Ideal gas behavior*. StatPearls Publishing.
6. Tenny, K. M., & Cooper, J. S. (2022). *Ideal gas behavior*. StatPearls Publishing.
   "For a gas to be "ideal," there are four governing assumptions:
   - The gas particles have negligible volume.
   - The gas particles are equally sized and do not have intermolecular forces (attraction or repulsion) with other gas particles.
   - The gas particles move randomly in agreement with Newton's Laws of Motion.

- The gas particles have perfect elastic collisions with no energy loss.

In reality, there are no ideal gases. Any gas particle possesses a volume within the system (a minute amount, but present nonetheless), which violates the first assumption. Additionally, gas particles can be of different sizes; for example, hydrogen gas is significantly smaller than xenon gas. Gases in a system do have intermolecular forces with neighboring gas particles, especially at low temperatures where the particles are not moving quickly and interact with each other. Even though gas particles can move randomly, they do not have perfect elastic collisions due to the conservation of energy and momentum within the system.

5

Throw It in the Average

Important Points Checklist

- What "throw it in the average" means, and why it is preferable to other approaches
- What is meant by "fluctuation is normal and expected"
- The danger of concluding a few polls in succession are indicative of a "trend"

Understanding political polls is not quite as simple as rolling dice, I don't want to lead you on. While it's a useful (and accurate) way to understand how the statistical "margin of error" works, along with some other probability concepts, poll numbers are not as simple as dice rolls. With dice rolls, every roll of the dice is basically the same, no matter who does the rolling.

Conversely, with political polls, different pollsters have different methodologies, different weights – even different sample sizes – that complicate taking all of them in a simple average, and calling that number "the" average.

Poll averaging, and the related "**aggregation**," can involve weighting polls by recency, perceived quality, among other things. Aggregating polls is not a novel idea I'm presenting to you, but one employed by multiple companies and forecasters. This approach is by no means flawless; the decision of which polls to include, how many polls to include, and if some polls should count for more than others is just scratching the surface.

"Throwing it in the Average" or looking at an individual poll that may have been very rigorously conducted only to use it as part of an average with others might feel overly simplistic or insufficient. After all, what if you mix "good" polls in with "bad" ones?

In that sense, it may at first feel a little contradictory of me to say that we should both have high expectations for what polls (plural) can tell us, and not spend too much time analyzing any individual poll. But this approach has a strong statistical basis.

In addition to the benefit of becoming less susceptible to the mediafication of poll data, and not overly scrutinizing what a single, snapshot, and possibly

 DOI: 10.1201/9781003389903-5

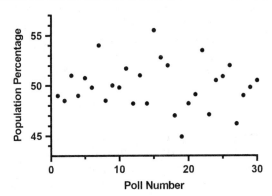

FIGURE 5.1
A simulation of 30 polls with a given population proportion of 50%, and a margin of error of plus-or-minus 4% are shown. Code can be found in endnotes.[1] Chart by the author.

outlier piece of information *means* (and talking about it for days or weeks), looking at the average of polls is better from an accuracy standpoint, too.

On the topic of "noise" in polls, and the importance of not overanalyzing one poll, a friend of mine had a neat idea to create a simulation in which a population's true value is 50%, and the margin of error is +/− 4%. That is, the true population value is *known with certainty*, 50%, and the simulation provides a lot of independent results from that population. That means, given the 4% margin of error (MOE), **each dot represents a result that could have been given by an ideal poll.**

I charted the results of that simulation in Figure 5.1.

Looking at this chart, you probably notice that the range of results is pretty wide and that few returned exactly "50%." A few were on the edge of the margin of error (46% and 54%) and a couple even outside of it.

Remember, looking at this plot of 30 polls, we *know* the true population percentage is 50%, a benefit we don't get in real life! And yet, if we were to look at most of these polls *individually*, it's unlikely that's what we would estimate.

In most real-life examples, the true population proportion can only be estimated, but this simulation demonstrates a powerful concept by effectively approaching the problem backwards.

> Instead of "given this poll data, what can we estimate about the population?", it asks "given this population, *what would poll data estimate about it?*"

While polls in real life don't enjoy the obvious benefit of "knowing the true population proportion," nor do we usually get to see what lots of polls conducted at the same time would produce, the concepts demonstrated by this simulation are next in line in terms of those crucial to understand. If a simulation in which the true population **is known** rarely returns that exact number, isn't it fair to say that even a very good human pollster, conducting a poll at this time, could have returned any of these 30 results? While they all meet the "reasonable estimate" standard we should have of polls, very few are precise.

The most important thing this demonstrates about poll data can be summarized as:

"Fluctuation is normal and expected."

Look at the range that these simulated, ideal polls produced for a known population proportion of 50%, with a known margin of error of +/– 4% and a 95% confidence level, and consider that next time the results of a single nonideal poll are being reported, or analyzed. Even polls conducted with remarkable precision can return different results, but if we think of polls as we should, a reasonable estimate, we allow ourselves to not become overly fixated on what one poll *means*.

Of course, in real life, we don't typically have the benefit of knowing the true population proportion when polls are conducted, and notably, unlike the simulation, results can change over time.

Set that aside for now and consider the fact that if 30 equally good polls, ideal or otherwise, were conducted at the same time on the same population, they might produce very different results. Fluctuation is normal and expected. But watch what happens when we introduce the "average" approach.

Very quickly, what looks like a noisy and wide spread of results – if we consider the average of them – produces a consistent and accurate one.

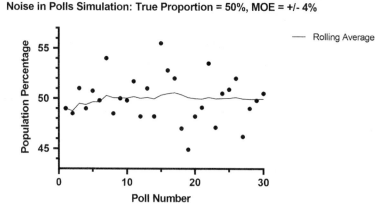

FIGURE 5.2

A simulation of 30 polls with a given population proportion of 50% and the margin of error of 4%, but with rolling average introduced. Chart by the author.

You might note a few clusters and patterns within this data such that applying the mediafication lens that prefers content to accuracy, they would likely be reported as showing "momentum" or a "trend." Yet, we know with certainty each of these polls were taken from a population proportion of 50%; another note to consider when poll momentum and trends are discussed.

Right around now is the time where my discussions tend to devolve, and rush, into specific political examples of times that a poll result (or even a polling average) "missed" the result by a large amount. That's how it is reported in the media, explained and analyzed by perceived experts, and even cited in research papers on the subject.

I've openly and intentionally avoided political poll analysis so far. If things feel simple for you to this point, your perseverance won't go unrewarded. If you've learned something, even better. By considering only political polls, as most people do, the fundamentals of what polls tell us are overlooked.

And in case I sound like a poll apologist, to the contrary, both in theory and in practice, polls can be and have been wrong. In fact, given that we know the true population proportion in my simulation above was 50%, within the margin of error or not, each and every one of those dots has an absolute error equal to exactly the amount it deviates from 50%. Even the poll average can be called wrong by the same calculation.

But in the simulation, we *know* the true population proportion at the time of each poll. We do not know the true population proportion at the time of each poll in most real-life instances – and especially not in political polls. This creates a conflict that – whether intentional or accidental – leads to prominent analysts and experts to make some egregious mistakes pertaining to if, how, why, and "how much" the polls were wrong.

Those egregious mistakes lead many others – in both the public and among other experts – to believe that poll data is in "crisis,"[2] tell the public "don't trust the polls,"[3] and more directly, that polls have been "wrong"[4] by much more than they actually were, with little to no ability to support the claim.

I'll address each of those beliefs and more, but one step at a time.

I think almost everyone who is interested enough in polls to look at them and talk about them, and certainly anyone willing to read books about them, understands, or is capable of understanding these core concepts of polls covered so far:

- Polls are reasonable estimates, they're not perfect.
- It's necessary to understand what an ideal poll measures, even if only in theory, before trying to calculate what nonideal polls measure.
- Lots of polls of good quality, taken together in an average, will give you a better estimate in the long run than any one poll.
- The error in a poll is the difference between the reported result and the true value for the same population.

Notes

1. Code is in R
   # Set the number of draws
   draws <- 30
   # Simulate draws from a binomial distribution and divide by 594
   results <- round(rbinom(draws, 594, 0.5)/594, digits = 4)
   # Plot the results on a scatter plot
   plot(1:draws, results, xlab = "Poll number," ylab = "Proportion of Support,"
   main = "Results of Ideal Polls with known 50% Proportion of Support").
2. Graham, D. A. (2020, November 4). *The polling crisis is a catastrophe for American democracy*. The Atlantic. https://www.theatlantic.com/ideas/archive/2020/11/polling-catastrophe/616986/
3. Chalabi, M. (2016, January 27). *Don't trust the polls: The systemic issues that make voter surveys unreliable*. The Guardian. https://www.theguardian.com/us-news/datablog/2016/jan/27/dont-trust-the-polls-the-systemic-issues-that-make-voter-surveys-unreliable
4. Too many examples to cite.

6

What Makes a Poll "Wrong" Part 2

Important Points Checklist

- What a "target population" is, and why it's important to define
- The role frame error can play in "total error"
- What "weighting" is, and why it's done
- The terms within the formula for "poll error"

The statistical margin of error is the most tangible number – and the only "error" number reported – with poll data. But as you know, the margin of error is not the only reason a poll could produce a number that differs from the population.

This is one reason election pollsters sometimes attempt to differentiate between registered voters and likely voters and include this distinction in their report. If a sample includes all registered voters, but you're trying to measure the preferences of people who will ultimately vote – your poll results with respect to the election could be skewed! After all, not all registered voters are likely to vote. It's reasonable to say that a registered voter who has never voted is probably less likely to vote in the next election than one who has voted in the past 10 elections.

To use a more direct example, imagine if a poll asked a huge, random sample of 16-year-olds who they would vote for. Considering they're not eligible to vote, how useful is that data regarding the target population of "actual voters?" Not very useful!

To that end, it's understandably important for a poll sample to reflect the **target population**, the specific population you want to know something about. The ways pollsters qualify who "likely voters" are can vary – it is not an exact science. But how well they can estimate that is another factor that contributes to poll accuracy – and pollster quality.[1]

It's so important, in fact, that there's a name for erroneously including or excluding members from a poll sample who aren't part of your target population: **frame error**.[2]

Including a bunch of 16-year-olds in your "likely voter" election poll sample is an "erroneous inclusion" frame error, since they're not eligible to vote.

DOI: 10.1201/9781003389903-6

Excluding someone from your "likely voter" sample who has voted in each of the past 10 elections and says they will definitely vote this year is an "erroneous exclusion" frame error. Frame errors are not usually this egregious in modern polling (and these are not the only types of frame error) but qualifying who among your sample are likely voters is another tough job pollsters must navigate – and can contribute to poll error.

By saying who they believe are "likely voters," pollsters are making their best approximation of both voter turnout and voter demographics, and weighting their raw poll data accordingly.

Frame error, the previously discussed nonresponse error, and the statistical margin of error – while still not the only factors that contribute to total poll error, are almost certainly – as it relates to most professional polls, political polls included – the three largest potential contributors.

So, let's update the definition.

> **Poll error = Statistical Margin of Error + Nonresponse Error + Frame Error + Other Error**

and

> **Poll error = Reported Poll Value – True Population Value**.

In total:

> **Reported Poll Value – True Population Value = Statistical Margin of Error + Nonresponse Error + Frame Error + Other Error**.

Again, there are lots of terms, but none of them are too complex. All I'm doing with these definitions is breaking down the factors that can contribute to a poll not matching the true population value. Being able to "plug in numbers" to this formula, while it might sound simple when I put it that way, is exactly how it can be broken down.

> For a valid definition, all the factors that *cause* poll error must equal the poll error, measured or observed – the whole must be equal to the sum of its parts.

Understanding this will provide insight to how and why I can say with such confidence the existing analysis related to poll error – and what polls tell us – is wrong.

One final topic to touch on before moving forward, that I've only mentioned briefly this point, is *weighting*.

In most polls, especially political ones, pollsters know that the people who respond to their questions aren't always a representative, truly random sample. In order to correct for this, they apply **weights** to their results.[3]

I previously used gender to give an example of how weights might work, but to consider a more relevant dog-household poll example:

If you know your town's addresses are 30% apartments, but the sample from your poll only had 25% apartment addresses, you might *weight* your data to correct for that discrepancy in hopes of getting an *even better estimate* of some characteristic – in this case whether or not they have a dog.

It seems reasonable to say that whether someone lives in a house or an apartment could impact the probability they have a dog.

You could even weight by marital status, age, and if the household has kids – these can all influence the likelihood of having a dog. If you'd like to dig deeper and think of even more examples that could impact the accuracy of a dog-household poll, you could weight for those too.

Excuse me, stranger, do you have a dog? Also, how old are you, do have kids, and are you married?

Sometimes, taking a poll is much more complex than just identifying a random sample, contacting them, and asking what you want to know – which is hard enough. In order to reliably weight your results, sometimes you need to know more.

All in all, just understand that weighting is intended to make poll results more accurate compared to the population of interest, regardless of sample size and response rate – but those weights are largely up to the pollster. The process is scientific, but it is not an exact science. Different pollsters weight on different factors, and even on a different *number* of factors.[4]

For the dog-household polls, some pollsters could weight their results to account for the number of apartments versus houses, others by how many children in the household, others by age – and some could weight for all of these factors. Even ideal polls with a random sample and 100% response rate should consider weighting, for the simple fact that even random samples are not always perfectly representative.

In my opinion, though I don't think it's too controversial, pollsters should be allowed to include or exclude any factors they choose from their weights – *as long as* they are transparent with these methodologies and acknowledge the possibility that their methodology could increase their poll's error (as well as reduce it).

Reasonable people can debate the best methodologies for weighting, but unless you want to *do* polls and forecasts, understanding these weights is not particularly important to understand how to *read* polls.

Why?

The results reported by a pollster are after they weight them. If their weights are good, bad, or otherwise, their *reported* results are largely reflective of the quality of their poll.

Regarding alleged political polling failures of the past, analysts famously argue whether people lied to pollsters, or that the samples weren't weighted properly, or unlikely voters voted in high numbers, or maybe people with certain voting preferences simply tended not to respond to polls. All of these are possible explanations, but as for *how much* each of these factors contributed to the error, the only honest answer is, "don't know."

Identifying and quantifying how much each variable contributes to a poll's wrongness – if it is rightfully called "wrong" at all – is a complex task. Fortunately, for both educational and practical purposes, there are ways around this.

By eliminating those much harder to quantify, "other" sources of error, we can understand at a fundamental level what polls do, how they should be read, what can cause polls to be "off" – as well as what can't and doesn't.

With that, I circle back to ideal polls, a concept I first tried to develop using data from simulations, for the simple fact that simulations are very quick to conduct, and the "rules" or parameters can all be known with certainty.

This is where the "100 years of pushback" I talked about in the preface starts to come into play.

Most people, including experts who regularly use simulations and models in their work, say that simulations such as the "fluctuation is normal and expected" example – though it proves my point about ideal polls – aren't practical or relevant to "real polls." In short, when first presented with it, it seems everyone believes that the ideal poll standard isn't useful and is impossible in practice.

To which I'd respond: that's why you read the book!

It's very useful, and it's not impossible. We can, very easily, create and even conduct such an ideal poll. It doesn't require the ability to code, knowledge of advanced statistics, nor even much time.

In doing this, we can develop a better understanding of what polls tell us.

The central question to every discussion of polling – especially political polling – were the polls *right*? How much were they *off*?

That leads me to what I believe is the most important unaddressed topic in this field today: how we define and calculate poll error, and poll accuracy.

The definitions and formulas I have provided so far regarding poll error are not controversial. But earlier, I noted that the field has a "wild" disagreement about what the "reported poll value" and "true population value" are.

These disagreements are based on a plain misunderstanding of how polls as a statistical instrument are intended to be used. I believe those issues can and should be fixed. I will start by addressing "true population value."

Notes

1. Rosentiel, T. (2012, August 29). *Determining who is a "likely voter."* Pew Research Center. https://www.pewresearch.org/2012/08/29/ask-the-expert-determining-who-is-a-likely-voter/

2. Grossmann, W. (2020, March 25). *Frame error—CROS—European Commission.* CROS. https://cros-legacy.ec.europa.eu/content/frame-error_en

3. *MOE.* Langer Research Associates. (2022). https://www.langerresearch.com/moe/

4. Kennedy, C. (2020, August 5). *Key things to know about election polling in the United States.* Pew Research Center. https://www.pewresearch.org/short-reads/2020/08/05/key-things-to-know-about-election-polling-in-the-united-states/

7

Introducing: Simultaneous Census, Present Polls, and Plan Polls

Important Points Checklist

- The relationship of a census to a poll
- What a simultaneous census is, and why it matters for poll accuracy measurements
- The difference between present polls and plan polls, and why they should be differentiated

Naturally, a census – where the entirety of a population is asked the same question(s) – is very hard to do, and impractical in most cases. I can't imagine a scenario where I needed to be so accurate with my "households with dogs" question that I required a census.

But indulge me. I promise I'm going somewhere.

If I take a poll of a major city with a sample of 600 and report that 300/600 households (or 50%) have a dog, how can I possibly know how accurate my poll is, or its error?

For the sake of this exercise, assume that I have taken an ideal poll.

We know the definition of poll error is the difference between reported results and true population value. Is there any way to know for certain the *true population value*?

The only way to know for certain how accurate this poll was, the only way to know the true population value, is to take a true census: in this case, to ask all 100,000 households. Maybe such a feat could be accomplished by enlisting the help of a few thousand friends.

If the person in your census doesn't answer the first time, come back (or call back) later. It's a census, after all – rigorous.

Now is where things get interesting. Assume in your rigorous census, you're able to eventually get a response from every single household. 100%.

DOI: 10.1201/9781003389903-7

Can I calculate poll error now, by comparing the 50% reported in the poll to the percentage of dog-households obtained in the census? Whatever the discrepancy – if any – is the error, right?

Not so fast.

When you took your poll, with a random sample of 600 from a population of 100,000, that specific population – and by extension, the sample you asked – was based on the household data you collected *at that time.*

In the time it takes to do a true census of 100,000 households, whether a few weeks, months, or years, things can change. In the time between the poll and the census, the city might build a new dog park encouraging residents to get dogs, or a major housing association might implement new restrictions on pet ownership, causing a substantive decline in households with dogs.

Even without any of that, many of the residents of the city at the time of the poll may have moved away (thus potentially part of the poll and not the census), and many non-residents at the time of the poll may have moved to the city (not possibly part of the poll, but part of the census).

Do registered voter households with dogs move to the city in the same proportion they move out of it?

You can't say with any level of certainty – and certainly not the one that would withstand any scientific scrutiny – that the underlying population from your poll is the same as the one in your census taken later.

Remember, if the underlying population of the census is different from that of the original poll, we can't rightly attribute the entirety of the discrepancy between the census and the poll as the poll's error. We can't do that for the same reason we can't apply the data from your nearest major city to your next-nearest major city and assume all the numbers – margin of error and all – are the same. Nor could we assume "households with pets" data from the United States should apply equally to the United Kingdom.

You could argue the latter example is a bit more extreme, but you couldn't argue that each apply the same underlying reasoning: random samples taken from a specific population can't be said to apply (with the same margin of error [MOE] and confidence level) to a different population.

Which is to say, a census alone, as long as it is taken from *nearly* the same population can get us close to the answer, doesn't quite resolve "how do we calculate a poll's error?"

In the dog-household example, we know that the specific population we took the sample from was different from the specific population from which we took the census, but it is very hard to know exactly *how much* different.

So then, what does that poll data tell us?

The Simultaneous Census

The only valid way to measure any poll's error – the discrepancy between the reported result from the sample and the true value for that specific population – is not just a census, but a *simultaneous* census. As far as I know, there's no existing literature in the world of polling or statistics on this concept, but I hope it's pretty straightforward: if the thing a poll attempts to measure can change over time, a census taken some time later can't be considered an authoritative standard for the accuracy of the poll to be measured against. Nor can the margin of error given by the poll's sample size be judged according to how well it aligns with that future population's value.[1]

If I complete a poll of 600 households in your nearest major city today, it is not hard to explain why a census completed five years from now is not a proper or valid way to measure the accuracy of that poll: the population from which I took the random sample is not the same one from which the census was taken. We can apply that same reasoning to a census completed one year from the poll, one month from the poll, and one week from the poll.

There are changes in the population of interest between poll and census that must first be accounted for *before* "instrument error" or "poll error" can be calculated.

In the current, textbook definition of poll error is as follows:

Poll error = Reported Poll Value – True Population Value.

It is *assumed to be understood* that the "true population value" must come from the **same population** as the "reported poll value." But based on their characterizations of what the "true population value" is, this is clearly not understood, even among the field's experts. As I'll show and explain in more detail later, present methods substitute a *future value* for "true population value," despite the fact that this future value measures a different population than the poll did.

Given the fact that it is both a more accurate representation of what polls measure and that "true population value" is a source of confusion for both experts and the public, I believe "true population value," as it relates to the definition of polling and statistical applications, should be referred to as "simultaneous census."

A simultaneous census is a census that is conducted at the same time (simultaneous) of the entirety of the same population (census) from which a poll's random sample was derived – *and asks the same question(s) with the same options.*

If a census is not simultaneous, and the population of interest can change, that non-simultaneity *alone* can cause a difference between a poll's reported value and the true population value, but is *not attributable* to any poll error – nonresponse, frame, margin of error, or "other." Likewise, if a poll asks a *different question* than the census, the discrepancy between responses to different questions cannot be attributed entirely to any poll error.

It's true that any census, even if it's not simultaneous, offers a great deal of insight into the accuracy of a poll. Even a census taken a year after a poll is a much better way to approximate the poll's accuracy than no census at all.

But at some point, whether weeks or months or years later, the approximation can become less than approximate.

While a simultaneous census might be the only "perfect" way to measure poll error, it makes sense that a census completed very close to the poll, whether a matter of days or weeks, can provide a very close approximation to simultaneous.

Given a census of dog-households taken a year after some polls on the subject, a rigorous researcher might try to account for some of the factors that could have caused people in that city to get or not get dogs in that timeframe. If this city implemented a new program to offer free pet adoptions, while it might not cause dog ownership rates to go up substantially, it would be lazy to simply assume this change had no effect.

This simultaneous census concept, though it may feel abstract right now, is also vital to understand as it relates to polls because it allows us to continue to keep the variable of interest – poll error – isolated from factors beyond its scope.

As with the example above, there are factors that can cause "households in your nearest major city with a dog" to change over time.

In this sense, it's important to remember that polls as a tool are intended to give us insight into a population *right now*.

In the dog-household poll example, the difference between "poll" and "census" is narrow. In political applications, they are not.

To illustrate, and draw a comparison to how political polls are currently judged:

If in your poll, a household responds "no, I don't have a dog," you accurately record it as such, and then some weeks later when you're conducting your census, you return to that same household thinking it's just a formality – but when you ask them, you're greeted by a barking puppy, along with a grinning "yes, we have a dog."

Is It fair to say your poll was *wrong?* Comparing your poll to your census, was there a poll error?

It seems uncontroversial to say that, no, if someone reported not having a dog in your poll, and sometime later had one, it's not accurate to say that the poll was wrong. People who don't have dogs sometimes get dogs at a later time; this is not something a poll should be expected to capture. A poll disagreeing with the census in some way, even if it appears significant, cannot always be attributed entirely and solely poll error.

The current, accepted definitions of what makes a poll "wrong" would say that, yes, that is a poll error. The definitions say that the poll's failure to predict that this household would get a dog soon after not having one is a "poll error."

That's not correct. This is why it's important to understand how a poll works in ideal situations before trying to analyze it in nonideal ones – otherwise you'll misassign errors.

When analyzing a poll, you must understand that the poll's results can only be generalized to the specific population you polled, at the time you polled them, for the same question that was asked.

This is the value provided by understanding a simultaneous census.

Poll error = Reported Poll Value – Simultaneous Census Value.

This definition is no different from the current one, except for the fact that it specifies the standard by which polls, as a statistical tool, should be measured: how close is the reported result to the whole target population *right now*?

And this is only scratching the surface of what I mean by experts not understanding how to accurately answer the question "what do polls tell us?"

All Polls Aren't the Same

To this point, I've avoided using political polls in my examples, without really explaining why, other than that they're the most difficult class of polls. True, political polls rely on people truthfully divulging personal information to strangers, but the challenge is much greater than that.

Polls can be divided into two categories:

- Polls about things you did, currently do, or things about yourself (which I will refer to as **present polls,** since they're about an existing state, not concerning the future)
- Polls about future plans (which I will refer to as **plan polls**)[2]

These polls are fundamentally different in two important ways:

If you ask someone about *future plans,* you have to allow for the possibility that some in your population – maybe very many of them – are undecided. No matter how nice you are, no matter how much they want to tell you, and no matter how certain they are to eventually choose one of the options, they might genuinely not know which option they will choose. They can also change their mind between when you ask and some future time.

While if I ask you about something you *did*, or something about yourself *currently*, "I don't know" or "undecided" isn't a valid option. Nor can you change your mind about your current or past state.

The main differences between a present poll and a plan poll is that it is not possible to be undecided in a present poll, and it's not possible to change your mind in a present poll.

Some present poll examples:

- Do you have a dog?
- Do you smoke?
- Are you married?
- Did you vote in the most recent election?
- What's your favorite color?
- How many children do you have?

Whether it's a "yes/no" or open-ended question, you can't be undecided or change your mind about any of those things! And if you do change your mind in some future date, it doesn't have an impact on what the researcher asked, and what they wanted to know – the same cannot be said for plan polls.

In both present polls and plan polls, people can refuse to respond, and they can "prefer not to say." However, *only* in a plan poll is a response of "I don't know yet" (undecided) valid, and only in a plan poll does someone changing their mind at a later date impact the data.

To illustrate, consider the following "plan poll" examples:

- Do you plan to get a dog this year?
- Do you plan to vote in the upcoming election?
- Who do you plan to vote for?
- What will you have for lunch tomorrow?

All of these questions have the possibility that someone genuinely doesn't know (undecided), *and* that someone states with a high level of confidence their future plans but changes their mind. As my wife can happily attest, my response to whether I *plan* to get a dog this year could change from negative to affirmative at some point; plans can change!

Please read each of the plan poll questions, and consider if someone could reasonably change their mind about those future plans – or be undecided.

Those factors do not exist in present polls.

Current analysts and researchers do not differentiate between what should be considered distinct categories of polls, they evaluate them with the same standards; that's a problem.

Plan polls – of which political polls are a huge subset – frequently have a lot of people who are undecided or "don't know" what they will do. Within the category of political polls, our minds typically gravitate toward the polls regarding who will lead our country, but there are also polls taken on local elections, whose candidates are usually lesser-known. As a consequence, the number of voters who are undecided is typically much higher – even among those who are certain they will eventually vote.

Likewise, some of those people who do state a preference or plan may change their mind after the poll.

How those "don't knows" and "mind changers" are accounted for in the calculation of poll error is not a trivial issue.

At its core, what you *plan to do* and what you *did* are different questions because they have different possible responses.

Applying the simultaneous census standard – better put, understanding that the simultaneous census is the proper standard to measure poll accuracy – would start analysts, researchers, and even the public on a path to better understand poll data.

Not content with doing more simulations, I opted to conduct real polls to demonstrate this concept.

Notes

1. Observing that election results don't fall within a poll's "margin of error," as many experts have and do, is a textbook example of not understanding what the margin of error in a poll applies to, and why the Simultaneous Census standard is needed. A poll's observation includes undecideds. The margin of error given by the poll applies to all of its observations, at the time of the observation, including undecideds. Not a future value. Direct examples of this will be given later.

2. Questions concerning the "past" and "present" fit under the umbrella of "present polls" as I have defined them because both mind-changing and "don't know" are not factors in them.

8

What's for Lunch?

DOI: 10.1201/9781003389903-8

Important Points Checklist

- How "What's for Lunch?" applies to political polls
- Why replicability is important for scientific research
- What the margin of error applies to plan polls
- What is meant by "apples to apples" in this chapter's context

Up until now, we've been inching closer to tackling political polls directly. Now we're sprinting toward them. What I've written about so far shouldn't be taken lightly: without a sufficient understanding of concepts like "ideal polls" and a "simultaneous census" and, importantly, the difference between a present poll and plan poll, the next step will be unproductive.

That next step includes a transition from present polls to plan polls.

The dog-household poll was a present poll. When you ask someone if they have a dog, the options are either "yes" or "no." You can't be undecided. A present poll such as this can be replicated easily – whether you ask co-workers or classmates about pet ownership, their favorite color, or a very long list of options. Since you're asking about a present state, if their status or preference changes in the future, that "mind-changing" doesn't impact what you wanted to know.

One creative way to work around this "present poll" problem when taking a census is to not ask "do you have a dog?" in the poll and "do you have a dog?" in the census, but instead "do you have a dog?" in the poll, and "did you have a dog on [insert date of poll]?"

While how well someone remembers can introduce sources of error, it's much more accurate than asking "do you have a dog?" in both poll and census, if the census is not simultaneous.

Admittedly, in present poll applications, while the simultaneous census concept still applies, the differences between simultaneous census and "within a few weeks" census are probably very small. The same cannot be said for plan polls.

When you ask someone what they *plan* to do in the future – such as what they plan to have for lunch the following Friday – you can *try* and limit them to two options: "Pack lunch/bring from home" or "Order/purchase lunch." But if they say "I'm not sure yet" or "I don't know," you must accept that as a valid answer. They genuinely might not know.

Creative pollsters sometimes ask, "if the election were today/tomorrow …" as a twist on "who do you plan to vote for?" but in a plan poll, you can't trick time. That's not to say there aren't better ways to ask certain questions (defining "better" here as, "eliciting a truthful response"), but the best answer someone can possibly give you about what they plan to do in the future is what they currently plan to do. And, even if the elections were tomorrow, they can still be undecided.

This introduction of "undecideds" creates an entirely new problem for calculating poll error; one that the traditionally accepted definitions miss, and you'll understand why it's wrong by the end of this chapter.

This time, instead of a hypothetical scenario, I enlisted the help of two local elementary schools to take a poll. Actual data from actual people.

Population: Staff of two local schools, 91 individuals (N = 91).

Sample size: 80 ($n = 80$).

Characteristic: Lunch plans on a day in the near future: pack/bring from home versus order/purchase.

Random sample: Names of all 91 staff members were entered into a spreadsheet, numbered 1–91. A random number generator was used to identify 80 members for the poll.

A sample size of 80 was chosen to produce a margin of error comparable to political polls: around 4%.

By taking a random sample from these two schools regarding their lunch plans on Friday, my goal is to be able to estimate what the population of 91 staff members will have for lunch on Friday.

Perhaps a little tongue-in-cheek at this point, I concluded that all staff members were "likely lunchers" – that all of them were likely to have lunch on Friday. This is an important distinction, for the same reason "registered" and "likely" voters, as mentioned earlier, are important to distinguish.

After gaining permission to conduct the poll, I had to do a little work in advance. I received permission to conduct the poll on a Tuesday and thought about how much time would be reasonable to allow for a plan poll such as this. The upcoming Thursday being the Thanksgiving holiday complicated things. I considered asking about "next Friday," but at the risk of being misunderstood regarding which Friday I was asking about (a form of **specification error**), I opted to conduct the poll on the following Monday.

With the timeline in place, I considered the possibility that asking them just one question "What do you plan to have for lunch on Friday?" would impact my poll results in some way because the participants in the poll did not know I would be returning on Friday to record their actual lunch choice – the census. If they knew the intent of the poll, there was some chance that could inadvertently influence the outcome.

But I also promised the poll would only take about a minute.

So I opted to ask three questions instead of just one.

Question 1: What is your typical commute time to school? (A) Less than 15 minutes, (B) 15–30 minutes, (C) 30+ minutes.

Question 2: On Friday, for lunch, do you plan to (A) Pack/bring lunch from home, (B) Order/purchase lunch?

Note: In this question, I did not offer "undecided" as an option, but if they responded with "I don't know yet" (or some variation), I noted that in the poll result as "undecided." If they used a qualifier like "probably" such as "probably pack," I recorded it as "pack." Like many pollsters, I didn't make a "topline" distinction between someone who said "I always pack my lunch" and "I'll probably pack my lunch on Friday."

Question 3: During the holidays, I typically (A) Travel outside the state, (B) Stay home/Travel within the state, (C) Visit/travel both within and outside of the state.

I opted for the old school paper-and-clipboard, "in-person" poll technique, with one key goal in mind: reach a 100% response rate.

I achieved a 100% response rate.

I surveyed all 80 individuals I had previously identified to be part of my random sample.

Here are the results to the "What's for Lunch?" question:

Option	Response (%)
Pack/bring lunch from home	75
Order/purchase lunch	15
Undecided	10

This data has no weights applied, and I opted to use this as my "reported" result.

Seventy-five percent (60 individuals) responded "Pack/bring lunch from home."

Fifteen percent (12 individuals) responded "Order/purchase lunch."

Ten percent (eight individuals) responded with some variation of "undecided."

I had hoped for a somewhat closer number of individuals to choose from the two options offered, to more closely resemble a political poll. Maybe next time I'll think of such a question. Maybe you can conduct such a poll. But my hopes aren't what's important here.

What's important is that I've conducted an ideal poll. With real people, collecting real data, I can report a 100% response rate with a margin of error of approximately +/– 4% and 95% confidence.

Now, you might be thinking to yourself that it's "silly" to conduct a poll of 80 individuals to know something about a population of 91. Conducting the poll this way requires me to accept a 4% margin of error, when with just a little more effort, I could have completed my census of 91 individuals, and had zero margin of error.

I'll admit, in practice, it is a little silly.

But with the lessons it can teach, "for science," this is not silly. In fact, what this poll does for how polls can be understood is more valuable than any poll Gallup or any pollster in history has ever conducted.

First, "What's for Lunch?" is a real poll that had zero nonresponse, something often dismissed as impossible or purely theoretical. Such a poll could also be conducted on a much larger sample size, and an individual or small group could easily accomplish this in an afternoon.

Second, unlike most political polls, this type of poll can be conducted and replicated by any interested researcher from amateur to expert; replicability is a foundational principle of science.[1] Now, this is not to say political polls are not scientific, most are, but it is to say that if the content of my findings and/or conclusions it assists in drawing are inaccurate, then it should be very easy to show how and why.

Third, possibly most importantly if only for the fact that it's true but hard to understand: refer to the charts presented in Chapter 2 showing the margin of error calculation for polls of different population sizes. As it pertains to what polls are intended to measure, they *mean* the same thing.

The only thing this experiment does is manipulate the sample size and population numbers to be more manageable while maintaining ideal poll standards.

There is no substantive mathematical difference between a poll that contains a random sample of 80 for a population of 91 and a poll that contains a random sample of 600 for a population of millions.[2]

My poll said:

Pack/bring lunch from home: 75% +/– 4%[3]

Order/purchase lunch: 15% +/– 4%[4]

Undecided: 10% +/– 4%[5]

If you insist, you could imagine those options as political candidates, or "support/against" options on a referendum. What this poll said about the lunch preferences of my population is no different from what a political poll says about voter preference.

So let's take the first step in analyzing this plan poll.

Does this poll say that *after lunch on Friday*, the result should fall within this range – with about 95% confidence?

It does not.

I only express what this poll data *doesn't* mean before explaining what it does mean because it's a common misconception perpetrated by experts in the field, and therefore misunderstood by the public and media; it's a mistake that needs to be corrected.

I'll elaborate.

Does this poll say, predict, or in any way assume that approximately 10% of individuals in the population will be undecided about what to have for lunch on Friday … after they have lunch on Friday?

Of course not. When I spell it out this way, people start to see why this "compare the poll to the future result" standard for accuracy is flawed. But unfortunately, the experts the public relies on to provide statistically valid analysis have, in this instance, failed.

The reason this poll does not predict that approximately 10% of people will be undecided about what to have for lunch on Friday, after lunch on Friday, beyond the logical contradiction, is one you hopefully – having read the chapters leading up to this – already understand.

The margin of error given by the sample size can only apply to a simultaneous census. Even if the census isn't quite simultaneous, the margin of error can only be interpreted as applying to each answer, based on their preferences at the time of the poll.

What this poll tells us – like all polls – is that if we were to conduct a simultaneous census, our results should fall within the margin of error as frequently as that confidence interval specifies. It has nothing to do with what they eventually have for lunch. And the difference between our poll's reported results and that simultaneous census is our poll's error.

I didn't conduct "What's for Lunch?" because I was particularly concerned about lunch preferences, believe it or not.

I did it because it was a fun way to illustrate a knowable, testable, and repeatable concept related to polls that, because of the way political polls contaminate our understanding, both experts and the public miss or dismiss as abstract or purely hypothetical.

So here's the big question:

For the data that I collected, with a 100% response rate, what does that 4% margin of error apply to? Does it apply to what they *eventually* decide to do for lunch on Friday?

Or can it only apply to their preferences right now?

Answer:

What this poll says is that if I completed my census simultaneously (whether in this specific case by asking 11 more people, or in other statistically equivalent applications by asking thousands or millions more) then each option should fall within the range given by the random sample 95% of the time.

The poll estimates, with 95% confidence, that if a census were taken simultaneously to the poll, the results would fall within:

Pack/bring lunch from home: 75% +/– 4%[6]

Order/purchase lunch: 15% +/– 4%[7]

Undecided: 10% +/– 4%[8]

And that's the end of what this poll data says, and what it means. Period. End of analysis.

"What do you *plan* to eat for lunch on Friday?" and, after lunch on Friday, "what *did* you eat for lunch?" are not the same question. While they may seem similar *enough* on the surface because they're on the same topic, they are very different.

The same can be said for "Who do you *plan* to vote for?" as measured by a poll, and "Who *did* you vote for?" as measured by election results.

In the plan poll, "who do you plan to vote for?" you can be undecided and/ or change your mind. In the present poll, "who did you vote for?" you can't be.

The margin of error given by a poll also applies to the number of people who say they are undecided. If you take a plan poll, and some people say they are undecided, that *means* if you had taken a simultaneous census, then approximately the same number of people would have also said they are undecided. Ignoring this group, or assuming their eventual preference, is not a valid application of statistics.

Understanding the importance of differentiating between present polls and plan polls is another necessary step for analysts and a teaching point for learners. If the underlying variables in your analysis are not the same (such as, a poll with undecideds and a census without them), then making a direct comparison that fails to calibrate for this variable change is not valid. Here, the term "comparing apples to apples" really should be taken into account.

The margin of error given by the sample size, in addition to only applying if simultaneous, must also be a census: **everyone in the target population must be asked the same question(s) and given the same options**.

Applying this, if we ask one group who they *plan* to vote for, and approximately 10% are undecided, then some time later we ask that same group who *did* they vote for, when undecided is no longer an option, how can we account for the fact that we're asking different questions with different options?

The current methods don't because they treat plan polls like "who do you plan to vote for?" and present polls like "who did you vote for?" by the same standards, even though they have different possible responses.

Before putting forward some possible solutions to this problem, I should first introduce and explain how existing methods try to tackle this "apples to apples" problem: by pretending it isn't one. This will address the field's use of conflicting "reported poll values."

Notes

1. Plesser, H. E. (2018). Reproducibility vs. replicability: A brief history of a confused terminology. *Frontiers in Neuroinformatics, 11*, 76. https://doi.org/10.3389/
   fninf.2017.00076

2. A "finite population correction" is required when the sample size makes up a substantial portion of the population as in "What's for Lunch?". While this is a "mathematical difference," I don't think it could be argued as substantive. The margin of error in these examples is nearly identical, and that is the variable of interest that this experiment intends to isolate.

3. As the proportion of support moves away from 50%, the margin of error becomes smaller. Given that political polls only ever (to my knowledge) report this "50%" proportion as their MOE, I opted to use that value for all options here. I believe it would be unnecessarily complicated and possibly confusing at this point ("everything you need to know ... without oversimplifying") to report more precise MOEs than pollsters do. I would support pollsters reporting more precise margins of error for their poll's values, but in this chapter – and for the rest of the book – I am only trying to offer a way to interpret results as they are currently reported, including for those less mathematically astute.

$z = 1.96, p = 0.75, N = 91, n = 80$

$MOE = 0.849/25.584 \times 100 = 3.317\%$

4. See note 3.

$z = 1.96, p = 0.15, N = 91, n = 80$

$MOE = 0.7/25.584 \times 100 = 2.736\%$

5. See note 3.

$z = 1.96, p = 0.10, N = 91, n = 80$

$MOE = 0.588/25.584 \times 100 = 2.298\%$

6. See note 3.

7. See note 4.

8. See note 5.

9

The Fallacy of Margin (Spread) Analysis

Important Points Checklist

- What the "spread" in a poll refers to
- What the "margin" in a poll refers to – and how it differs from margin of error
- The difference between observations and predictions, and why it matters for poll data
- The assumption made by the Spread Method regarding undecideds' eventual choice
- The value of "isolating the variable of interest" as it relates to testing currently used methods

While the difference between two numbers in a poll is often and rightly called its "margin," I want to avoid the possible confusion of "margin" and "margin of error." I'll instead refer to this by another accepted and commonly used term: **spread**.

The spread is the difference between the top two results in a poll (excluding undecideds).

Let's first adopt the practice of what I'll call the **Spread Method** of poll error analysis, citing FiveThirtyEight for the fact that they're the most influential modern adopters of it (I cannot say for certain where it originated) and analyze my "What's for Lunch?" poll based on its spread.

> In our opinion, the best way to measure a poll's accuracy is to look at its absolute **error** — i.e., the difference between a poll's margin and the actual margin of the election.[1]

What my poll said, to report it accurately,[2] was approximately:

Pack/bring lunch from home: 75% +/− 4%

Order/purchase lunch: 15% +/− 4%

Undecided: 10% +/− 4%

DOI: 10.1201/9781003389903-9

I put the margin of error with each of these results to illustrate what a poll tells us – and tries to tell us.

However, what most US media outlets, analysts, and pollsters *report* this poll says is:

Pack/bring lunch from home: 75% (+60)

Order/purchase lunch: 15%

with "+60" referring to the "spread" between the top two options. Typically, the margin of error is included in a footnote.

Polls are often characterized by how much the leading candidate is "up by," so for this example, you might hear or see it reported simply as "Pack/bring lunch is up 60 points" with no mention of the numbers underlying that calculation.

+60. The difference between the two options is 60. Pack lunch is "up by 60." The school newspaper ran with this. What does this mean for the cafeteria?

Ok, no mediafication here, let's put our "informed consumer" eyes back on. Why don't these reported numbers add up to 100%? What happened to the other 10%? Do people undecided about what they're going to have for lunch on Friday typically just not have lunch on Friday?

These seemingly rhetorical questions might come off as "cheeky," but if I'm nice enough about it, it can contribute to a meaningful discussion and understanding.

The poll data, reported as

Pack/bring lunch 75% (+60)

Order/purchase 15%

ignores the existence of undecideds, and the margin of error. While characterizing the lead as "up by 60" is valid on its face, does this data predict that, on Friday after lunch, the number of people who pack/bring lunch from home will be, or should be, +60?

Believe it or not, most every respected analyst and statistician who conducts poll research would say that, yes, the accuracy of this poll can be rated on how close the eventual result is to "pack/bring lunch" winning by 60. This is literally the definition used by FiveThirtyEight and most American analysts. It isn't a new phenomenon; this is how they've always done it.

In their own words, FiveThirtyEight said this approach "assessed how accurate" polls were by how good they are at "predicting the margin separating the two leading candidates in each race."[3]

This is a systemic problem that is based in fallacious reasoning.

What this poll says, and what every poll ever taken says, is that if you conducted a *simultaneous census of the same population,* the respective results given by the poll would (with 95% confidence) be within the margin of error – with undecideds included. And that's it.

Polls can only make observations and estimates: people make predictions.

According to the Spread Method of calculating poll error, a poll's error is characterized as the difference between the spread of the poll and the spread of the result.

In calling the difference between the spread of the poll and spread of the result an "error," **the *assumption* of spread analysis is that even if a poll were provably perfect, undecideds *must eventually split* evenly to the top two candidates; otherwise, the poll was inaccurate.**

In our 75–15–10 poll, with 10% undecided, the only way to maintain a +60 spread is for the 10% undecided to distribute 5% and 5%, which would give an 80–20 result.

The only way someone could conclude it's the poll's job to predict what "don't knows" will do is if they don't know what the "true value" given by a poll refers to. This is clearly the case right now in this field. Fortunately, we can conduct lots of easily replicated experiments to test the simultaneous census standard if it is for some reason controversial.

If the people who were undecided in my poll – or any poll – eventually choose the same option, or decide in any manner that is contrary to what an analyst or expert assumed they would, that is not relevant to the poll's error, properly defined.

It is not controversial nor is it debatable to say that assuming what undecideds will eventually do is a function of a prediction or predictor, not a poll or pollster.

And prediction error is very different from poll error.

Taking what should be described as a poor assumption (which I've generously called a "prediction error") and blaming "poll error" is *the primary cause* of diminished public confidence in polls because it is based on a misrepresentation and misunderstanding of how polls work.

"Why Polls Were Mostly Wrong"

reads one headline from Scientific American. The article utilizes the spread method to reach that conclusion.[4]

"Here's why political polling is no more than statistical sophistry"

reads another headline, this one from Fortune.[5] The examples cited in the article to support this claim, of which there are many, also refer to the spread of the polls and poll averages compared to the eventual result.

A book much longer than this one could be printed just from headlines of articles describing how "wrong" polls have been based on spread and spread alone.

The media – whether they are themselves analysts or not – have unquestioningly adopted the same flawed assumptions as the experts; after all, why would they have any reason to question them?

The industry's perceived experts (who incorporate their assumptions to the definition poll error) say a poll or "the polls" were "wrong" because the spread of poll(s) didn't match the spread of the result.

In doing so, they're saying in no uncertain terms that *their assumption* about how undecideds should split *cannot possibly be incorrect*. Media outlets themselves propagate this fallacious assumption utilized by the experts, and as such the public is powerless to understand *that* it's wrong – much less how and why.

That's where I hope to help. These things are testable and knowable. Substituting assumptions where we can have data isn't just lazy, it's bad science.

If the eventual result of the "What's for Lunch?" census *is* +60, does that mean the poll had zero error, statistical margin of error or otherwise? Can we know that with any level of certainty?

If the eventual result *is not* +60, does that mean the poll had an error equal to the difference between the spreads? What if the result spread is far greater or smaller than +60, and not explainable by the margin of error alone?

This is the problem with the spread method's definition of poll error, and why it is a fallacy:

If we can say with certainty in easily replicated experiments that the only source of error in a given plan poll is the margin of error – at least, that the "other" variables are negligible – but that method's formula *consistently produces results far greater than the underlying sources of error*; then, those definitions fail to reliably measure poll error.

In other words, if the output of the "Spread Method" says the error is different from the measurable "sum of all sources of error," then it is wrong. This is the problem created when you don't isolate the variable of interest and don't have an ideal standard to compare to: you're left with multiple variables influencing the result, are incapable of saying where the error came from – and can be left with a huge error in your "error" calculations.

To critique the potential error in "What's for Lunch?" by breaking "poll error" down to its component parts, before getting to the results:

What are the sources of error in a poll with a 100% response rate other than the margin of error? Response rate isn't the only factor that can contribute to poll error, but what else could there be? Could they account for a poll-to-election spread discrepancy of 5 or 10?

The fact that this definition of poll error can so easily fail (both theoretically and in practice) but still be used and accepted uncritically by the industry's experts is why you may have noticed my tone ranges from disappointment to astonishment about these issues.

While spread analysis is problematic largely because it is so pervasive, it is not the only industry-used measure of poll error.

Notes

1. Rakich, N. (2023, March 10). *The polls were historically accurate in 2022.* FiveThirtyEight. https://fivethirtyeight.com/features/2022-election-polling-accuracy/
2. See note 3 in Chapter 8 for more precise MOE values.
3. Silver, N. (2010, November 5). *Rasmussen polls were biased and inaccurate; Quinnipiac, Surveyusa performed strongly.* FiveThirtyEight. https://fivethirtyeight.com/features/rasmussen-polls-were-biased-and-inaccurate-quinnipiac-surveyusa-performed-strongly/
4. Dickie, G. (2020, November 13). *Why polls were mostly wrong.* Scientific American. https://www.scientificamerican.com/article/why-polls-were-mostly-wrong/
5. Sonnenfeld, J., & Tian, S. (2022, November 16). *Pollsters got it wrong in 2018, 2020, and 2022. Here's why political polling is no more than statistical sophistry.* Fortune. https://fortune.com/2022/11/16/pollsters-got-it-wrong-2018-2020-elections-statistical-sophistry-accuracy-sonnenfeld-tian/

10

The Fallacy of Proportion Analysis

Important Points Checklist

- The primary method currently used to evaluate polls in the United States and outside of it
- How the "literature" version of the Proportional Method calculates poll error
- How "reported values" differ for the same data in the Spread versus Proportional Methods
- The assumption made by the Proportional Method regarding undecideds
- What the Blended Method is, and its purpose
- The effect of data's proximity on the simultaneous census standard
- How asking questions with different options impacts "error" calculation
- Why polls are done if they're not predictions

While the Spread Method reaches some problematic conclusions from its flawed assumptions, research published in the *Journal of the American Statistical Association* refers to a different method as "standard in the literature."[1] I later learned that this method is often used to report and evaluate poll data outside of the United States.

That method, which I'll call the **Proportional Method**, just like the Spread Method considers only the two principal options or "two-party" in reference to politics and disregards undecideds and any other or "third-party" options in calculating the "poll accuracy" number.

The proportion, reported as a percentage, is calculated by subtracting the undecided and "third-party" options from 100 to create a new denominator, and using each of the "two-party" options as the numerator.

To use the "What's for Lunch?" results as an example, the Proportional Method would not report the respective responses as 75%, 15%, and 10%, nor even 75%–15%.

Before reporting any results, this method would subtract the 10% undecided and create a new denominator: 100 minus 10, so 90.

Then, for each of the two poll numbers, find the percentage out of 90:

$$75/90 = 83.33\%$$

$$15/90 = 16.66\%$$

This result, depending on the rounding preferences of the source, would be reported as something like:

Pack/Bring Lunch: 83%

Order/Purchase Lunch: 17%

Here is where things start to get really bad – both from a data literacy perspective and for the public.

The Proportional Method – which is very different from the Spread Method in what it asserts is the poll's reported value – also often reports results with the spread!

That's right. Depending on the method of choice for that pollster, analyst, media outlet, or as established by your country's tradition – the same results can be reported in different ways.

The *same exact poll data* could be reported via the Spread Method as:

Pack/Bring Lunch 75% (+60)

Order/Purchase Lunch 15%

Or via the Proportional Method as:

Pack/Bring Lunch 83% (+66)

Order/Purchase Lunch 17%

Think ahead: For readers who are already astute in margin of error calculations, consider how (and if) the margin of error given by the sample size applies when results are reported this way.

If you see the problem with how polls are characterized and mischaracterized – whether in regard to error or the plain reported numbers – then we're getting somewhere.

Regarding the Proportional Method's characterization of poll error, like "proportional" implies, it says that the **result** proportion of the "two-party" number should match the **poll** proportion of "two-party" number. It defines the difference between one candidate's poll proportion and their result proportion as the poll error.

Properly reported,[2] my poll said:

Pack/bring lunch from home: 75% +/− 4%

Order/purchase lunch: 15% +/− 4%

Undecided: 10% +/− 4%

But via the Proportional Method, many analysts would report that my poll said:

Pack/bring lunch from home: 83%

Order/purchase lunch: 17%

By the Proportional Method, even if the poll had zero error, the only way for the 75–15–10 poll to end in a 83–17 result is for undecideds to distribute proportionally to the decideds (8.3–1.7, or approximately 8–2).

The assumption of proportion analysis is that even if a poll were perfect, undecideds *must* split proportionally to decideds, or there was a poll error.

As you can see, there is direct and irreconcilable conflict between the two methods for calculating poll error. If the eventual outcome of my lunch poll were 80–20, the *Spread Method* would classify the poll as having zero error (+60 spread in poll and +60 spread in result), while the Proportional Method would say it had a three-point error (leader proportion was 83% in poll vs. 80% in result).

On the other hand, if the eventual outcome were 85–15, both would report the poll as having an error. While the Proportional Method would characterize it as a two-point error (leader with 83% in poll vs. 85% in result), the Spread Method would characterize it as a catastrophic 10-point error (+60 spread in poll vs. +70 spread in result).

Yes, if the poll had a spread of 60 and the result has a spread of 70, the Spread Method would say there was an enormous 10-point error. From whence came this error? Could it be that the poll was accurate and the undecideds just strongly favored one option? No, it's the *poll* that must have had an error.

So which one is better, you might ask? Neither! Wrong is wrong. Both methods make assumptions and assert that if their assumptions are not matched, then the poll – not their assumptions – must be wrong.

The Blended Method: To Illustrate the Unfounded Assumptions Used by Experts

If I invented a definition of poll error called the **Blended Method** that called the average of the Spread Method and Proportional Method the Totally Real

Definition of poll error, it would be just as valid, even though I just made it up. There is zero scientific or statistical superiority to any established method over the Blended Method – because they're unscientific calculations that are based on a misunderstanding of what polls do.

The only reasonable argument I have encountered regarding these varying definitions of "poll error" is that they may have useful applications as it pertains to building forecasts – and maybe they do. And maybe my Blended Method does. It doesn't matter.

When it comes to providing an accurate definition of poll error and measuring poll accuracy, neither of these commonly accepted definitions (nor the one I just made up) are valid. Polls are not predictions, no matter how much we want them to be.

All of these definitions are wrong and problematic in their own ways. Regardless of which is closer to the eventual result in the "What's for Lunch?" example (or for every poll and election ever conducted), there is only one accurate definition for poll error. And it requires a comparison with a simultaneous census.

Again, what this poll question *tells us* is that if I conducted a ***simultaneous census*** with my poll, (with 95% confidence) the results[3] would be approximately as follows:

Pack/bring lunch from home: 75% +/– 4%

Order/purchase lunch: 15% +/– 4%

Undecided: 10% +/– 4%

That's it.

What does this poll tell us about the future preference of the 10% undecided? Nothing. It only tells us that they're *currently* undecided, and there are approximately 10% of them. If all of the people who are currently undecided end up packing lunch, if none of them end up packing lunch, and if they split 50/50 or 80/20, the poll doesn't tell us – it doesn't *try* to tell us.

To say a poll was wrong because it didn't tell us what the undecideds would eventually do is as illogical as … saying a poll was wrong because it didn't tell us what undecideds would eventually do. I'm out of analogies here, friends. We've reached the ground floor.

That's not to say polls can't be wrong. It's to say that **if we call a poll or "the polls" wrong, we must do so for valid reasons.**

Regarding the issue of poll error, all that I hope to establish – assisted by the understanding of a simultaneous census – is that not only *can't* a poll predict what (or who) undecideds will ultimately choose: it doesn't *try* to.

Is this simultaneous census definition of poll error that says you must ask the same question with the same options to the same population at the same time too strict?

No. It's correct. Regarding the application of what polls do, the data they provide, up to and including the maths underlying the statistical margin of error itself: this is what polls tell us.

Now, if you consider only human subjects and the logical near impossibility of conducting a census literally simultaneously with a poll, you might contend this standard, while correct, doesn't have much real-world application. For one, it does still have real-world application regarding understanding how polls work – more on that later – but for the sake of detail, let's talk about application.

As discussed earlier, the "households with dogs" poll could differ from a census taken some time later for reasons unrelated to poll accuracy. But in a present poll, those differences are probably small in a timeframe of weeks or even months: the options (yes/no, in this case) do not change.

Nonetheless, the following still applies: **the closer a census is to simultaneous, the more precise your measurement of poll error can be.**

That being said, if the only problem with the current, accepted, and expert-used definitions of poll error was that the census is taken a few days or weeks after the poll, while useful to understand that calculation is imprecise, I wouldn't go writing a book about it.

The problem is that the current definitions say that the election – where a group is asked **a totally different question** from the poll – is the "census" which polls should be measured against. And that the difference between them is equal to the poll's error, pretending their assumptions can't be wrong. So here we are.

The traditional, still industry standard of poll analysis, plainly does not recognize "who do you plan to vote for?" and "who did you vote for?" as different questions. There are major logical and statistical problems created by not understanding this difference, some of which I've already presented, others which I'll elaborate on ahead.

In order to call the difference between a poll and census an error (accepting the fact that the census may not be simultaneous, but a reasonably close time-frame provides a very good approximation), you have to ask both groups the same question, with the same options.

Here is an easy application of the "different options ... can't claim any difference is an error" concept:

Poll a random sample of students from a population of a class or classes, appropriate for your desired margin of error.

Ask them: Do you like the color blue, red, or green better? Record the results.

Then take a *census* of the class(es) and ask them if they prefer blue or red: **no more "green" as an option** in the census, even though it was an option in the poll.

To use a simple example, say the class is evenly split between blue and red – 40% say blue is their favorite color of the choices given and 40% say red – with the remaining 20% citing green.

Can we say the error of this poll is measured by how closely the census – which doesn't have "green" as an option – comes to being 50% and 50% for blue and red?

Both the Spread and Proportional Methods would say, "yes."

The correct answer, however, is "no."

If you have students, ask them to identify why. Most students will provide both the correct answer and valid reasoning – with many recognizing the faulty assumption made.

The two questions had different options; they're measuring different things. Just because the questions are related doesn't mean we can pretend they're the same.

"Green" is just as valid of an answer in this poll as "I don't know yet" is in a political one.

All this example does is take what is inherent to a plan poll (can be undecided before some date, can't be undecided after it) and apply the same concept to a present poll: removing an option many people may have chosen.

If I had included such an experiment here, and called the blue-to-red spread or proportion as equal to the poll's error, inserting my own assumption about how "greens" must have decided, it would have been immediately flagged by reviewers of this book as statistically invalid (or perhaps illogical, wrong, or some less nice adjective).

Yet, this is how poll analysts are allowed to proceed uncritically in the treatment of undecideds.

I suspect you'll find that this "color preference" experiment will produce results "outside the margin of error" quite often, if you elect for the Spread, Proportional, or Blended Method.

But by the Simultaneous Census standard, it will fall within the margin of error exactly as often as it should.

My shouldn't-be-controversial statement is that a poll including green as an option cannot be compared "apples to apples" to a census that *doesn't* include green as an option and have the difference in the spread or proportion of the "primary two" options called the poll's error.

Well, not if you want your calculations to be valid.

The poll tells you nothing about the *eventual preferences* of people who preferred green in the initial poll. The poll doesn't *try to tell you* anything about the eventual preferences of people who preferred green in the initial poll.

Green was an option, and they chose green. The fact that green would eventually no longer be an option (whether they know it or not) doesn't change the simultaneous census application.

Figuring out the percentage of green-preferers who will eventually prefer blue or red if/when green is no longer an option is a wonderful thing to study, if you're interested in informing a prediction about your population's color preferences.

But no matter what you *think* people who prefer green will choose when green is no longer an option, that is still *your* assumption, not the poll's. That assumption introduces a source of error that the poll didn't have, and one that is not inherent to the poll instrument.

Similarly, whether you are team "50/50" or team "proportional" as it relates to the best assumption for undecideds – or team "blended" – that's still *your* assumption, not the poll's.

By inserting their own assumptions (or "traditionally accepted" assumptions) into their definitions of poll error, experts, and by extension the media that report it, have misguided people into believing polls try to do something that they don't.

And this is where things, I hope, start to come together if they haven't already.

Just because the "spread" or "proportion" of blue-red preferers matches the result doesn't mean there was *no* poll error. And just because the spread or proportion doesn't match doesn't mean there *was*.

Want to analyze the poll's sample to see if it was truly random? Wonderful. Want to scrutinize the poll's report to make sure they properly weighted their reported results for gender, declared major, or whatever other variables might impact someone's color preference? Great. Don't care much about these things? Neither do most people. That's okay, leave it to the experts.

The research regarding what can or has caused actual poll error matters a great deal, and improving methods is both good and necessary. But improving methods requires properly defining terms and understanding what polls try to measure in the first place.

Criticizing polls for not agreeing with the eventual result in some way, asserting that *this alone* is evidence of their "wrongness," is not just evidence that experts don't understand how polls work, it is proof. No one who understands what the data given by a poll means would say or do this.

The attempt to characterize poll error, even approximately, based on the results from different questions with different options causes experts to make assertions regarding poll error that are not statistically defensible. Questions that should be straightforward, such as, "what do polls tell us?" aren't just not being addressed by the experts in the field, *they're being actively undermined*.

I'll preface this by reminding everyone that this is not personal. That being said:

The way perceived experts talk about polls and poll error today is having the same effect as the tobacco company's hired doctors in the 1930s. The public is being misinformed. The "error bars" the public assigns to these things – "do polls provide reasonable estimates?" like "does smoking cause lung cancer?" are being widened. Unlike the 1930s, I do not believe the entirety of the misinformation can be attributed to financial incentives, though that might be part of it; I think it's more likely that most experts have just never bothered to be critical enough of any "traditionally accepted" definitions, even if they're easily shown to be wrong.

Beyond the public, this misinformation undermines the progress that could be made academically related to how polls should be conducted, and how to improve methods. Even the ability to better approximate poll error would be an enormous step in the right direction.

If your poll must allow for undecideds, and you want to calculate poll error, your census must also allow for undecideds. If your census does not have (or allow for) undecideds, you must account for them. To say anything else, as both existing definitions of "poll error" do, is a fundamental misunderstanding of how polls work.

So why do polls, if not to make a prediction?

This is the most common question I'm asked whether I'm "teaching" or "preaching" about the need to stop treating polls as predictions. And it's a great one.

As discussed in earlier chapters – and I think we agreed – if tasked with predicting what percentage of people would pack/bring lunch from home versus what percentage would order/buy lunch, or the color preferences of a class, or if a household has a dog, it's much better to have some poll data to inform the prediction than just guess.

The key word here: *inform* the prediction.

In the present poll example "do you have a dog?", if my reported poll results say that approximately 50% of people responded "yes," then I can very reasonably use 50% as my prediction for the whole population. The poll and census questions are the same – and the options, yes or no, are the same.

When I am asking a "plan poll" question, like the "What's for Lunch?" poll, and as happens in political polls, it's something very different. There's a really big step between poll and prediction. That "big step" is related to how much time there will be between the poll and event, and how many undecideds there are in the plan poll. The more time there is, and the more undecideds there are, the bigger the metaphorical step that has to be taken between poll and prediction.

In plan polls, properly reported, the two main options almost never add up to 100%. Even in plan polls with more than two options, all of the "decided" options combined rarely add up to 100%.

But a prediction *must* add up to 100% because the eventual result will always equal 100%. That means the *predictor must* account for the undecided.

Can the predictor say they think the undecideds will divide 50/50 between the pack and buy lunch options? Sure!

Can the predictor say they think the undecideds will split 80/20, proportional to the decided members of the poll? Entirely reasonable.

But a prediction requires – at the absolute minimum – making an assumption regarding the undecideds, so the numbers add up to 100%, *and* making an assumption about how many people, if any, will change their minds between the poll and result. If those assumptions are ultimately incorrect, that does not mean the poll was or must have been *wrong*.

It means the assumption, prediction, or forecast was wrong.

Another "thinking ahead" challenge for poll and stats veterans is as follows: can you think of any way to test whether an assumption about what undecideds would eventually do was right or wrong?

The lunch poll – and any plan poll, including political ones – do not predict, *nor attempt to predict*, the eventual preference of undecideds.

This is a limitation of polls as a tool. Understanding the limitations of a scientific or statistical instrument such as a poll must be considered when calculating and assigning error.

The current, accepted, and used definitions of poll error – one assuming undecideds should split evenly to the major candidates, the other assuming undecideds should split proportionally to the decideds – are not describing poll error, as they purport to. These are assumptions whose discrepancy can only be considered a *prediction error, forecast error,* or simply a bad assumption.

They're not just wrong in the sense that there are a few days or weeks between the poll and census; we can accept that as a close enough approximation, usually (though I've never encountered a serious researcher in any field, especially statistics, who reached the conclusion "this is close enough, usually" and didn't try to improve it).

Most egregiously, they're wrong in the sense that their assumption regarding what undecideds should do – could, *by itself,* cause them to say *a poll* is "wrong" when that measure of "wrongness" has nothing to do with the data given or reported by the poll.

Notes

1. Shirani-Mehr, H., Rothschild, D., Goel, S., & Gelman, A. (2018). Disentangling bias and variance in election polls. *Journal of the American Statistical Association, 113*(522), 607–614. https://doi.org/10.1080/01621459.2018.1448823
2. See note 3 in Chapter 8 for more precise MOE values.
3. See note 3 in Chapter 8 for more precise MOE values.

11

Instrument Error: Weighted Results and Literal Weight

Important Points Checklist

- How polls and scales are related
- The difference between "user error" and "instrument error"
- Observations versus predictions and the consequences of blending them
- The meaning of "herding" and why it exists
- The role "changing attitudes" can have on poll data, and why it matters

Ideal polls, check. Simultaneous census, check. Plan polls versus present polls, and how the existing methods evaluate poll accuracy, check. The next step is to consider the role pollsters themselves play in the data collection process, factors they can and can't weight for, how to measure the accuracy of a tool when the thing it's measuring will almost certainly change over time, and to consider the limitations of your tool or instrument in research.

Polls are a statistical tool with a specific and valuable purpose: to provide a good estimate about a population without needing to ask that entire population.

Just like if you're using a scale to measure someone's weight, or a stopwatch to measure how much time it takes for someone to run some distance, you should understand the limitations of any tool you use and incorporate that into your analysis, and characterization of that tool's "error." A person who blames a tool for producing an inaccurate result when some of that inaccuracy – perhaps very much of it – has nothing to do with the tool, is a problem that should raise concerns regarding the reliability of that person's analysis.

There are a number of factors that *can,* as with a poll, cause a scale or stopwatch to produce inaccurate values. Some of those are rightly classified as instrument errors, some of them are not.

DOI: 10.1201/9781003389903-11

In the case of a scale, it doesn't matter if it's mechanical or digital, or whether your preferred mass unit is pounds, kilograms, or stones, scales have a specific purpose: measure how much force is exerted on it. That's it. When someone stands on a scale, that force is converted to a number – a weight.

If you jump on a scale, you might get an inaccurate reading compared to your true body weight: not the scale's fault. If you place a scale on a soft surface, which absorbs some of the force where you stand, you might get an inaccurate reading compared to your true body weight: also, not the scale's fault.

If you stand on the edge of a scale, or put all your weight on one side, you might get an inaccurate reading compared to your true body weight: not the scale's fault.

I actually tested some of these factors, such that my home scale produced results off by as little as 5% (standing on the edge) and as much as 60% (placing the scale on carpet) compared to the reading it gave taken on a hard surface just seconds earlier.

> Is it a **"scale error"** because the scale produced results "off by" up to 60%? No. If you use an instrument – or analyze the data it gives – in a way that it is not intended to be used, the kindest description of that would be "user error."

This is why it is important to specify what an ideal poll is, so that you understand how that tool functions in an ideal environment – and can properly calibrate for nonideal ones.

You might find me a bit unqualified to comment on a scale's accuracy if I placed a scale on carpet, it gave a reading far off my true body weight, and I said, "someone needs to *fix* this scale."

If I weighed myself while standing on a scale I had first placed on carpet and compared that to the measurement taken at my doctor's office a week later, it would be unreasonable – though quite funny – to conclude that the difference between "home scale" and "doctor scale" is entirely attributable to "scale error."

Scales are a tool not intended to be used on carpet, and the measurement was taken on carpet. If you assign the entirety of the discrepancy between the two values as "scale error," then it's reasonable to conclude that you either don't know how scales are supposed to work (e.g. not on carpet) or simply don't care to account for what should be, to anyone serious about finding how "wrong" the scale was, an obvious factor in the overall discrepancy.

Whether placed on carpet or concrete, a scale can be wrong. But how much of the discrepancy is attributable to misusing the tool and how much is attributable to the scale's overestimation or underestimation of the true weight are separate factors.

To say its error is equal to the difference between home result and doctor result *alone* isn't reflective of the tool's accuracy.

In order to accurately determine how much error a tool had, you must be able to account for the factors that do (and don't) contribute to any observed discrepancy between measurement and eventual result.

Figuring out whether a 2-, 20-, or 120-pound measured weight difference between "home scale" and "doctor scale" is entirely attributable to scale error will help us better understand poll error.

So far, I've mostly talked about ways the scale can produce results that aren't reflective of my true weight, if I did not understand how a scale is supposed to be used. But there's more than that.

Consider also: is it possible to gain or lose a few pounds in a week, just through diet and/or hydration status? Is your weight changing over the course of a week a "scale error"? Because we *understand* what scales try to tell us, we wouldn't assume or assert that our scale giving different readings a day or week or even a few hours apart is entirely attributable to the scale's error.

To put it all together, a good analyst must account for *all* the factors that could cause the observation given by an instrument to differ from some future result – both in how close to an ideal environment that instrument was used, and if what it tries to measure could change over time.

A scale can't control for the fact that someone used it incorrectly, or that weight can change in a week. It's just a scale.

And now we're getting into how to properly analyze what a *tool* tells us, calculate that tool's error and the importance of accounting for all the variables that truly contribute to *instrument error.*

Let's Weigh in

Imagine there's a huge financial (or bragging rights, if you prefer) incentive in predicting what my weight will be when I visit the doctor next week. Don't worry, you won't hurt my feelings, I want you to win.

Would you rather just look at me and guess, or use the results from multiple scales to inform your prediction? Okay, that's too easy.

But there's a catch.

There will be three different scales, each flawed in some way: some of those ways are known, and others are unknown.

- **Scale one** is placed on a hard, flat surface 10 days in advance of my appointment, but I try to put all my weight on my left leg.
- **Scale two** is placed on carpet six days in advance. I stand on it normally.
- **Scale three** is placed on carpet three days in advance, and I try to put all my weight on my left leg.

Anything else you'd like to know to inform your prediction? Well, these scale companies have some incentive to get my weight right too.

How about some *weighted* (ha) data?

The company that runs "scale one," I'll call them Scalester 1, noticed a disproportionate oversampling coming from the "left" and believe they have corrected for this in their reported, weighted data.

Scalester 2 knows their methodology that puts the scale on carpet is flawed, but they've always done it that way and don't want to risk changing methods. They have a good idea of how to weight for this in their reported data.

Scalester 3 is new but has learned from Scalesters 1 and 2 and applies that knowledge to correct for both known flaws in their reported number.

Question One: Is it fair to judge a company's results based on their reported, weighted data, even if we know their data collection method was imperfect?

Yes, that's fair. Imperfect methods, whether from an imperfect sampling procedure or an imperfect scale reading, are typically inevitable. Whether due to weighing yourself on carpet or accounting for nonresponse, how a company weights their results, and how much they weight different factors is up to them. For that reason, as long as the company is transparent with their methodology (e.g. not just guessing), it's entirely fair to judge a company's error (scale or poll) based on their **reported result**.

And now to get into the good stuff.

Question Two: Should a company attempt to weight their results to account for the time between measurement and result?

That is, should "Scalester 1" weight their results to account for the fact that my doctor's appointment isn't for another 10 days, and report what they think my weight will be in 10 days? Should "Scalester 3," just 3 days before my appointment, try to account for this?

Or should they just report their observation?

This question gets right to the core of how instrument error is measured versus how it should be measured, and another of many issues within this field that has been allowed to survive uncritically.

Knowing that the underlying methods used to generate the measurement may be flawed in some way, it's considered not only standard but *critical* to calculate and report weighted results, and rightfully so. Just like in the scale examples, we may not be able to say *exactly* how much our raw (unweighted) number is "off," but we can say with a high level of certainty that they are off to some extent. No problem.

Everyone would agree, I hope, that the goal of any measurement – scale or poll – is to produce a result as accurate as possible. The scale reading that is produced with the scale sitting on carpet is highly unlikely to be accurate.

But accurate *compared to what?*

The standard for accuracy for any tool, the standard for "error" of any instrument, as I've outlined throughout and a concept that desperately needs to be taught and reinforced – in both the public and among experts – is the simultaneous census. Or in this case, a simultaneous weighing.

When I talk about scale data, it seems obvious; with polls, it's clearly not.

The only way to measure (with ~100% accuracy) the various scale accuracies would be to weigh myself using the company's methodology of choice, with them reporting their "weighted" result, **simultaneously** to me getting weighed at the doctor's office.

Having accepted the (for now) logical impossibility of such a test, I think it's fair to say the difference between the result reported by the "Scalesters" with their respective methodologies and weights, compared to being weighed at the doctor within a few seconds, minutes, or hours, is close enough to the actual error. Not much time for my weight to change, in that instance.

But I'm not getting weighed within seconds, minutes, or hours. The "result" won't be recorded for days. The Scalester's accuracy will be judged largely – if not solely – on their closeness to the *eventual result*. So, should the companies attempt to weight their reported results for how much time is left to the event of interest?

And here it is:

By utilizing definitions that say the difference between reported result and eventual result is entirely attributable to **instrument error**, the analysts, statisticians, and experts are saying: yes, as pollsters and Scalesters, you **should** weight your results to try and account for the time between your observation and the eventual event.

See a problem with that?

Being boxed into "traditional" definitions that are flawed in various ways, the field's experts are asking pollsters whose tools are designed only to observe things to instead provide a forecast. Not to **collect data** and tell us what it says about "right now" – which is the only thing polls and scales as tools are intended tell us – but to **predict** what that data will tell us after some time.

The distinction between observation and forecast is becoming nonexistent.

Consequently, most non-US pollsters have adopted variations of the Proportional Method in which they assume, guess, or forecast what they think undecideds will eventually do – and then report that data which includes those assumptions **as if it were the poll**.

It's unclear if this "blending" of poll and forecast has started in the United States.

But back to the Scalesters, put yourselves in the shoes of Scalester 3 who conducted a test three days before my doctor's appointment. Their methodology was flawed in two ways: the scale was placed on carpet *and* I tried to put all my weight on my left leg.

They have to apply weights based on multiple variables to figure out my "true weight." But they know that. They're newcomers to the Scalester world

Reported Weight Over Time

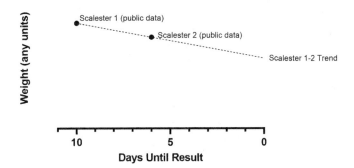

FIGURE 11.1
A trend showing the public results from Scalesters 1 and 2 along with a trend leading up to the result "weigh-in." Chart by the author.

but are very good with data; they're confident they can do better than existing methods.

But being a new Scalester, this report could make or break their reputation for years to come, maybe forever.

They do have one huge advantage, however: they're the closest to the event. By nature of being closest to the recording of "eventual result," they can minimize the impact of my possible weight fluctuation between measurement and result.

But if you know how your company's error will be judged, that's not your only advantage. They also have access to previous pollster's reported results.

My weight 10 days ago (as reported by Scalester 1) compared to my weight 6 days ago (as reported by Scalester 2) showed a downward trend, as noted in Figure 11.1.

So Scalester 3, weighing me 3 days later – would not be surprised to see that trend followed, become flat, or even go the opposite way. They're human and recognize patterns, but also understand that 2 or 3 readings is hardly a trend. They've read my book and understand that **fluctuation is normal and expected** (in both body weight and polls).

While they might personally have guesses about what my weight will do, they're good pollsters, and only want to report what their data says.

And then they weighted their data and noticed something they couldn't dismiss. Not only had my weight loss trend continued, it sped up. Clearly, like a politician running for office, I had a goal: I wanted to have the best possible performance.

The top dotted line in Figure 11.2 represents the original "trend" based on Scalester 1 and Scalester 2's public data, while the lower line represents the new "trend" accessible only to Scalester 3, including their not-yet-public data.

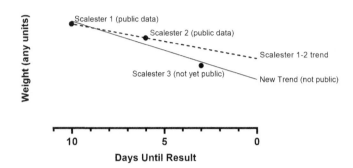

FIGURE 11.2
A chart showing Scalester 3's reported results along with a new trendline, along with the original trendline from Scalesters 1 and 2. Chart by the author.

My final weigh-in, the result against which each Scalester's accuracy will be judged, will be taken three days later.

So, not yet having released their results to the public, with three more days before my result "weigh-in," should Scalester 3 simply report their findings based on the data they just took, or should they put their metaphorical thumb on the scale?

Their result doesn't line up with the trend from the other Scalesters – should they "nudge" their number closer to the Scalester 1–2 trendline for fear of having an outlier result and being derided as "wrong?"

Or, on the other side, do they have enough confidence in the trend of their new findings to "nudge" it closer to the new trendline, to get lots of credit if it ends up being "right?"

Or should they just report the data as they observed it?

The industry of "poll error" analysts and experts tells them they'll be judged based on how closely their reported results match the *eventual results*. So I'll leave it up to you to decide what Scalester 3 should do, and if this scenario of me getting weighed has any application to pollsters in real life – and if this internal conflict might be present in the poll industry today as elections approach.

What Would You Do?

Here's my take, for those interested, and the only "just for the experts" section I'll be writing. This is for pollsters, analysts, statisticians, and people doing work in this field:

Given constantly expanding access to forecasts, prediction markets, and easy access to results of other pollsters, there's a systemic risk of poll data becoming less and less independent, to the point that having 20 polls that appear independent might be no better than having two that are actually independent. Nate Silver said he had proof of this **herding** phenomenon, in his article "Here's Proof Some Pollsters Are Putting A Thumb On The Scale" in which pollsters report numbers closer to other pollsters for fear of being "wrong" as early as 2014.[1]

Similarities in the "thumb on the scale" idiom are coincidental, though I believe mine is funnier.

He and others at FiveThirtyEight clearly did some great work to identify and quantify this "herding" effect, and more work has been done on it since.

To the analysts and forecasters who understandably worry about what this might do to the quality and reliability of poll data in the future: look in the mirror. Think about what role *you* may be playing in this troublesome (and potentially, for the poll data industry, catastrophic) trend. Critique and condemn pollsters for the right reasons, work with them to fix methodology and weights as needed … and fix *your* definitions of poll error to better measure actual error.

Personally, I gained an entirely new level of respect for otherwise reputable pollsters with a long history who report outlier results knowing it could impact their reputation. I'm probably *more*, not less, inclined to trust such a pollster in the future – even if their reported number was farther from the actual result than others. And if you're thinking, "well yes, based on their methodology, reputation …"

Consider if this pollster were new and *had no reputation*. This one election will be their only reputation for at least the next several years, and a bad "miss" could take decades to correct. What's their incentive?

They have incentive to herd. You give it to them.

In a panel where Jonathan Brown, founder of Sextant Strategies & Research, was asked about the herding effect and whether pollsters are concerned about their reported data not being true to what they had collected, he said, "It's better to be wrong with everyone."[2]

That's the state of the poll industry today, and it's not just the pollsters who are to blame. They know how their poll accuracy will be judged, and regardless of what their actual observations say, many may conclude, "it's better to be wrong with everyone."

A savvy person with a budget who wants to make a name for themselves in polling and/or forecasting could conduct polls of varying quality, apply weights of varying validity – with the explicit goal of trying to predict the actual result – and according to the existing definitions of "poll error" that are currently used, *they would be entirely justified in doing so.*

The best pollsters wouldn't be the ones who had the best methods, collected the best data, or even applied the best weighting techniques. It would be those who put their thumbs on the scale in the way the experts say they should and tried to build the best forecasts.

That's the future (if not present) of the polling industry unless the definitions of "error" are fixed.

Back to the content.

Changing Minds

You may have noticed my Scalester example. While it mostly focused on weight (mine) and weights (adjustments to raw data), there was the added caveat that my weight was changing over time. How can real-life pollsters adjust for that? Should they?

> The questions "what does this scale say my weight is?" versus "what does this scale say my weight will be in a week?" are quite different. One reflects what the tool actually tries to tell us, while the other is projecting our desires onto the tool.

You can ask the scale whatever you want, but it's only designed to give you one answer. No matter how nicely you ask, the absolute best it can do is give your weight right now.

Polls, in that sense, are no different.

You can ask someone who they're going to vote for (or what they're going to have for lunch). The best they can possibly do is give you the answer that is true for them right now. The highest standard we can possibly have of scales, or polls, is that they give us an accurate reading *right now*.

Introduce humans to the mix – humans whose job it is to read or calculate what the tools tell us – you get an entirely different problem.

If you start imposing the *desire to know the future* on a tool that doesn't try to (and can't) tell you the future – and confuse your misplaced desire to know the future for a "scale error" or "poll error" – then your understanding of how that tool works is insufficient, and your calculations of "error" will range from mildly inaccurate to catastrophically wrong.

> Being unable to predict the future is not a weakness of any of these tools. Nor is it a feature. It's just a reality. The definitions should be fixed to agree with reality.

We have tools that, at their very best, can give us a reasonable estimate about something right now, and humans want to use them to predict the future. This is not as big of a problem as it might sound; these tools are wonderful for helping us inform our predictions! So much so, I dedicated part of my book to this fact. But unfortunately, for pollsters, it's not that simple.

One large group of people, pollsters, uses a tool to collect and report data. Then at least two other groups created at least two different sets of rules (the Spread Method and the Proportional Method) that impose their own perception on what your "reported data" was, compared to the eventual "actual result" (which they also have different methods of calculating), and this is the standard for accuracy against which your tools are judged.

If that sounds highly unscientific, that's because it is.

And it just so happens that the group collecting the data knows the (unscientific) criteria on which their reported data will be judged, thus creating the vicious dilemma illustrated in the Scalester example: report the data as it is, or as you think it will be? Or, to borrow from an earlier expression I used to express my dismay about inaccurate, made-up definitions – try and blend the two?

Judging pollsters based on their closeness to the eventual result creates an overwhelming incentive to produce not poll data but blended poll-predictions (polldictions, or pollcasts), and this directly undermines the reliability *both* of those reports and threatens the reliability and trustworthiness of the industry as a whole.

It's up to analysts, statisticians, and experts to fix their definitions and calculations. Once pollsters understand we just want your observations, not your predictions, we might be able to correct the direction the political poll industry is going, if it isn't too far gone already.

The best possible data a *plan poll* can give, since we're trying to know about a *future* result, is data to *inform a prediction*. How we use that data – how many undecideds there are and what they'll eventually decide, and even how many might eventually **change their minds** – is a function of an assumption, prediction, or a forecast, *not a poll*.

In the "What's for Lunch?" poll, if someone responded that they planned to purchase/order their lunch – because that's what they usually do – but (a couple of days later) happened to have leftovers from a large meal the night before, and packed that for lunch, thus contradicting their original response – was the poll wrong? The current definitions of poll error would say, illogically, yes – a person changing their mind is a poll error; the poll was wrong.

It's not as though mind-changing, sometimes referred to as "**changing attitudes,**" is not a variable unknown or of no concern to researchers. By only using polls very close to elections in their calculations, they consider changing attitudes to be of minimal impact. Fair enough. It's probably rather rare that someone changes their mind about who to vote for, or what they'll have for lunch that same week.

Likewise, it's probably rather rare that someone's weight changes in a week. So?

Something being "rather rare" does not justify saying we don't need to account for it in our calculation of error, ever. It also doesn't mean that small changes can be assumed as zero.

If I happen to get very ill the week leading up to my doctor's appointment, it's fair to say my weight might change more than would otherwise be expected, and not negligibly. Even if I'm just on a healthy diet-and-exercise plan, my weight could change in a short period of time.

If my weight changed in a week, according to the current definition of instrument error, the scale reading a week ago was wrong, and therefore, that scale probably needs fixed – or the Scalester needs to fix their methods. If someone changed their mind about what they'll have for lunch on Friday, according to the current definition, the poll taken a few days earlier was wrong. What was wrong about it? What is there to fix?

The misplaced assumption that asking a tool to tell you the future, if it fails to do so, means the *tool* was wrong has more ramifications than you might think. And this simple, illogical starting point is the first domino of how we get to where we are today.

It seems silly to think about standing on a scale and asking it "what will I weigh next week?" and calling the difference between your weight today and next week a scale error; you asked it about next week, after all. **Yet that's precisely the current standard experts use to analyze polls**.

It doesn't matter if you ask people what they plan to do tomorrow, or next week, or next month: the best they can do is answer about their plans right now.

If in the last week before an election, a political candidate is implicated in a major scandal, it seems fair to consider that in this instance, *just maybe*, the effect of changing attitudes isn't negligible, and certainly not "zero."

But going by either currently used definition:

We must assume the number of people changing their minds close to an election will be 0.

If that is later shown to be untrue, do the analysts concede there was an assumption error on their part?

Nope, still the *polls* that were wrong. Poll error. In addition to being incorrect, that characterization is harmful for public understanding of polls.

As you understand, there are plenty of *real* reasons polls *can* be wrong, but disagreeing with someone's assumptions, even an expert's assumptions, *isn't one of them*. This is why I advocate for a "show your work" approach. If changing attitudes can impact the difference between poll and result, you should account for that.

As you also understand, changing attitudes is not the only step between poll to result. There's another, usually much bigger step: **account for undecideds**.

The Spread Method and Proportional Method's respective "poll error" definitions have very different philosophies regarding how they account for what undecideds "should" do, but both philosophies are flawed for the same reasons. In these methods, they're projecting their *desire* to predict something about the future onto polls – something polls can't and don't try to do.

You can apply whatever statistical measures of accuracy that you want: Root Mean Square Error, Mean Absolute Error, Blended Root Mean Square

Absolute Error, it doesn't matter. If your inputs are based on flawed assumptions, your analysis will still be just as wrong as any other.

The ability to input numbers into a formula doesn't matter if you don't know what the output means.

Both the Spread and Proportional Methods make assumptions they assume can't be wrong, and they approach the problem backwards.

Instead of assuming error is *unknown* and trying to quantify it, analysts assume error is *known* (by their respective Spread Method and Proportional Method definitions) and say the sum "total" of errors must somehow fit that number.

Regardless of your feelings toward poll reliability in general, polls are by far the best tool we have for informing predictions; polls will always be superior to anyone's best guess. But polls are not themselves predictions, as the current definitions of poll error assume.

If you can't separate the errors of your assumptions from actual poll errors, it's time to start over.

In the process of calculating poll error, by making an assumption about changing attitudes *and* what they think the poll says undecideds "should" do, the existing definitions are not comparing a poll to a simultaneous census, nor poll to census, nor "observation" to "true value."

The Spread Method and Proportional Method definitions compare their different perceptions of what the observation was, plus some assumptions, to the eventual result – which they also have different perceptions for – and conclude any discrepancy can only be a poll error.

There's a better way, and we can use "What's for Lunch?" to help find it.

Notes

1. Silver, N. (2014, November 14). *Here's proof some pollsters are putting a thumb on the scale.* FiveThirtyEight. https://fivethirtyeight.com/features/heres-proof-some-pollsters-are-putting-a-thumb-on-the-scale/
2. Saunders, D. J. (2016, November 12). *How herding blinded polls to Trump's big win.* San Francisco Chronicle. https://www.sfchronicle.com/opinion/saunders/article/How-herding-blinded-polls-to-Trump-s-big-win-10609374.php

12

What's (Actually) for Lunch?

Important Points Checklist

- The "broken down" version of the poll error formula
- "What's for Lunch?" eventual results compared to poll question
- The ability to conduct polls like this yourself, if you want
- How "What's for Lunch?" accuracy should be judged
- How this poll's margin of error applies to political polls

It's early afternoon on Friday, and my friends are happy to see me! Okay, that's an assumption.

Nonetheless, I'm back with my pen, paper, and clipboard to tabulate some results. Instead of a random sample, I'm now asking a census, 91 school staff.

While we're mostly interested in the whole "lunch" thing, for the sake of completeness, I went ahead and asked the appropriate variation of the same three questions from the original poll:

Question 1: What was your commute time to school today? (A) Less than 15 minutes, (B) 15–30 minutes, (C) 30+ minutes

Question 2: For lunch, today, did you (A) Pack/bring lunch from home, (B) Order/purchase lunch

Question 3: During the most recent holiday, I (A) Traveled outside the state, (B) Stayed home/Traveled within the state, (C) Visited/traveled both within and outside of the state.

While I'm sure you're eager to read about how long the staff's commutes are, let's consider one final time the largest potential errors from this poll before analyzing the results.

If this poll included individuals who were not likely to eat lunch – or perhaps never eat lunch – that would be an example of a frame error. Likewise, if I had mistakenly polled staff members from a school other than those I identified in my target population, that would be a frame error.

DOI: 10.1201/9781003389903-12

Frame errors, like nonresponse error, in most large-sample polls are difficult to quantify *precisely* but are the two primary contributors, plus the statistical margin of error, that can contribute to poll error.

In other words, the formulas for calculating poll error can be expressed in two different ways:

1. The difference between the reported result and the simultaneous census result
2. The sum of all the factors that *caused the difference* between the reported result and simultaneous census result

The same terms can be expressed as:

1. Poll Error = Reported Poll Value – Simultaneous Census Value,

and

2. Poll Error = Statistical Margin of Error + Nonresponse Error + Frame Error + Other Errors.

Any disagreement between the Reported Poll Value and the Simultaneous Census Value must be caused by some combination of these errors.

So, in its longest form, we can break the definition down to:

Reported Poll Value – Simultaneous Census Value = Statistical Margin of Error + Nonresponse Error + Frame Error + Other Errors.

The statistical margin of error for this poll is known: approximately +/– 4%.

Nonresponse Error? Well, everyone responded – can't attribute any error to nonresponse for this poll.

As for frame error, it's much easier to identify who among a school staff population will eat lunch on Friday than it is to figure out who among a registered voter population will eventually vote. Are political polls more susceptible to frame error than a "What's for Lunch?" poll? Absolutely.

But that's by design. That's not to say that frame error is not possible in a "What's for Lunch?" poll, that's to remind everyone that if we are going to call something an error, we must be able to quantify it, even if approximately. And a not-so-subtle reminder that when you're thinking about polls and poll error, that political polls are not the only types of polls.

So what else is there that could cause an error? "Other error."

This is not to be dismissive of the fact that other factors can't cause a poll error, they can. One potential source of "other error" in most polls would be the question itself: if the poll question is "leading" in some way, or the respondent doesn't understand it.

In total, I'm not asserting that it's *impossible* that my poll had some source of error. Just that there aren't many strong candidates.

In the original poll question, I *did not* include "I don't know yet" as an upfront option. I offered them the two options, and only if they said they were "undecided" (or some variation) did I count it. Furthermore, if someone's answer included a modifier like "probably" before their response, such as "I'll probably pack," I counted this the same as a direct "pack/bring from home" answer.

There's no "perfect" way to ask a question about what someone plans to do. If I had offered "I don't know yet" or "undecided" as an option in my original poll, it's fair to say at least many, but probably more, would have selected that option.

I could have given "undecided" as an option in my in-person poll question; then, if someone said they're undecided, follow up with a question with which option they would lean toward.

I could *have not given* undecided as an option, and even if someone volunteers that they're undecided, *still* ask them about a preference they lean toward.

Indeed, many reputable pollsters employ these approaches, and only report "undecided" if a respondent expresses neither a preference *nor* a "lean."

All of these methods have different advantages or justification.

While you can consider the pros and cons of each of these methods, this is one reason a diversity of pollsters is beneficial: if lots of pollsters are asking the same question(s) in the same way, and if there's an error (like, how many are *actually* undecided, or how many are *actually* likely voters?), then those pollsters are all susceptible to the same errors.

But there was only one pollster in "What's for Lunch?".

Here is what everyone in the "What's for Lunch" population eventually did:

Pack/bring lunch: 84.6%

Order/purchase lunch: 15.4%

And a reminder of the poll data, with approximate margin of error:

Pack/bring lunch: 75% +/− 4%

Order/purchase lunch: 15% +/− 4%

Undecided: 10% +/− 4%

The result had a "spread" of 69.2, while the poll only had a spread of 60. As it relates to "nonresponse error" and "frame error," thanks to a 100% response rate and 100% of likely lunchers having lunch, I think it's fair to say my poll has ~0 error in those categories.

The statistical margin of error is unavoidable for any poll; it's about 4% for this one.

Is there any kind of "other error" my poll may have had? Is it possible some respondents didn't understand the question? With humans asking and answering questions – and a human tallying the results – it might be impossible to say there's *no possibility* for *any other error*, but there's not a strong case that other errors could be substantive.

Here, there is a huge discrepancy between the poll and the result, if you only look at spread, but the margin of error is the only realistic source of error.

Either this poll – with results that existing methods would claim are far outside the "margin of error" – is an anomaly or it shows polls aren't reliable, or …

It provides tangible, reproducible proof that demonstrates the comparison of "poll to eventual result" is based in a flawed understanding of how polls work.

I Have a Secret

On the same day I conducted my poll, I conducted a (nearly) simultaneous census by asking 11 more people the same question. Surprise!

Not only do polls only try to tell us someone's current position, but the statistical margin of error given by the sample size can also only apply *to those results*, those responses, and that population.

The original poll said that with 95% confidence, a simultaneous census would produce a result in approximately this range,[1] applying the margin of error:

Pack/bring lunch: 71% – 79%

Order/purchase lunch: 11% – 19%

Undecided: 6% – 14%

That's what this poll actually tells us. **This is what all polls actually tell us.** A likely range of results for a population right now.

You might find this range unimpressive for a poll of 80 to a population of 91 – a fair critique.

But apply what we learned here to what this means for plan polls with a sample size of 300 for a population of 600. And plan polls with a sample of 400 for a population of 1,000. And plan polls with a sample of 600 for a population of 100,000, or a million and beyond.

All of these have nearly identical margins of error, and the same Simultaneous Census applications apply.

This kind of experiment is easily reproducible.

For those interested, if you have access to a population of any of these sizes, you can try it yourself: take a random sample of an appropriate size for the population of interest. You can use my "What's for Lunch?" example of a plan poll, or come up with your own.

Importantly, you (the researcher) have to decide how and when you're going to take the census. Eventually, simultaneously, or both?

With access to a few hundred or more individuals, you could give all of them – a census – some plan poll.

Consider a plan poll like "What do you plan to have for lunch on Friday?" given to a census of 500 individuals on paper. Their responses to that poll can be completed and placed into a large bin.

If you're handy enough to export the results into a spreadsheet, or conduct the poll online, that works too.

But if you're serious about the research factor, after having collected all of those responses, you're only allowed to access a *random sample* of the responses – not look at all of them.

Each time you take a random sample from that bin or spreadsheet of responses, you're taking a "poll."

Access a random sample appropriate for your desired margin of error.

You can take as many polls as you would like in order to inform a prediction of what the whole population said; it's up to you.

Importantly, do you think the error in those polls you took would be properly measured by comparing them to the simultaneous census results, which you have in the bin, or comparing them to the "eventual results" which you could collect later?

Comparing polls to the simultaneous census would be the proper way to measure poll error. You can know with certainty how much error each poll had long before the result, because polls – even plan polls – don't try to measure future results.

As I will, you can also compare your polls to a simultaneous census (the correct "true value") and to the eventual result (the currently used, but incorrect "true value") – to observe how much error not using the simultaneous census standard can introduce.

If you conduct such experiments, you'll probably note that your various polls produce different amounts of poll error, even if you use the exact same methods.

Even with a true simultaneous census of results in a bin to compare your poll's error to, the statistical margin of error alone guarantees the inescapable truth of polling: **fluctuation is normal and expected.**

But fluctuation does not mean useless.

I want to hear about your polls! Seriously. Share them with me.

I'll share my simultaneous census results a little later, after first covering how the Spread and Proportional Methods would characterize my "What's for Lunch?" poll, now that there are results to compare them to.

Note

1 See notes 3–5 in Chapter 8.

13

Real Polls + Bad Math = Fake Errors

Important Points Checklist

- How the margin of error is currently misunderstood in the field
- The appropriate application of margin of error
- What assumptions the Spread and Proportional Methods make about "Changing Attitudes"
- What is meant by "cramming" and why current standards encourage it

Anyone who tries to characterize a plan poll's error based solely on its poll-versus-result spread or proportion is wrong. Moreover, anyone who classifies a poll as "outside the margin of error" based solely on the poll-to-election result is also wrong. This understanding is not insignificant or inconsequential, it's a mistake still made by experts today.

At Berkeley Haas, their study "Public Election Polls Are 95% Confident but Only 60% Accurate" was interpreted to mean that you should not be "nearly as confident as the pollsters claim" in election polls.[1]

The fact that non-experts and the media would have diminished confidence in polls from this study is not an accident: the experts said you shouldn't be as confident as the pollsters claim.

There is one big problem with that. Here's the data they used to reach that conclusion:

> So, for example, a poll finding that 50% of 800 likely voters favor a particular candidate in an upcoming election, then a 3.5% margin of error would imply a confidence interval predicting 95% chance that the candidate will receive between 46.5% and 53.5% of the vote in the actual election.[2]

This is not correct. It would imply they would receive between 46.5% and 53.5% in a simultaneous census of the same population, given the same questions, *allowing for undecideds*. The margin of error given by the sample size doesn't apply to different questions with different options.

DOI: 10.1201/9781003389903-13

To take their example to its conclusion, if a two-candidate election poll with 3.5% margin of error showed:

Candidate A: 50%

Candidate B: 40%

Undecided: 10%

Does this poll suggest, predict, or in any way conclude that Candidate A must eventually receive 46.5%–53.5% and that Candidate B must eventually receive 36.5%–43.5%, or the poll is "outside the margin of error"?

Their definition says it does. Let's test it.

Even at the positive extremes, 53.5% and 43.5%, these two-way election numbers only add up to 97%. Election results will always add up to 100%.

Their method, due to its flawed assumptions, would conclude that any election with a moderately high number of undecideds will *always* produce results "outside the margin of error." This is not what the margin of error is.

There's *no possible combination*, given 10% undecided in an election with only two choices (and very few combinations given more than two) that at least one of them *doesn't* gain more than 3.5% compared to their poll number. With a fundamental understanding of what polls do, it shouldn't be surprising that only 60% of election polls produce a number within 3.5% of the eventual result: undecideds must eventually decide, and there are often a lot of undecideds. This fact has nothing to do with the poll's margin of error.

The same can be said for "What's for Lunch?" in which the decided preferences were split 75%–15%. Since the eventual result must add up to 100%, there's no possible combination by which at least one of them doesn't end up "outside the margin of error given by the poll," if you make the mistake of thinking the numbers and margin of error from the poll apply to the eventual result.

In polls with undecideds, a result different from the poll number *should* happen quite frequently, and it has nothing to do with the poll's accuracy. The margin of error in a poll applies to the number of undecideds, too. Not the eventual result.

Misusing the most fundamental statistic of the field – the margin of error – does not inspire confidence that experts understand how polls work.

In a similar blunder, Nate Silver's 2014 article "How the FiveThirtyEight Senate Forecast Model Works" shows that in their database of polls from the final three weeks of campaigns since 1998, "the actual results fell outside the poll's reported margin of error almost 25 percent of the time."[3]

The method he used to arrive at this "25 percent" figure is based on some great work by Charles Franklin, who provided formulas to calculate the margin of error differences between polls.[4]

Between polls.

Silver took Franklin's accurate and useful calculations for how to compare *polls to other polls* and used them to compare *polls to elections*. This led him to wrongly conclude that the results fell outside the margin of error "almost 25 percent of the time." Franklin was very clear in his work that the presence of undecideds impacts the calculation for error between polls and took this into account. Silver ignored this fact and used the formula to compare polls (with undecideds) to elections (with none).[5]

The ability to plug numbers into a formula isn't useful if you don't understand what the numbers mean.

Polls are not elections. They have different variables. You can't compare them as such.

If university researchers and one of the best-known forecasters in the world don't know how to interpret poll-to-election data, how can the public or media be expected to?

The above examples are two from a list of thousands that make the same mistake, some more egregiously than others.

The fact that the "true value" a poll tries to measure *is not* the eventual result is clearly not understood by many researchers and experts actively publishing in the field; this demonstrates that there is an urgent need to teach the simultaneous census. Understanding what a simultaneous census is – and that this is what is meant by the "true value" in statistics – would rectify this mistake and allow them to do better research.

On that note, below is (for the last time, probably) a side-by-side of the poll data and result, as they should be properly reported and understood.

The poll data

Pack/bring lunch: 75% +/– 4%[6]

Order/purchase lunch: 15% +/– 4%

Undecided: 10% +/– 4%

Eventual result

Pack/bring lunch: 84.6%

Order/purchase lunch: 15.4%

What the Spread Method Says

The Spread Method isn't concerned with undecideds, nor particularly concerned with the margin of error. It only cares about the difference between the top two "decided" numbers: in this case, 75 and 15. The spread is 60, so it can be characterized as "+60."

So what *does* the Spread Method have to say about the margin of error? Does +/– 4% apply to it?

We can try applying the margin of error to the spread: 60% +/– 4%.

Does this method say the final spread, according to the margin of error, should be somewhere between 56% and 64%? That seems too narrow.

What if we apply the +/– 4% to *both* of the results (75% and 15%) to get 71%–79% and 11%–19%, and then take the spread? That makes more sense!

So, given the two extremes, that gives us a spread range of 71%–19% (+52) to 79%–11% (+68). Does the Spread Method say the final spread should be between 52 and 68, or it's outside the margin of error?

Well, note that at both positive extremes here, 79% and 19%, the numbers still don't add up to 100%.

While the "60% accurate" study *misunderstood* what the margin of error means, other respected researchers just *ignore* what the margin of error says because they prefer to look at the spread.

> The margin of error given by the sample size only applies to the poll at the time of the poll, and to all the polls' options: a simultaneous census. The margin of error given by a poll has nothing to do with the eventual result.

But what the Spread Method does is not up to me, it says what it says.
The poll result said "+60."
The final result was 84.6%–15.4%. That is a "+69.2" spread.
The Spread Method says my poll had a 9.2-point error.

What the Proportional Method Says

Like the Spread Method, the Proportional Method is not concerned with undecideds or third parties. It's especially dismissive of the margin of error.
In this case, this method says my poll said:

Pack/bring lunch: 83%

Order/purchase lunch: 17%

This would also be reported as "+66" by many pollsters, analysts, and experts.

In response to my work – which demonstrates a knowable, provable, but apparently unpublished fact that the margin of error given by a poll's sample size and confidence interval does not pertain to the eventual result – I've been pointed to ways to calculate the margin of error after accounting for a range of possible undecided preferences, through models that seem perfectly reasonable. This is a fine area of research, but irrelevant to the more pressing issue.

When tobacco companies funded research on the impact of genetics and pharmacology on lung cancer prevalence, that was real research. But it didn't answer or address the question: "What is the primary cause of lung cancer?"

Similarly, calculating the margin of error for the various methods – potential or actual – doesn't answer or address the question: "What was the cause of poll error, and how much error was there *in this specific instance?*"

If Proportional Method or Spread Method users wish to calculate a new margin of error for their methods based on historical or experimental findings, fine, have at it. But now you're calculating *the method's* error, not the poll's.[7]

Conflating what a poll says with what these methods and their assumptions say it says again does not inspire confidence that these issues are understood and will be corrected.

If the eventual results don't match these reported results, that speaks to the accuracy of *the method's assumptions*, not the poll's.

In addition to what these assumptions mean for dismissing the margin of error given by the poll's sample size, **think about how problematic it is to report a poll with lots of undecideds as having no undecideds.**

Consider what would happen if there were an even higher number of undecided voters, and those undecided voters disproportionately supported the candidate with *fewer* decided voters. Not that that could ever happen, right? (And if it does, that's a poll error, please don't blame our bad assumptions.)

Forget all that and celebrate for now because the Proportional Method says my poll's error is the difference between one option's poll results and the eventual result: 84.6%–83%. Very impressive, only 1.6-point error!

Where's the Error?

Ignoring, for now, that the two accepted methods for calculating poll error say that there was 1.6-point poll error (Proportional Method) and a 9.2-point poll error (Spread Method) and that this is a huge discrepancy regarding how much error the "What's for Lunch?" poll had – let's try to quantify some things. Or, "show your work."

In order to accurately call something an error, as I hope has been established, the "census" or "true value" you compare your *reported* data to must be *simultaneous*. If it is not simultaneous, there are variables (not caused by instrument error) *that must first be accounted for before any claims of error can be made.*

In the absence of a simultaneous census, instrument error – like poll error – can only ever be known approximately.

In the "Scalester" example, the Scalesters all have flawed methodologies: scale placed on carpet and/or person being weighed stands with weight on one leg. This addition was necessary for the analogy comparing scales to polls because polls will always have a notable margin of error, no matter how "ideal" they are.

However, there was only *one* variable that needed to be accounted for to approximate instrument error for the Scalesters – weight change over time.

If I lost two or five pounds in a week, a realistic scenario for someone with a weight loss goal, saying a scale (or Scalester) was "wrong" *because* their reported value was "off" from the doctor-measured value a week later, is not correct. This is not what is meant in statistics by "true value" in the definition of error.

In order to calculate scale error, you must account for (or at minimum, allow for) the possibility that weight has changed over time. The same is true for poll data compared to election data.

Any assertion of "there was this much poll error" should be subject to the same scrutiny and criticism as the polls themselves – perhaps, due to their perceived authority, even more.

A poll provides data with the admission that it could be wrong: margin of error and more. Poll error and "accuracy" calculators are not as candid.

Frankly, given that both methods for calculating poll error are largely determined by that method's assumptions, it would be equally as valid for a *pollster* to say every poll that disagrees with their poll has an error. Not just an error, an error equal to the magnitude by which it disagrees with their poll.

If you think that would be a conceited assertion on the part of the *pollster*, then why are the differing definitions of poll error, whose conflicting and easily falsified definitions that assume their assumptions can't be wrong given such authority?

It's because the people who use these definitions do not understand what poll data means, their words and work prove it.

But a simple fact: weight change is not always negligible over hours, days, or weeks. Just because it often is, even if it usually is, doesn't mean it always is. If you want your "error" calculation to be accurate, you must account for it.

Likewise, there's no reason to accept (if we care about accurate definitions) that voter "mind-changing" is negligible over the course of a week – or 3 weeks – as is the standard cutoff for "poll error" calculations by users of both the Spread Method[8] and Proportional Method.[9]

Just because it often is, even usually, doesn't mean it always is. So what do these methods do to account for this? Nothing. Whether you take your poll one day or 21 days before the election, you're judged on the same standard. If there's a good reason to believe voters might have changed their minds in those 21 days? Doesn't matter. It's the *polls* that were wrong.

If undecided voters eventually decide in a way that doesn't agree with the method's assumptions, you should account for that. Neither of the current methods allow for that.

The standards being *wrong* should be enough, but that's not the only problem.

In addition to the previously mentioned problem herding creates for the trustworthiness and accuracy of poll data, and the incentive created for pollsters to try and report their numbers as a forecast, there's *another* problem, which is more of a disincentive: if you know you're going to be judged solely on the closeness to the *eventual result*, would you rather take measurements weeks in advance or wait until the last possible minute?

"**Cramming**" – or doing most of your data collection as late as possible – creates another, avoidable problem caused by the accepted poll error definitions.

Knowing that body weight, like voter sentiment, might change over time – having lots of data over a long period of time would be useful for forecasters (and the public, for that matter) to separate "trends" from "noise."

Properly read – and properly judged – scales and polls are the best possible tool for those respective jobs. But for what? To be told how "wrong" your three-week-old reported value was because it didn't match the spread (or proportion) of the eventual result?

A scale, if properly read, can only give us our weight *right now*. Asking a scale what you'll weigh in a week, or three weeks, doesn't magically change what the scale can tell you.

Polls are no different.

But plan polls – political or otherwise – don't just have changing attitudes to account for.

There are *two* major variables of interest that impact the difference between the value reported in a plan poll compared to the eventual result: changing attitudes *and* undecideds.

Nonresponse error and frame error – among others – are potential errors inherent to the poll instrument and are something pollsters can and do try to weight for.

But then, the analysts who calculate poll error say changing attitudes not being zero, and undecideds not voting how their method assumes they should, *is also a poll error*, even though the poll can't and doesn't try to control for that.

The only way that conclusion would make sense is if you believe polls and pollsters should be able to predict how undecideds will eventually vote, how many people will change their mind, and who they will eventually vote for.

In other words, the *industry-used* definition of poll error puts "changing attitudes" and "how undecideds eventually vote" *on the wrong side of the equation.* Changing attitudes and undecided ratio must be accounted for *before* poll error can be calculated. It is step one. It is also not a small step, nor should it be optional.

The current "poll error" definitions require believing one of two things, the first of which is contradictory, the second of which is bad science:

1. **Pollsters** *should not* **try to predict the eventual results, but also pollster's accuracy should be measured by how close they are to the eventual results. (Spread Method)**
2. **Pollsters** *should* **try to predict eventual results and instead of reporting what their poll observed, they should report their forecast in place of the poll. (Proportional Method)**

So let's fix the definition.

Notes

1. Counts, L. (2021, January 7). *Election polls are 95% confident but only 60% accurate, Berkeley Haas Study finds. Berkeley Haas.* https://newsroom.haas.berkeley.edu/research/election-polls-are-95-confident-but-only-60-accurate-berkeley-haas-study-finds/
2. Kotak, A., & Moore, D. A. (2020). Public election polls are 95% confident but only 60% accurate. *Behavioral Science & Policy, 8*(2), 1–12. https://doi.org/10.31234/osf.io/rj643
3. Silver, N. (2014, September 17). *How the FiveThirtyEight senate forecast model works.* FiveThirtyEight. https://fivethirtyeight.com/features/how-the-fivethirtyeight-senate-forecast-model-works/
4. Franklin, C. H. (2007). The 'margin of error' for differences in polls. *ABC News.* https://abcnews.go.com/images/PollingUnit/MOEFranklin.pdf
5. As undecideds and third-party responses increase (as $p1 + p2$ moves away from 1.0, using Franklin's terminology), the standard error for the difference between candidates decreases: but this is for comparing polls to other polls. Elections are a different calculation. With a higher number of polled undecideds – and to an extent in US elections, third-party support – the range of possible outcomes within the margin of error *increases.* The consequence of this is that in elections where "third-party + undecided" is higher, a smaller standard error is observed by this formula because it only intends to compare polls to other polls, leading to an improperly narrow expectation for the range of outcomes within the margin of error for the election. This is a more technical "apples to apples" problem.

6. See notes 3–5in Chapter 8 for MOEs reported here.

7. I think this also gets into the difference between credibility intervals and confidence intervals. A topic better discussed elsewhere, or here: https://aapor.org/statements/understanding-a-credibility-interval-and-how-it-differs-from-the-margin-of-sampling-error-in-a-public-opinion-poll/

8. Mehta, D., Radcliffe, M., & Shan, D. (2022, January 9). *Polls policy and FAQs*. FiveThirtyEight. https://fivethirtyeight.com/features/polls-policy-and-faqs/

9. Shirani-Mehr, H., Rothschild, D., Goel, S., & Gelman, A. (2018). Disentangling bias and variance in election polls. *Journal of the American Statistical Association, 113*(522), 607–614. https://doi.org/10.1080/01621459.2018.1448823

14

My Simultaneous Census

DOI: 10.1201/9781003389903-14

Important Points Checklist

- Changing Attitudes and Undecided Ratio's impact on poll-to-election result
- The difference between poll error and forecast error
- How to calculate poll error when a simultaneous census is and isn't possible

Since polls can only measure a current state or provide a "snapshot," the best definition of poll accuracy would require the poll to be compared to a simultaneous census.

However, a simultaneous census is not *normally* available. The best we can do with political data is compare the poll numbers to the eventual result. But there are factors *not caused by poll error* that can impact the discrepancy between poll number and eventual result.

1. **Changing Attitudes**. People can change their mind for any reason or for no reason. Failure to foretell the future isn't the poll's fault, therefore cannot be a poll error. Since Changing Attitudes can result in a gain or loss of support after a poll is taken, this number can be positive or negative.

2. **Undecided Ratio**. If undecided voters all vote for the same option, split 50/50, or any such combination in between (including in elections with more than two options), since this is not something a poll tries to measure, it cannot be included in the category of "poll error." This ratio accounts for both the percentage of undecided voters who ultimately voted for that candidate, *and* how many there were. Since undecided voters eventually choosing an option can only add to an eventual vote total, the undecided ratio can never be negative.

DOI: 10.1201/9781003389903-14

So, to correct for known variables that can impact the discrepancy between observation and eventual result, but are not properly classified as poll errors, here's the definition:

Eventual Result = Simultaneous Census to Poll + / − *Changing Attitudes* *+Undecided Ratio.*

In the absence of asking a census of voters "did you change your mind?" and "were you undecided x days before the election, and for whom did you ultimately vote?", these numbers can only be approximated. And even with a census, the reliability of voter's self-reported mind-changing and undecided-ness might not be totally reliable.

One way this could be addressed is by following up with your pre-event random sample and asking that same group what they did after the event – and comparing their responses from before to after. While your random sample could only offer an approximation of mind-changing and how unde-cideds decided, it's a lot more valid than the current standard: guessing or assuming.

In any event, **the weighting process used by pollsters, properly done, is only intended to estimate the Simultaneous Census value, *not* the eventual result.**

By design, polls cannot account for future changing attitudes, or eventual preferences of undecideds. That's not a weakness of a poll, nor it is a feature. It's just a reality. Weighting methods, therefore, *should not* attempt to account for changing attitudes and undecideds – because that would be a forecast, not a poll.

Attempting to make a forecast from one poll, and reporting the forecast as if it were the poll – as companies that use the Proportional Method (and variations of it) do – creates many problems, some of which I've covered, others I'll discuss more later, and it also confuses the public.

In fact, FiveThirtyEight expressly *excludes* polls from their database that have "blended" their data in this way. They use the analogy of "using some-one else's barbeque sauce as an ingredient in your own barbeque sauce."[1]

What they mean by this is that they don't want their poll averages to be influenced by forecasts – just polls – a reasonable position, if you ignore how they say poll accuracy should be judged.

Poll accuracy – and pollster reputation – is presently judged on its close-ness to the eventual result, and weighting methods should – and rightfully do – try to account for some of the things that might cause their poll's raw data to disagree with the simultaneous census value, such as nonresponse.

But polls and pollsters **cannot and should not** try to weight their data to account for people who might change their minds, or how that pollster thinks undecideds might eventually decide. That step goes far beyond the function of a poll, and introduces sources of error that have little to nothing to do with their poll data, or the quality of their poll. Even if a pollster has a good reason to believe, based on data they've collected, that those who say they're undecided will eventually split 60/40 or 80/20 in favor of one option, that's not their job. Their job is to report what they observed.

It's a **forecast's job** to account for changing attitudes and undecided splits, not a poll's.

But, it is said and accepted that if your poll doesn't agree with the *eventual result* in the way the analyst believes it should, that's how much *error* the poll had.

The solution to this isn't to allow pollsters to weight for changing attitudes and undecided splits – though given the current environment, they would be justified in doing so. The solution is to fix how we – the public, pollsters, forecasters, and experts – judge a poll's accuracy.

If you're trying to forecast something about the future, and you're incorrect, that's not a poll error, that's a forecast error.

We can say with absolute certainty (owing to their conflicting assumptions) that either the Spread Method or Proportional Method, in every single election, will always have some amount of error. That amount may be small, or it may be large, but it's nonzero and could be larger than poll error itself in many cases.

But these methods blame the polls for the entirety of "error," their faulty assumptions included. If you make assumptions about what a tool *should* measure, but the tool doesn't try to measure what you say it should, you will *never* be able to properly evaluate that tool's accuracy.

> Measuring the accuracy of a tool not intended to be predictive by how predictive it is, is the epitome of bad science.

Not only do these incorrect assumptions cause experts to misjudge a poll's *accuracy*, they also cause experts to misjudge a poll's *utility*, or how useful they are. While polls are neither predictions nor forecasts, they can offer a great deal more insight than experts currently seem to understand. It would be quite ironic if their invalid "poll-to-result equals error" metric caused them to miss some valuable information that polls *can* provide related to *informing a prediction* of the eventual result.

But for now.

If Your Forecast Was Wrong, Then That's a Forecast Error

Here is the definition of poll error, given what polls, as a tool, are intended to do:

> Reported Poll Value – Simultaneous Census Value
> = Statistical Margin of Error + Nonresponse Error
> + Frame Error + Other Errors.

With a Simultaneous Census being theoretically impossible (at least in political elections), it's both fair and necessary to incorporate changing attitudes and undecideds into our definition. No, a poll conducted three weeks or three days before an election does not necessarily mean changing attitudes is negligible or zero. **You can make that assumption if you want, but that's your assumption, not the poll's.**

And a poll certainly doesn't tell us how undecideds will eventually decide.

Building on the simultaneous census standard, the complete poll error formula that accounts for **when a simultaneous census is not possible, as in political applications:**

$$\left(\textbf{Reported Poll Value} + / - \textbf{Changing Attitudes} + \textbf{Undecided Ratio}\right)$$
$$- \textbf{Eventual Result} = \textbf{Statistical Margin of Error} + \textbf{Nonresponse Error}$$
$$+ \textbf{Frame Error} + \textbf{Other Errors}.$$

In order to know how accurate a poll's reported value actually was, these variables must be accounted for – not assumed to be known.

For example, if a pollster's raw data, before weighting, has one candidate at 48%, but their weighting techniques determined their sample (due to nonresponse, frame, or "other") overestimates that candidate's support, the reported poll value could be lower than that raw number – say, 46%. That reported poll value *is not a prediction of the eventual result* and makes no attempt to correct for future changing attitudes or undecideds, nor should it. It's an effort to estimate what a simultaneous census of the target population would say, what it observes about the population right now. After the election, data can be collected to approximate Changing Attitudes and Undecided Ratio, which can then be compared to the eventual result as an estimate of total poll error, as seen in the formula above.

The "What's for Lunch?" Simultaneous Census

Here's a quick recap of what happened.

I identified 80 individuals at random from a population of 91. I reported those results earlier (75%–15%–10%) – and this poll comes with approximately a 4% margin of error.

After asking all 80 individuals I had identified at random and noting those results, I went back and asked the remaining 11. I collected data from the entire population of interest about what they planned to have for lunch and asked them the same questions at almost exactly the same time.

So, which of the following gives the best standard for calculating my poll's actual error:

   a. The near-simultaneous census

   b. Spread Method

   c. Proportional Method

   d. Blended Method

If you answered "the near-simultaneous census," you are correct.

My census allowed for the remaining individuals to be undecided, just like those in the poll.

My poll's margin of error – like all polls – applies to each result (including third-party and undecideds) at the time of that poll.

If a proclaimed third-party voter changes their mind the day before an election because they're afraid their vote won't make a difference, or is voting strategically? Not a poll error. Undecided voters favoring one option over another? Not a poll error. Someone who said they were going to pack their lunch, but an hour later, they made plans to go out to lunch with a few other teachers? Not a poll error.

If you want to make assumptions about what third-party voters will eventually do, you can. Assumptions about what undecided voters will eventually do? You can. But it doesn't matter if those assumptions are entirely defensible or completely made up – those assumptions are yours, not the poll's.

If an assumption is wrong, that's an assumption error. If a forecast is wrong, that's a forecast error. If a poll is wrong, that's a poll error. If an assumption or forecast is wrong – not a poll error.

Now that we're all on the same page, here are the near-simultaneous census results:

Pack/bring lunch: 69/91 (75.8%)

Order/purchase lunch: 14/91 (15.4%)

Undecided: 8/91 (8.8%)

Applying the **absolute error** formula appropriately (measured result – actual result) and comparing the results of the poll to the near-simultaneous census give us the following:

Regarding the group that said "Pack/bring lunch," my poll had a 0.8-point error: 75% in the poll, 75.8% in the census.

My poll's "Order/purchase lunch" group had a 0.4-point error: 15% in the poll, 15.4% in the census.

And though my poll "overestimated" undecideds by 1.2%, we could still say it had a 1.2-point error: 10% in the poll, and 8.8% in the census.

All of these results are well within the margin of error established in the poll. This is what the "margin of error" means in a poll.

And properly read, this is what the poll tried to tell us. **This is what *all* polls try to tell us.**

Note

1. Mehta, D., Radcliffe, M., & Shan, D. (2022, January 9). *Polls policy and FAQs.* FiveThirtyEight. https://fivethirtyeight.com/features/polls-policy-and-faqs/

15

Introducing: Adjusted Poll Values

Important Points Checklist

- What "most off" and mean absolute deviation (MAD) are and what they apply to
- The lack of accountability created by current "error" definitions
- The role elections play in Simultaneous Census calculations
- How Changing Attitudes and Undecided Ratio can be estimated
- Why methodology is important when considering error

Having understood that a simultaneous census is the *proper standard* to compare polls to, not the eventual result – the "What's for Lunch?" poll still wasn't perfect.

How can we describe and measure the accuracy of this poll?

One method to rate this poll's error would be to use the largest discrepancy of any value it measured, the "most off" method. In this case, the undecideds contained a 1.2-point error.

Another acceptable method would be a MAD approach, which would say

$$\frac{\text{Sum absolute value of errors}}{\text{Number of groups}}$$

Which for this poll-to-simultaneous census would be

$$\frac{0.4 + 0.8 + 1.2}{3} = 0.8.$$

I'm not asserting that the "most off" approach or "MAD" approach are the "best" ways to measure poll accuracy. While they may be, I suspect some other methods, given simultaneous census data, would be something statisticians might be better equipped to evaluate. But unless and until the experts recognize how and why their current calculations are flawed, they won't be able to fix them.

DOI: 10.1201/9781003389903-15

Speaking of flawed: the Proportional Method said my poll had a 1.6-point error, while the Spread Method said it was off by 9.2.

Meanwhile, the "most off" method said my poll had a 1.2-point error, while the MAD method said 0.8.

If you think it's unfair that I get to compare my data to the *simultaneous census value*, and the best the presently used methods could do is the *eventual result*, remember:

Every single poll ever taken has a simultaneous census value. Just because that value happens to be unknown in most of those cases is irrelevant. That simultaneous census value is not only the best possible result a poll can give you, but also the result a poll tries to give you. Comparing a poll to an eventual result, without accounting for people changing their minds or how undecideds voted, introduces **forecast error** or an assumption error[1] in the calculation of poll error.

"What's for Lunch?" is a replicable plan poll that demonstrates both methods fail to accurately measure poll error.

How much forecast error did the Spread and Proportional Methods attribute to my poll, that should have been attributed to their flawed assumptions?

Show Your Work

Given the fact that I conducted a near-simultaneous census, achieved a 100% response rate, and everyone I identified as "likely lunchers" ultimately "lunched" I would say the only realistic source of error in my poll is the margin of error itself.

For the sake of brevity, and to be as hard as possible on my own poll, I'll use the "most off" method to calculate my "What's for Lunch?" poll's error. It was "most off" about undecided.

The definition:

> Reported Poll Value – Simultaneous Census to poll Value
> = Statistical Margin of Error + Nonresponse Error
>   + Frame Error + Other Errors.

And replacing these values with the known values:

$$10\% - 8.8\% = (-4\% \text{ to } 4\%) + \sim 0\% + \sim 0\% + \sim 0\%.$$

The reason I reiterated here that the statistical margin of error can be positive or negative (in addition to the hopefully obvious fact that this is the definition) will come into play a little later. While I don't think it's possible my poll had greater than 0% nonresponse, 0% frame error, or 0% "other" error, I noted them as "approximate" in case there's something I neglected to include.

From Whence Came This Error?

The Proportional Method says:

$$\text{Poll error} = 1.6\%$$

$$1.6\% = (-4\% \text{ to } 4\%) + \sim 0\% + \sim 0\% + \sim 0\%$$

On its face, a 1.6% error is pretty good and entirely reasonable given what we know about the quality of the poll and the simultaneous census. But it would be unfair to let this discrepancy slide.

Even using the "most off" method in my near-simultaneous census, 1.2%, how do we get to 1.6%? It has to come from somewhere. Nonresponse, Frame, somewhere else? In this case, it comes from the method's assumption error. While it might seem like splitting hairs in this instance, 1.2 versus 1.6 are more than 28% different. It'd be like calling a 3-point error greater than a 4-point error. When you attribute the entirety of a discrepancy to "poll error," this is not trivial. And as you'll see, this notable assumption error still exists despite the fact that, in this instance, the method that made a *good* assumption about what undecideds would do!

And going by MAD, the Proportional Method contributed as much error (0.8 points) as the poll itself had.

The Proportional Method contains, *at least*, a 0.4-point error in its characterization of my poll. Just like a lazy pollster who has a sample size of 3 but ends up close to the "right" answer, flawed methods should not be given undue credit.

Nonetheless, in this case, the Proportional Method ended up much closer to the actual error obtained in the near-simultaneous census than the Spread Method did.

As for the Spread Method, it said:

$$\text{Poll error} = 9.2\%$$

$$9.2\% = (-4\% \text{ to } 4\%) + \sim 0\% + \sim 0\% + \sim 0\%$$

There's no work to be shown here other than throwing the "Spread" formula out and starting over. If the numbers can be this far off the near-simultaneous census, your formula is wrong.

The very imprecise Spread Method that assumes undecideds will split 50/50 to the major candidates is allowed to exist because in competitive elections – where the most polling is done – a 50/50 undecided split is, to most, indistinguishable from a 65/35 one. Three-point poll error, 5-point poll error, who cares, it's all the poll's fault.

Flawed methods have no incentive to fix themselves when the methods' flawedness can be blamed on something else.

So is it possible, in a competitive election with higher-than-usual undecideds, plus a lot of those undecideds eventually supporting one candidate, that this formula would drastically overestimate poll error?

Don't skip ahead. There are still a few things to cover first, but yes – this will come up in Chapter 23, where I'll cover Donald Trump's surprising presidential win in 2016.

For the Proportional Method and Spread Method to be fixed first requires the experts who propagate them to acknowledge that the methods need fixed. By ascribing to those methods, they duck accountability, using polls as scapegoats when their method-of-choice's assumptions are wrong; when those assumptions are wrong, they just bake that wrongness into how wrong "the polls" were.

In other words, they call any perceived discrepancies between poll-and-result a "poll error," regardless of the source of the discrepancy. Their methods' assumptions, by circular reasoning, can't be wrong.

If they're serious about quantifying poll error in a way that truly measures poll error, and doesn't duck accountability by saying "if my assumptions were wrong, well, that's the poll's fault – see you next time," then let "What's for Lunch?" be the first step in that direction.

Let's talk about what being serious about quantifying poll error might look like.

In Summation: What Really Happened

I did a poll and shared those results. I then went back a few days later and found the eventual results, by asking a census of the population of interest what they had for lunch. I concealed from you the fact that I had conducted a near-simultaneous census at the time of my poll.

Given that **what polls try to tell us pertains to a simultaneous census, not some future result** – this data was important for me to – eventually – share with you.

Yes, that is to say what the staff ultimately did has nothing to do with how accurate my poll was – or how accurate any poll is. I knew within 10 minutes of leaving, when I tallied the simultaneous census results, that the poll was very accurate.

The difference is, in real life, we don't typically have the benefit of a simultaneous (or near-simultaneous) census. For political data, **elections are the nearest-simultaneous near-census we can get**. But understanding that polls try to tell us about a simultaneous census, not an eventual result, we must accept that the poll-to-result calculation by itself is an invalid way to measure poll error.

By understanding *which variables contribute to poll error and which don't*, which is the foundation provided by the ideal poll concept, we can better approximate poll accuracy; by better approximating poll accuracy; we can use polls to make more informed predictions.

This is why I didn't jump immediately into political polls, or even plan polls. Every example I've given to this point – and the few more that remain before we talk politics specifically – is with a purpose: to tackle one poll-related topic at a time. Jumping straight into political polls, as expert misunderstandings of them clearly attest, risks avoidable mistakes.

When I took my "results" census on Friday in which I asked everyone what they had for lunch that day, I also compared the responses from the 80 individuals in my poll sample (what they had for lunch) to their response in the original poll (what they said they *planned* to have for lunch).

That uncovered some interesting – and important – information.

Of the 72 decided people I polled, six had changed their mind. Is six a lot? That depends.

Would 6/72 decided people (8.3%) or 6/80 total (7.5%) be a large percentage of people to change their mind in a political poll? Maybe. Regarding what they'll eventually have for lunch? I don't think so.

But what *I think* doesn't really matter. What matters is what is accurate.

All six didn't change their mind in the same direction, though.

Four changed their mind from "pack/bring lunch" to "order/purchase lunch."

Two defected the opposite way: from "order/purchase lunch" to "pack/bring lunch."

Compared to the original poll that said:

Pack/bring lunch: 60/80 (75%)

Order/purchase lunch: 12/80 (15%)

Undecided: 8/80 (10%)

We now know the *decideds* from that original, random sample eventually chose:

Pack/bring lunch: 58/80 (72.5%)

Order/purchase lunch: 14/80 (17.5%)

Note: here, there are no margins of error because I'm talking specifically about the poll's sample. The margin of error is only introduced when the poll results are generalized to the population. Undecideds are not excluded from the denominator because their eventual decisions cannot be assumed: note that one option experienced a "net loss" in this step, compared to the poll.

If you remember what the "final results" were, you might be able to predict where this is going.

Of the people who said they were undecided in the poll, 8/8 (100%) ultimately packed their lunch.

It's important to note that a previously decided person changing their mind is a separate step – and a separate calculation – from how undecideds eventually decide. An undecided person doesn't "change their mind" in the same way that a person who states a preference does: an undecided person "makes a decision."[2]

As for the fact that, in my poll, 100% of undecideds did the same thing: that's a lot. Is that realistic in a political circumstance? Not really. Does what I think matter? Again, no; what matters is what's accurate.

Collecting post-event data from the same sample – and applying it to pre-event data – gives us a better way to calculate poll error.

In the absence of a simultaneous census, the experts in the field – and by extension the media and public – have mistaken the eventual result for the "true value." Whether they do this knowingly or subconsciously, I'm not sure – but wrong is wrong.

Regardless of what any expert says about poll error, it can only ever be estimated. If they disagree, I'd request that they tell me exactly how much error can be attributed to the statistical margin of error for any given event, compared to nonresponse error, compared to frame error.

By characterizing poll error with a precise number, in addition to misinforming the public about what polls do, they've assigned an undue level of confidence and precision to their own assumptions.

The "post-event sample" proposal would require collecting data from people who were originally part of the poll's random sample, and while that data would come with a margin of error of its own, this approach would provide a far better approximation than the current methods that simply rely on assumptions.

Everyone, I hope, already agrees that poll data is better than assuming or guessing – but when it comes to poll error, for some reason, experts prefer to assume.

Moreover, in adopting a calculation that properly allows adjustments for the inevitable change over time that comes with any plan poll, pollsters have more incentive to conduct polls earlier, and analysts have a tool to check potentially unscrupulous pollsters who release polls with motives other than accuracy.

Adjusted Poll Methods

Here's how I'd approach the problem:

Adjusted Poll Value = Reported Poll Value + / – *Changing Attitudes Since Poll* + *Undecided Ratio Since Poll.*

Forget for a minute my inability to resist conducting the simultaneous census from my sample – while it proved an important point, it is of course not possible in political poll applications.

Let's say all we know is reported poll result, eventual result, and the data we collected from the post-event poll of our random sample.

Here was the result, expressed as a percentage:

Pack/bring lunch: 84.6%

Order/purchase lunch: 15.4%

Here was the original poll data, expressed only as a percentage:

Pack/bring lunch: 75%

Order/purchase lunch: 15%

Undecided: 10%

Finally, after the event, we found out the decided group in the original poll ultimately chose:

Pack/bring lunch: 72.5%

Order/purchase lunch: 17.5%

Given the net impact of changing attitudes between the poll and eventual result, we can **adjust** our poll data to account for those who were decided at the time of the poll, compared to how that same, specific population (decided at the time of the poll) eventually decided.

"Pack/bring lunch" underperformed their poll number among those who were decided at the time of the poll, while "Order/purchase lunch" overperformed their poll number.

Changing attitudes accounted for a 2.5% discrepancy between poll and eventual result.

Existing methods would assign this 2.5% discrepancy to poll error. This is incorrect.

Now that we have adjusted poll numbers based on what *decideds did*, taking into account this mind-changing, we can look at what the *undecideds did*.

Remember, the existing methods make *assumptions* regarding what undecideds eventually did. I don't think my proposal to quantify what they actually did, instead of assuming, should be especially controversial.

Undecideds in this poll accounted for 10%. The "ratio" portion in the "Undecided Ratio" formula means that up to "10%" can be assigned to each option, according to how those undecideds chose. For a visual, after adjusting

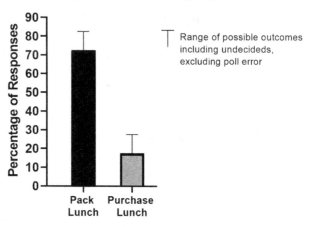

FIGURE 15.1
A chart that shows the results of how previously decided individuals ultimately decided, thus accounting for Changing Attitudes. The bars on top of each option are the maximum percentage each option can receive after accounting for undecideds. Chart by the author.

for changing attitudes, the bars in Figure 15.1 represent the *maximum* number of undecideds each side can receive.

After an event – election or lunch – there are no undecideds.

That 10% is "up for grabs," and it has to go somewhere. We can either assume we know what undecideds will do (or did), or collect data to better estimate it; you know which side I prefer.

If in my poll that had 10% undecideds, I later found that 50% chose "pack" and 50% chose "purchase" then the ratio would be 50%. In that case, 50% of the 10% total number of undecideds, 5%, would be added to each adjusted result.

Likewise, if they had split 80% for "pack" and 20% for "purchase," I would assign 80% of 10% (8%) to "pack" and 20% of 10% (2%) to "purchase."

In my post-event survey of those who stated they were undecided in my original sample, I found that this group eventually opted 100% for "pack/ bring lunch."

That means 100% of the 10% previously undecided should be apportioned to the option they chose for the "pack/bring lunch" option. Figure 15.2 shows the impact of both Changing Attitudes and Undecided Ratio.

Here are the complete adjusted poll results, accounting for both changing attitudes and undecideds.

Using the formula:

Adjusted Poll Value = Reported Poll Value + / – *Changing Attitudes + Undecided Ratio.*

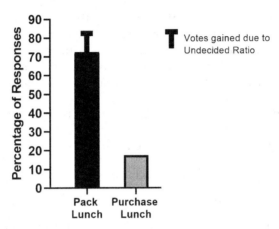

FIGURE 15.2
A chart that accounts for both Changing Attitudes and the Undecided Ratio in the "What's for Lunch?" poll. Chart by the author.

Adjusted poll results: pack/bring lunch

$$75\% - 2.5\% + 10\% = 82.5\%$$

Adjusted poll results: order/purchase lunch

$$15\% + 2.5\% + 0\% = 17.5\%$$

Remember, the poll's random sample isn't always perfectly reflective of the population – which is why there's a margin of error in the first place. But that doesn't mean knowing if/how many changed their minds, or especially how undecideds eventually decided, isn't important. It's the most valuable data you can have, aside from the original poll itself.

Considering changing attitudes and eventual undecided preference are things polls do not try to measure, and since the eventual result is the nearest data to a simultaneous census that is typically available, poll error calculations should be *adjusted* to account for this.

In the absence of a simultaneous census, Adjusted Poll Value is a better basis for poll error than Reported Poll Value. Neither the Spread Method nor the Proportional Method *should* be used. MAD and other methods are much better, but the following illustrates what their Adjusted methods would show.

The Adjusted Spread Method

The Spread Method – which compares the poll's spread to eventual result spread – originally said this poll had a 9.2-point error.

The **Adjusted Spread Method**, which incorporates data collected after the event to provide an estimate of how changing attitudes and undecideds impacted the eventual result, would say the following:

Adjusted Spread Poll Error ≈ Adjusted Poll Spread – Eventual Result Spread.

This method would have calculated:

$$\text{Pack/bring lunch: } 75\% - 2.5\% + 10\% = 82.5\%$$

$$\text{Order/purchase lunch side: } 15\% + 2.5\% + 0\% = 17.5\%$$

The Adjusted Spread Method would read this adjusted poll's value as "82.5 − 17.5," or "+65."

The Spread Method had a +60 spread. After accounting for changing attitudes and undecideds, it did not have a "+60" spread but an adjusted "+65" spread.

Compared to the eventual results, "+69.2," the Adjusted Spread Method would say this poll was "off by" the difference between 65 and 69.2, a 4.2-point error.

This 4.2-point error calculation is much closer to reasonable than the original "off by 9.2-points" characterization, though still contains a large forecast error, given what we know about the simultaneous census.

To be clear, I am not advocating for this method to be used. Spread is a terrible way to characterize a poll and its accuracy and should be done away with entirely.

I'm only demonstrating that if you were to twist someone's arm and force them to use "spread" to estimate poll error, there's a better way. You must at least account for the variable change between observation and event (changing attitudes and undecided) before attributing the entirety of the discrepancy to poll error. The current formulas do not.

The Spread Method assumes 0% changing attitudes when it was actually 2.5%. It assumed 5% undecided for each option when it was 10% and 0%. Its flawed assumptions resulted in an approximately 7.5-point forecast error, in absolute terms.

The Adjusted Proportional Method

With the same, necessary adjustments, let's see how the **Adjusted Proportional Method** – the Proportional Method after accounting for changing attitudes and undecided preference – would have evaluated poll error in this instance.

Adjusted Proportional Poll Error ≈ Adjusted Poll Proportion – Eventual Result Proportion.

This method originally characterized the proportions as:

Pack/bring lunch 75%/90% (83.33%) and 15%/90% (16.66%).

But changing attitudes reduced that number from 75% to 72.5%, only for undecideds to bring it back up to 82.5%.

With the eventual result of 84.6%, the Proportional Method would have characterized this as:

Pack/bring lunch: 82.5% – 84.6% = –2.1%

Order/purchase lunch: 17.5% – 15.4% = 2.1%

The Adjusted Proportional Method would have estimated a 2.1-point error. As it happens, its original estimate of 1.6 was closer to the true value given what we know about the simultaneous census.

But why? Since we've identified all of the variables of interest, we can "show our work."

The Proportional Method, like the Spread Method, assumes 0 changing attitudes.

Its saving feature in this instance was its assumption that undecideds would break proportional to decideds (8.33/10 by the poll results, 8.25/10 by the adjusted poll results).

While it was wrong about its 0 changing attitudes assumption (–2.5 instead of 0) and also wrong about its undecided assumption (~8.3/10 instead of 10/10), these effects nicely **cancelled each other out**: –2.5 and +1.7 left it with a net forecast error of only about 0.8.

Now we're into the *real analysis* of why "closer to result" doesn't always mean "better." I don't think it's better to be lucky – having two errors nearly cancel each other out – than good.

This method for estimating poll error, without improved standards, would be rewarded for making two errors that happened to nearly cancel each other out, even though it's just as likely those two errors would compound each other.

The best way to judge polls would be against a simultaneous or near-simultaneous census that allows individuals to choose from the same options. Experts, analysts, media, and public alike all understand that "undecided" will eventually not be an option. But polls don't know that. Polls can't tell us what that group will eventually decide when "undecided" is no longer an option. In the absence of a simultaneous census, as happens with elections, Adjusted Methods should be implemented to give better estimations of poll accuracy, and methods such as "most off" or MAD should be adopted in place of Spread or Proportion.

Working Backwards

One option that has some promise, in addition to asking members of a previous sample after the event, would be to conduct a post-event poll of a new random sample of a now-known population: actual voters. Asking actual voters when they decided, and for whom they ultimately voted, would give researchers a new piece to the puzzle for calculating poll error.

On top of that, there's the benefit of figuring out whether a given poll's method would have identified that voter as a "Likely Voter," which could offer an estimate of frame error as well.

Working backwards, if someone reports having decided for whom to vote on election day, versus a week before, versus a month before, that data can be used to judge a poll's accuracy according to when it was conducted in the same way I did with my near-simultaneous census.

Collecting this type of data wouldn't be unprecedented. **Callback studies** are sometimes used in other surveys[3] and have been used sparsely in polling, but as far as I can tell, there's little data collected in this manner *after* elections compared to the vast amounts before it.

Polls asking actual voters *when they decided* already exist,[4] but polls asking people who they voted for as they leave their polling place, **exit polls**, are sometimes maligned for not providing a truly random sample. In an era where more people than ever are voting early or not in-person, these methods become more unrepresentative.[5]

Nonetheless, a poll of actual voters asking when they decided might allow researchers to say something like, "a week before the election, the results were approximately 45% Candidate A, 45% Candidate B, and 10% Undecided."

And that is the most appropriate – if it can be detected accurately – standard to judge poll accuracy against.

The biggest benefit this method would have are its "fail safes" such as comparing your actual voter poll sample to actual voter results – if 60% of your *sample* reports having voted for Candidate A, but only 50% of voters actually did, you know your sample is imperfect – whereas with plan poll data, you have to estimate. You can use your post-event poll data compared to the now known election result as a much more objective way to weight the data. Likewise, for subpopulations of how many men and women reported voting for each candidate in the poll compared to the known election results, and so on.

A post-election poll is much easier to conduct and weight because it is a present poll, not a plan poll, and it has the huge benefit of a known standard to compare to: the election itself.

Asking someone who they plan to vote for weeks or even a day in advance has the potential of some being undecided. Pollsters must approximate "likely voters" and lots more. But a poll of "who did you vote for?" is much easier to weight because your target population is now "actual voters", and the election results are known.

While the "fail safe" detection is a huge benefit most political polls don't have, and an unquestioned improvement on existing methods, it's not that I have zero reservations about this approach. I'm not sure how accurately people can be counted on to remember when exactly they decided, among other things – especially now that they know the election result.

The two options I briefly discussed here, Adjusted Poll Values and "Working Backwards," aren't intended to be an exhaustive list of options that I think would "fix" the poll error calculation problem but rather an illustration of the existing flaws and a step toward better ways to calculating it.

As long as everyone can agree that the current methods for calculating poll error are flawed, and could be and should be improved, that's sufficient progress for now.

To this point, I've expressed and hopefully substantiated the problems with the current poll error calculations, and that those responsible for those calculations should take both accountability and action. As of today, the best we can do as informed consumers of data is look at polls ourselves, *ignore* the "analysis" centered on "spread," and the proportional results that pretend there are no undecideds.

I'll give you some valuable tools to do that later.

Maybe the adjusted poll formulas I gave above resonated with you, maybe you ignored them because formulas aren't your thing.

A word of reassurance: if you are not personally interested in, or capable of, correcting the currently used definitions for and applications of poll error, don't sweat it. Even if you have no interest in calculating poll error yourself, an understanding of how poll error is *currently* characterized, versus how it should be characterized, will allow you to understand the underlying reasoning when we get to the more central topic of "how to read polls."

In the Proportional Method example above, two erroneous assumptions (overestimating Changing Attitudes but underestimating the Undecided Ratio) happened to nearly cancel each other out; I don't think calculations such as those should be rewarded. We should classify both of those errors properly: assumption errors or forecast errors.

While polls shouldn't be immune to criticism, that criticism should be valid: separate a method's erroneous assumptions from poll error. I don't want you to lose sight of the fact that it's not *just* the numbers from a poll or forecast that should be subject to criticism.

The lazy student who took a poll with a sample size of 3, despite his poll having 0 error, should still have their methodology criticized and corrected in the future. Polls with flawed methodology should not be – as some analysts believe – given a "pass" because it may have a track record for accuracy. This is true whether your opinion of their accuracy is whether you grade that accuracy as elite, "average," or otherwise.[6]

Just as I don't think poll error calculation methods that cancel each other out should be ignored, as happened in the Proportional Method "What's for Lunch?" case, nor do I think that polls whose errors happen to cancel each other out should be praised.

Notes

1. Specifically, a systematic theoretical error that will always underestimate the potential impact of undecideds on eventual results.

   Kirkup, L., & Frenkel, R. (2006). Systematic errors. *An introduction to uncertainty in measurement: Using the GUM (guide to the expression of uncertainty in measurement)* (pp. 83–96). Cambridge University Press. https://doi.org/10.1017/CBO9780511755538.008

2. Changing Attitudes between poll and result will always result in a "net-zero" and can be positive or negative for any option. Undecided Ratio can only be positive because "undecideds" will always, eventually, be 0%, and they can only add to a "decided" option's total.

   True, someone who says they are currently "decided" can later change their mind to "undecided" between poll and election – attitudes (or who someone plans to vote for) can change multiple times between poll and election.

   But the poll being a snapshot (i.e. the simultaneous census standard) means that we can only detect the population's current preference.

   That is to say, considering only poll-to-election (not, e.g., poll-to-poll) Changing Attitudes must be net-zero.

3. Davila, E. P., Zhao, W., Byrne, M., Webb, M., Huang, Y., Arheart, K., Dietz, N., Caban-Martinez, A., Parker, D., Lee, D. J.. (2009). Correlates of smoking quit attempts: Florida Tobacco Callback Survey, 2007. *Tobacco Induced Diseases, 5*, 10. https://doi.org/10.1186/1617-9625-5-10

4. Asperin, A. M. (2021, September 15). *Nearly half of California voters decided who to support in recall election more than a month ago: Exit poll.* KRON4. https://www.kron4.com/news/politics/inside-california-politics/nearly-half-of-california-voters-decided-who-to-support-in-recall-election-before-today-exit-poll/

5. Cohn, N. (2014, November 4). *Exit polls: Why they so often mislead.* The New York Times. https://www.nytimes.com/2014/11/05/upshot/exit-polls-why-they-so-often-mislead.html

6. Silver, N. (2023, July 1). *Polling averages shouldn't be political litmus tests.* Silver Bulletin. https://www.natesilver.net/p/polling-averages-shouldnt-be-political

   "Look, there might be good reasons to exclude Rasmussen based on their methodology, although I'd note that their track record of polling accuracy is average, not poor."

16

Compensating Errors and Poll Masking

Important Points Checklist

- What compensating errors are, why they matter, and how they relate to poll data
- What a compounding error is
- How poll masking relates to compensating error, and why it should be accounted for
- The relationship of undecideds to a candidate's likelihood to "overperform" or "underperform" their poll number
- Why assuming the eventual preferences of undecideds is "known" is a bad approach

There's a term I've only previously seen used in accounting that can apply to both poll error calculation methods and to polls themselves: compensating error.

In accounting, a **compensating error** occurs when two or more errors offset each other resulting in a "net effect" of zero or near zero. For example, if I am owed $100 from two people, $50 each, but one person mistakenly pays me $60, and the other pays me only $40 – if I'm only aware of the fact that I have received $100, and not where the money came from, I might mistakenly believe both individuals paid what they owed me.[1]

It's possible that these errors would go unnoticed because no immediate red flags are raised. I suppose that's why this definition is part of accounting classes.

If accounting was your job, would you be comfortable saying that since you were owed a total of $100 and received a total of $100, that there was no error?

I would approach this differently, as would a good accountant who is serious about their job. I would characterize these as two $10 errors, try to figure out how and why these errors happened so that – maybe – they could be better prevented in the future. If $10 errors can go unnoticed just because, this time, they cancel each other out – who's to blame (or, I suppose, thank) when the $10 errors go in the same direction in the future?

DOI: 10.1201/9781003389903-16

If I only received $99 when I was owed $100, this would constitute **immediate evidence of an error.** Depending on your level of concern, you might conduct an analysis of where that error came from. So, where did it come from?

The most reasonable assumption is that one individual simply underpaid by $1. But do we know that for sure? Does it matter?

If we care about calculating error, it does. If it happens that one individual who paid us $55 mostly offset an individual who paid us $44, this does not constitute a $1 error. It constitutes *two separate errors* of "+5" and "−6." The fact that these errors, in this instance, happen to have nearly cancelled each other out does not mean the error(s) were small.

A "spread" accountant would say that there was almost no error in this scenario.

More to the point regarding compensating errors and improving the standards of analysis, which is worse: two errors of +$10 and −$10 or one error of −$2?

It depends on how "error" is defined.

An expectation about what the eventual result *"should be"* blinds you to the possibility of compensating errors.

The terminology used by the accounting text describes compensating error as a threat because it creates "lack of vigilance in error detection."[2]

By their nature, errors don't always cancel each other out. Given the fact that errors caused by any single factor in polls are typically (though not always) relatively small, it's likely that existing methods are only capable of detecting them if there are **compounding errors,** errors that accumulate in the same direction. By neglecting the possibility that polls that looked "right" could contain multiple errors because of their "poll versus eventual result" expectation, the calculators of "poll error" further demonstrate either inability or disinterest in truly quantifying poll accuracy and sources of error.

1. A poll that matches or nearly matches the eventual result isn't always without error.

2. A poll that reports values far from the eventual result doesn't always contain errors.

Both statements are theoretically, experimentally, and mathematically verifiable. They are indisputable. But by the current standards for calculating "poll error," these facts are considered both irrelevant and, paradoxically, incorrect.

As you know, errors can come from many different sources in a poll. But the fact that those errors can be "positive" *or* "negative" means that there is a chance of those errors compensating. How often compensating errors happen would be a wonderful topic of research for experts in this field – but it's a question they're not allowed to ask under the current definitions. If the spread of the poll is close to the spread of the election, they conclude the poll had little to no error, end of analysis.

This is supported by the fact that expert task forces are called to find out why the polls "missed" when the spread of an election is "off" by a lot, but if the spread is close, good job everyone. This is that "lack of vigilance" accounting textbooks talked about.

There are too many possible combinations of overcorrections, undercorrections, or uncorrected errors to name in poll data – but more vigilance would be an inevitable and welcome side effect of better poll accuracy definitions.

In the earlier accounting example, a +$10 error cancelled out a −$10 error. In polls, the margin of error could similarly cause a −2.5% poll error but be cancelled out by a nonresponse error of +2.5%. In this instance, the pollster's methods of correcting for nonresponse would go without scrutiny because their poll had "zero error."[3]

In my opinion, two errors of "+10" and "−10" that happen to cancel each other out should not be viewed more favorably than a single error of "−2." But the status quo in the polling industry insists that this be the case.

It's entirely possible that a poll contained error(s), but the error goes without scrutiny when it closely matches the eventual result.

Owed $100, received $100, therefore zero error! This is not a good approach.

While an analyst comparing poll to result only sees the reported number versus the eventual result, this problem can happen with the polls themselves, too.

If enough people change their minds, and/or a candidate doesn't sway many undecided voters, it is *possible* for them to underperform a poll number without any poll error; it does not constitute "immediate evidence of an error," but underperformance in most cases still represents a red flag.

This next part is very important because it incorporates a concept that I've hopefully explained well enough (a simultaneous census) and will be applied to the analysis of poll data, politics and otherwise, that follows:

The **observed poll number** in any plan poll is the number reported by the pollster, allowing for *all possible options*, including "undecided" and third parties; it is not the proportional adjustment that excludes undecideds and third parties, nor is it the spread that doesn't account for undecideds and third parties.

If a poll has a 4% margin of error, it should be interpreted as, approximately:

Option A: 45% +/− 4%[4]

Option B: 40% +/− 4%

Option C: 5% +/ − 4%

Undecided: 10% + / − 4%

This poll does not tell us:

"Option A + 5"

which is what the Spread Method would report and is used to form the entirety of its basis for calculating poll error. This poll also does not tell us:

Option A: 53%

Option B: 47%

which is what the Proportional Method would report and is used to form the entirety of its basis for calculating poll error. This type of reporting isn't only incorrect, it's actively spreading misinformation.

Again, if a poll has a 4% margin of error, it should be interpreted as likely to produce a simultaneous census of approximately:

Option A: 45% + / − 4%

Option B: 40% +/ − 4%

Option C: 5% +/ − 4%

Undecided: 10% +/ − 4%

This is an exhaustive list of how poll numbers should be interpreted if you're interested in doing so accurately.
Now that I've summarized how to view reported poll numbers, it's time to talk about the first step to analysis. While the simultaneous census value in most polling applications is unknown, it exists, and it is the basis by which we should measure poll accuracy.

It's very possible for a poll that reports "45%" to have a simultaneous census value of 43%. In that case, it overstated an option's support by 2%. Likewise, the simultaneous census value could actually be 47%, in which case the poll understated that option's support by 2%. Both are 2-point errors.

But understanding the impact of compensating errors allows us to make an important and valuable observation: *the direction of the error matters.*

With mind-changing being relatively uncommon in political polls (though certainly not 0), and undecided voters usually being more than a couple percent, even if a poll *overstates* a major candidate's simultaneous census value – it's unlikely they don't get *at least* that many undecided voters to "cover up" for that overstated simultaneous census value.

That's a lot of words, but it's not a complicated concept. Let me explain what I mean using the example above.

Let's assume that, somehow, we know Option A's simultaneous census value is actually 43%; the poll overstated them by 2%. This is within the poll's margin of error, but an overstatement (and poll error) nonetheless.

But eventually, those undecideds have to go somewhere. In order for Option A to underperform their poll number *observed* as 45%, even though it's *actually only* 43%, they would have to receive fewer than 2% of the undecided vote, *and* not receive any net benefit of changing attitudes. That is not an impossible feat, but given approximately 10% undecided, fairly classified as "unlikely."

Undecided voters potentially "covering up" a poll error is a phenomenon I think of as "masking."

Poll masking refers specifically to the fact that **undecideds alone will typically cause a candidate to outperform their observed poll number**, even if those poll numbers overstated a candidate's simultaneous census value. It's a phenomenon common enough to justify giving a name to, and an important concept to understand in order to accurately interpret poll data as it relates to eventual election results.

Poll masking is a special category of compensating, *but it is not a compensating error.* A compensating error is when two or more *errors* cancel each other out, like the margin of error and nonresponse error. While a poll understating a candidate's simultaneous census value is an error, the proportion of *how undecideds eventually decide is not.*

If you put together:

1. It's more likely that a candidate *outperforms* their poll number when there are more undecideds
2. It's more likely that a candidate *underperforms* their poll number when there are fewer undecideds

This brings us much closer to a more cohesive understanding of how polls work and why experts read them wrong.

It is, of course, not a particularly advanced concept to say "undecideds will eventually decide" or that "elections have 0% undecided, and polls don't," but that very logical truth – combined with the understanding, "how they ultimately decide *is not* something polls can, or try, to tell us" – would put you at or near the top in the world as it relates to understanding poll data today.

The status quo only considers poll versus eventual result, and given how it characterizes poll error does not seem to care about the underlying variables that comprise it. In the rare cases you can find that they do care about those underlying variables, they do not properly account for them; that's a problem for judging poll accuracy.[5]

Allocating Undecideds: Polls > Assumptions

If you're still hung up on the "simultaneous census value" being unknowable, at least for political polls, remember: not knowing it doesn't mean we should just guess at it.

Remember earlier when you were presented with the options for estimating some characteristic, and you were given the options to "guess" or "look at a poll"?

We agreed, I hope, that looking at a poll – or lots of polls – while imperfect, is certainly better than guessing or assuming.

Why can't we do that for how we allocate undecideds for our calculation of poll error as well? We can, and should.

By working backwards, instead of assuming poll error is known, we can incorporate real data – not assumptions – to better calculate each option's simultaneous census value at any given time.

To illustrate:

Consider an election with two candidates: Candidate A and Candidate B. We want to compare two competing polls: Poll 1 and Poll 2.

Assume we are able to know (or approximate, given poll data leading up to and after the election) two things:

1. There were a relatively large number of undecideds: say, 15%, up to a week before the election.

2. Undecideds ultimately favored Candidate B by a large amount: say, 65%–35%.

Disregarding changing attitudes for now, given what the Spread Method and Proportional Method assume undecideds should eventually do, you might see why their characterization of which polls are the most accurate would be very *inaccurate* in this scenario.

Take the example:

Poll 1

Candidate A: 45%

Candidate B: 40%

Third-party and Undecided: 15%

The Spread Method would say anything other than an eventual result of Candidate A +5 is caused by poll error and therefore estimates that each candidate should receive about 7.5% from undecideds.

The Proportional Method (ignoring undecideds) would say anything other than Candidate A receiving 45/85 (~52.9%) of the two-party vote share is caused by poll error. By extension, it also assumes that ~52.9% of undecideds should vote for Candidate A, or about 7.9% of 15%.

But knowing that approximately 65% of undecideds eventually voted for Candidate B, as post-election poll data could tell us, would allow us to much more accurately calculate poll error.

Both methods would give poor ratings to Poll 1's accuracy, for, absurdly, not having predicted that more undecideds would eventually vote for Candidate B.

I think we should incorporate known, even approximate, values about how undecideds eventually decided into our poll error calculation. The current methods do not.

Poll 2

Candidate A: 44%

Candidate B: 44%

Third-party and Undecided: 12%

Now consider Pollster 2, who has consistently reported higher numbers for Candidate B, and/or lower numbers for Candidate A than other polls. Having read about the problem of herding, you understand we should give this pollster the benefit of the doubt. At minimum, their polls deserve fair consideration, and to have their poll error judged by the same standards as other polls.

For this poll, both the Spread Method and Proportional Method assume that undecideds should split evenly; to them, for this poll, anything other than a very, very close final result is indicative of poll error. And a very close final result means a very accurate poll.

Final result

Candidate A: 47%

Candidate B: 48%

Both the Spread Method and Proportional Method would agree: Poll 2 was extremely accurate! Or was it?

The Spread Method and Proportional Method have no way to account for what undecideds eventually decide. They assume their assumptions are correct, and that any deviation from that assumption is a poll error.

Where is the other 5% in this final result, you might be wondering? Third-party voters. Another layer to why political polls are the hardest class of polls. But both the Spread Method and Proportional Method consider "undecided" and "third-party voter" to be equivalent for their calculations.

The Spread Method would characterize Poll 1 as:

Poll Spread	Result Spread	Poll Error
+5	−1	6

And the Spread Method would break down Poll 2 as

Poll Spread	Result Spread	Poll Error
+0	−1	1

But this definition doesn't incorporate what we know. We know that about 65% of undecideds eventually voted for Candidate B, and we should account for it.

If we know about 65% of undecideds eventually decided on Candidate B, and still assuming (for now) Changing Attitudes are ~0:

$$\text{Adjusted Poll Value} = \text{Reported Poll Value} + / - Changing \ Attitudes + Undecided \ Ratio.$$

The adjusted Poll 1 value for Candidate B, who received 65% of the 15% undecided, or 9.75%:

Poll Candidate B	Changing Attitudes	Undecided Ratio	Adjusted Poll
40%	0	9.75%	49.75%

And for Candidate A, who received 35% of the 15% undecided, or 5.25%:

Poll Candidate A	Changing Attitudes	Undecided Ratio	Adjusted Poll
45%	0	5.25%	50.25%

$$\text{Adjusted Spread Poll Error} \approx \text{Adjusted Poll Value Spread} - \text{Eventual Result Spread}.$$

In total, the Adjusted Spread for Poll 1 would say:

Adj. Poll Spread	Result Spread	Poll Error
Candidate A +0.5	Candidate A −1	≈1.5

For Poll 2, with its 12% undecided and 65% going for Candidate B, that's an additional 7.8%

Poll Candidate B	Changing Attitudes	Undecided Ratio	Adjusted Poll
44%	0	7.8%	51.8%

And for Candidate A, receiving 35% of the 12% undecided, or 4.2%

Poll Candidate A	Changing Attitudes	Undecided Ratio	Adjusted Poll
44%	0	4.2%	48.2%

So in total, the Adjusted Spread Method for Poll 2 would say:

Adj. Poll Spread	Result Spread	Poll Error
Candidate A −3.6	Candidate A −1	≈2.6

In Summation

Poll	Method	Poll Error
1	Spread	6
1	Adjusted Spread	≈1.5
2	Spread	1
2	Adjusted Spread	≈2.6

Neither Adjusted Spread in this case is terrible, but according to Adjusted Spread, Poll 1 is clearly better.

On the other hand, under the traditional Spread Method, not only would the opposite be reported – that Poll 2 was the more accurate one, with just a 1-point error – but Poll 1 (who committed the sin of not guessing how undecideds would eventually vote) is characterized as off by 6. Which poll do you think was more accurate, given what we know about undecideds?

Poll 2 was a beneficiary of masking. While their poll number overstated Candidate B's simultaneous census value, the fact that they received a higher percentage of the undecideds concealed this.

I don't think overstating a candidate's poll number should be rewarded with a higher "accuracy" rating just because the pollster was lucky with how undecideds eventually decided.

On to the Proportional Method. Remember, the Proportional Method is based only on two-party vote share, and we only need to consider one candidate's poll-versus result proportion to calculate poll error.

The Proportional Method would characterize Poll 1 as

Poll Proportion	Result Proportion	Poll Error
Candidate A: 52.9%	Candidate A: 49.5%	3.4

And the Proportional Method would break down Poll 2 as

Poll Proportion	Result Proportion	Poll Error
Candidate A: 50.0%	Candidate A: 49.5%	0.5

Again, by this method, Poll 2 is rated as much more accurate than Poll 1. Poll 2 grades out as nearly perfect, in fact.

But when we apply what we know about undecideds, as with the Spread Method, a different picture emerges.

Originally, the proportion calculation for Poll 1 was 45/85, but now that we can account for the 15% undecided, knowing that candidate A received 35% of them (or 5.25%), the proportion becomes: $45 + 5.25$ (50.25) and $85 + 15$ (100).

Poll 1's adjusted poll value for Candidate A says 50.25/100, or 50.25%.

Poll Candidate A	Changing Attitudes	Undecided Ratio	Adjusted Poll
45%	0	5.25%	50.25%

Given the eventual result proportion of 47/95 (49.5%), the adjusted Proportional Method would classify this Poll 1's error as:

Adjusted Proportional Poll Error ≈ One Candidate's Adjusted Poll Proportion − Same Candidate's Eventual Result Proportion.

Poll Proportion	Result Proportion	Poll Error
Candidate A: 50.25%	Candidate A: 49.5%	≈ 0.75

Poll 2's adjusted poll values, for Candidate A: 48.2/100, or 48.2%.

Poll Candidate A	Changing Attitudes	Undecided Ratio	Adjusted Poll
44%	0	4.2%	48.2%

The Adjusted Proportional Method would rate Poll 2's accuracy as:

Poll Proportion	Result Proportion	Poll Error
Candidate A: 48.2%	Candidate A: 49.5%	≈1.3

Again, while neither Adjusted Proportion rates either poll as awful, it rates Poll 1 as much better. That is in contrast to the Proportional Method, which said Poll 2 was much better, and Poll 1 was far worse.

To summarize the Proportional Method errors for this scenario:

Poll	Method	Poll Error
1	Proportion	3.4
1	Adjusted Proportion	≈0.75
2	Proportion	0.5
2	Adjusted Proportion	≈1.3

Given what we know about undecideds, which estimate of accuracy is better?

Finally, a side-by-side-by-side. Four different methods, two different polls – with eight different approximations of poll error – given the polls, results, and approximate Undecided Ratio, each method would report the following:

Poll	Method	Poll Error
1	Spread	6
1	Adjusted Spread	≈1.5
1	Proportion	3.4
1	Adjusted Proportion	≈0.75
2	Spread	1
2	Adjusted Spread	≈2.6
2	Proportion	0.5
2	Adjusted Proportion	≈1.3

Adjusted methods agree that Poll 1 was more accurate, but the traditional methods strongly favor Poll 2.

While these examples of unnamed pollsters may seem inconsequential, remember: their reputations and public perception of their accuracy are determined largely, if not entirely, by the traditional (not adjusted) definitions of poll error.

It should be highly troublesome to anyone who cares about accurate definitions that a moderately large number of undecideds mildly favoring one option could cause the characterization of which poll was most accurate to be very incorrect. Also troubling, consider a scenario in which most polls are similar to Poll 1. The unanticipated (and unaccounted for) undecided behavior would lead an analyst to believe "the polls" were inaccurate.

For those interested in updating and improving currently used poll error calculations, feel free to build on anything I've included: don't just accept it as "correct" or "complete."

And I'll restate, if being able to calculate poll error yourself is not of interest to you, that's okay too. This chapter provides another big "why" leading up to the more direct explanation of how polls should be read. Seeing how errors can cancel each other out, causing less accurate pollsters to be rated as more accurate, demonstrates the need for better methods.

The point I think applies for all readers, and will be applied moving forward: do you see what factors might cause these methods to evaluate an inaccurate poll as accurate, and vice versa? If nothing else, given the large differences in what these accepted methods call "errors" you should understand characterizing polls as "off by" some specific number, as existing analysts do, is a dishonest approach.

Check Your Progress

There's nothing inherently wrong with using hypothetical numbers to illustrate some important concepts about polls. I doubt you could find a mathematician who hasn't, at some point in their lives, calculated the time a hypothetical train traveling at some speed in some direction would reach its destination. Being able to calculate when trains will reach their destination is not sufficient to make one a mathematician, but it is a pretty basic requirement.

The hypothetical numbers I've used to this point, to illustrate various concepts, aren't intended to make you a mathematician but to give you all of the basic requirements needed to read political polls.

Veterans of political polls might have recognized some of the numbers I used as barely hypothetical at all.

Fortunately, as I promised, you do not need to be a math or stats wiz to understand how to read polls.

There's also nothing inherently wrong with conducting a poll on a sample of 80 for a population of 91. The underlying application is the same as a poll of 600 regarding a population of millions. Asking someone "What's for Lunch?" provides an apolitical topic to help understand how plan polls operate: what they tell us, and don't tell us.

By now, I'm guessing some readers are interested in conducting polls and collecting data of your own, while others just want to know "how should I read political polls?"

Finally, our interests converge.

Of those interested in conducting polls, you might not have access to a large enough population. Of those not interested in the formulas or calculations, having braved the "underlying reasoning" portion of the book, you're all now ready and qualified to move on to political polls.

Let's start with what that data might look like *if all the polls were ideal*. Is that even possible?

Yes.

Notes

1. Accounting Corner. (2023, May 5). *Compensating error.* https://accountingcorner. org/compensating-error/
2. Accounting Corner. (2023, May 5). *Compensating error.* https://accountingcorner. org/compensating-error/
3. The difference between net error and absolute error is not unimportant because a pollster with two small but compounding errors of "+1" would be viewed less favorably (and reported as less accurate) than a pollster with two larger but compensating errors. Even the simultaneous census standard can only measure net error, but that should be the starting point, not the end of the analysis.
4. As the proportion moves away from 0.5, the MOE decreases. I'm not getting into it here for reasons discussed previously. See notes 3–5 in Chapter 8. This applies to later references to these data as well.
5. Given the number of variables that comprise poll error, their varying weight of contribution, the uncertainty around those measurements, and the risk of compensating errors, I believe a propagation of error function with square terms is one (if not the only) appropriate technique for quantifying poll error. The current uses of root-mean-square error to poll-versus-election results indicate a belief that compensating errors can't exist or are negligible, and that the "true value" intended to be measured by the poll is the election.

17

Welcome to Mintucky

Important Points Checklist

- What confounding variables are, why they matter, and how they relate to poll error calculations
- The danger of using assumptions in place of data in research
- Why analogue research can be useful
- What the field of inferential statistics studies
- Some ways the number of undecideds in a poll can impact the final result
- The biggest step(s) in going from polls, to poll average, to prediction
- Why the Proportional Method's reporting can be described as "fraudulent"

Chapter 1 introduced "variables of interest" with the smoking and lung cancer example. Researchers understood a lot of other factors that could, and do, play a role in developing lung cancer besides smoking. Failure to consider the possibility that other causes might also contribute to lung cancer, even to a lesser extent, would have been a major oversight by researchers – and possibly undermine the trustworthiness of their findings.

A lung cancer researcher should account for the fact that coal miners who smoke probably don't have the same risk for lung cancer as teachers who smoke. Smoking and lung cancer might be the variables of interest for the researcher, but smoking isn't the only variable that can impact the eventual result.

In this case, if a researcher considers only smoking and lung cancer but fails to account for the fact that environmental hazards can also contribute to lung cancer, the environmental hazard represents a **confounding variable:** a variable outside of those being studied that may "distort or mask" the relationship between those being studied.[1]

By considering only smoking and lung cancer, researchers would be unable to answer questions of importance, like how much does working in

DOI: 10.1201/9781003389903-17

a coal mine contribute to the development of lung cancer, even if they don't smoke?

In fact, it's this understanding (and not accepting assumptions where data should be) that has allowed researchers to build on the smoking and lung cancer findings. One such finding is the fact that secondhand smoke is also a cause of lung cancer.[2] But it wasn't until *after* the relationship between smoking and lung cancer was understood that the research could be built upon.

If experts had compared the prevalence of lung cancer in people who smoke to the prevalence of lung cancer in people who don't, and then concluded that the discrepancy equaled the amount smoking contributes – without correcting or controlling for known confounders such as environmental hazards or age – that would have been some unreliable research. Worse yet, if everyone in the field had simply assumed that "other factors" were known, or zero, there would have been no further research.

Questions like "why do people who don't smoke sometimes get lung cancer, and why?" are the kind of questions that make sense only if you don't place assumptions where data should go.

In the case of existing poll error methods, the variables of interest are identified as "reported poll result" and "eventual election result."

But by considering only poll and eventual result, those methods are unable to answer questions of importance, like how much did the eventual preference of undecideds contribute to the discrepancy between the poll and eventual result?

And here's how the smoking and lung cancer analogy, the understanding of confounding variables, and the ability to do better research when those confounding variables are accounted for relate to poll data:

Why do elections with a higher number of undecideds tend to have much higher poll error, according to the current methods?

> This is the first question I asked in my research, and this question
> can only make sense once you stop using assumptions in places
> where you should use data.

The conclusion I've reached, substantiated, and will continue to provide evidence for: elections with higher undecideds tend to have much higher "poll error" because undecided voters represent a confounding variable to the currently used and accepted measures of poll accuracy.

> If a confounding factor ... is recognized, adjustments can be made in the
> study design or data analysis so that the effects of confounder would be
> removed from the final results.[3]

The Adjusted Methods I introduced earlier are neither arbitrary nor subjective. Changing attitudes and undecideds can, and do, for every election, distort the association between poll result and eventual result. The amount it distorts isn't always the same, but it's always there.

Misreporting confounding variables is one of the simplest reasons the numbers given by an analysis can be invalid.[4] Adjusting for them is a necessary practice in any science; ignoring confounding variables (or making up definitions that assume their value to be 0) doesn't make them go away.

With most polls, it's typically impossible to fully account for these confounding variables because a simultaneous census is impossible. Typically, but not always.

Up until now, the main reason I requested that you not skip ahead is that there are a lot of underlying concepts to understand before tackling political polls.

With "What's for Lunch?", and assistance from the near-simultaneous census, we were able to look at the drastically different conclusions the two leading "poll error" methods would have had for that poll's error. It was also shown how both methods were inaccurate in their measurements: one with an indefensibly poor calculation, the other with a large compensating error.

Now, there are no more assumptions, hypotheticals, or near simultaneity needed to demonstrate these concepts.

When I put together this experiment, my goal was to mimic the conditions of political elections and the polls that precede them in a way that anyone could replicate. I wanted to collect realistic data but also perform the research as an "observer," and not someone who already knew the answer.

This is a type of analogue experiment.

Analogue research allows investigators to "exercise tight control over the implementation" of the variables of interest, including confounding variables.[5]

Analogue research should not be considered a gold standard for scientific research. What it does, however, is replicate under controlled and repeatable conditions something similar to a situation that occurs in real life.

This analogue experiment will help to cement the foundation of what I've talked about to this point and build on it all in one place before advancing to real political polls.

The Setup

I bought thousands of mints, some red and some green. When they were delivered, I gave my wife – who totally volunteered to help me – two simple, but important instructions:

1. Put most of them – but not all of them – in a giant bin. Mix well.

I knew how many I had purchased (an equal number of red and green) and that would make for a frustrating election. It's okay that I knew there were (probably) an approximately equal amount of red and green in the bin, but there was no way for me to know exactly how many of each there were beforehand.

2. The big step. Obscure the contents of some of the mints – at least a couple hundred of them.

I gave no instructions regarding which mints to obscure.

The mints I purchased were wrapped in transparent plastic. To obscure the contents of the mints, she used aluminum foil.

Now, I have a giant bin of reds, greens, and foil-wrapped "undecideds" of unknown contents and quantities. I later started referring to these "undecideds" as "silvers" solely for the color and not at all in reference to the famous poll statistician of a similar name.

And as a nod to the state in which I was born and raised, and my love for terrible puns, I referred to my giant bin of mints as "Mintucky."

The stage was set.

To give a little humanity to the experiment, I even named the candidates on the ballot:

Ricky Red, the Republican

Grace Green, the Democrat

Mints with the same color as the candidate's name would represent a supporter of that candidate. Silvers would be counted as "undecided."

Unlike in "What's for Lunch?" where I sneakily conducted a simultaneous census without you knowing, I can promise that no such secrets will be kept this time. The data, in the order I report it here, is the same order and time I accessed it myself.

Ready to make a prediction regarding the eventual winner – and percentage of the vote received – for each candidate?

Or would you like some polls to inform your prediction?

For those interested in recreating this experiment, you can consider marbles; mints; beans; or any number of small, countable objects. Remember to include undecideds, if you'd like to mimic political data.[6]

Data Collection

My data collection consisted of mixing the bin, taking a few hundred mints from it at random, and counting them: red, green, and undecided "silvers."

After counting the mints, I replaced them in the bin and mixed. These random samples were my polls.

The sample sizes for my polls ranged from the 400s to the 700s. There were a lot of mints to count for each poll, but each gave a similar margin of error.

I was not content with simply reaching in and grabbing a few dozen handfuls each time I conducted a poll, so I decided to enlist the help of a few "pollsters," each with slightly different sampling techniques. Some poll companies more creatively named than others.

1. **Shovel**. Just a snow shovel. This pollster digs deep and takes a sample as vertically as possible from the bin, but not from the edges or the exact middle.
2. **Twovases**. I used two vases, one in each hand. These two pollsters working as one company sample only from the sides of the bin.
3. **Bowl Poll Co**. I called it a bowl, though I was later informed it's a colander, this pollster takes samples from as near to the middle of the bin as possible.
4. **Bevtub Polling**. A "beverage tub" – a large container of almost equal width to my giant bin – this pollster gets a very wide sample from the surface.
5. **Box**. Just a box. It used a "mixed-methods" approach taking samples from both the edges and the middle.

Do these differing sampling methods make a difference? Doubtful. But in giving each pollster unique characteristics, with each collecting their own data, we have polls (and eventually, poll averages) very closely resembling what you might see in a real election from various sources. If any of the methods do somehow prove to be better, maybe we can learn something from that, too.

I started taking "polls" about four weeks before my election, with more polls closer to "election day."

First Polls for Mintucky

Pollster number one is Shovel. I dug as far down into the Mintucky bin as I could and lifted a huge pile of mints onto my living room floor. Four hundred ninety-nine mints, it turned out. Some red, some green, and some silver.

Time to count.

Below is the first poll from Mintucky!

Mintucky Special Election			
Candidate		Poll (%)	
Ricky Red (R)		46	
Grace Green (D)		44	
Undecided		10	
Nov 25, 2022	*Pollster: Shovel*	*499 mints*	*Margin of Error +/– 4%*

Of course, given an unhealthy obsession with "spread," the actual reporting would probably look something like:
"Ricky Red up 2%" or "Ricky Red up 2 points in the polls."
With that, a few questions to test your understanding:

1. Does this poll *predict* that Red will win? Why or why not?
2. Does this poll say that Red *will win by 2*? Why or why not?
3. What does this poll tell us about how the undecideds will eventually decide?

Unfortunately, even experts often answer these basic questions incorrectly. In his book *Strength in Numbers: How Polls Work and Why We Need Them*, G. Elliott Morris claims:

> the Marquette poll ... predicted that the Democratic candidate for governor in 2018 ... would win the election by a one-point margin.[7]

It's unclear which Marquette poll Morris was referring to here because in their poll closest to the election, they reported a "tie" (47%–47%) and the one prior to that reported the Republican candidate – not the Democratic one – as having a "one point lead" (47%–46%).[8]

Regardless, to ignore the third-party candidate in this race (who was polling as high as 5% in the final two Marquette polls, but received under 1% of the eventual vote) and to also disregard a not-insignificant number of undecideds in such a close race perfectly encompass the fallacious reasoning demanded by the Spread Method.

Hopefully, at this point, you can easily spot these common misunderstandings while reading along. The Shovel poll did not and does not "predict" that the Republican candidate for Mintucky would win by a 2-point margin, nor did the 2018 Marquette poll "predict" a 1-point margin for the Democratic candidate in Wisconsin, or for anyone. The applications of what inferential statistics teach for each of these scenarios are no different.

Unfortunately, for now, it's up to you as an informed consumer of poll data to understand these common misconceptions and avoid them. I hope the media and experts will correct their characterizations.

But to review, answering the questions above:

1. No. Polls do not make predictions about "who will win," that's called a forecast. Polls are neither predictions nor forecasts; polls are, at their absolute best, observations about a simultaneous census: what is true right now.

2. No. Even with zero undecideds, the margin of error alone would make any claim about what a poll says regarding "how much" someone would win by misleading at best – and there are lots of undecideds.

3. Nothing. A poll can't, nor does it try to, tell us how undecideds will eventually decide.

Nonetheless, even with just one poll, I'm going to violate my sworn oath as a Stats Guy and ask you to make a prediction from it, using Figure 17.1. We'll compare this prediction to your final prediction after seeing all of the polls, and then to the eventual result.

Remember, although there's a third option in the poll of "undecided," this is a two-way race. There are only two options on the ballot: Red or Green. In the end, Red and Green will account for 100% of the vote.

What do you think Ricky Red's final vote share will be, and how confident are you in that prediction? You can "slide" each of these options to a higher or lower number.

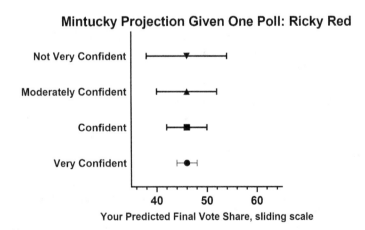

FIGURE 17.1
A chart that shows sample options with varying levels of confidence for Ricky Red's eventual vote share. Chart by the author.

What this poll tries to tell us is that the current composition (i.e. a simultaneous census) of the bin is as follows:

- Ricky Red 46% +/– 4%
- Grace Green 44% +/– 4%
- Undecided 10% +/– 4%[9]

And that's it.

This is a valuable piece of data for informing a prediction of the eventual result, but it is not in itself a prediction of the eventual result; anyone who states or believes otherwise is wrong.

It is certainly possible that this poll overstates or understates Ricky Red's simultaneous census value, which would rightly be called a poll error. But if you think undecideds will split 50:50, or proportional to decideds, or any other way (or you're unsure) – that's all part of your prediction – nothing to do with the poll's accuracy.

Given that we know there are only two candidates, undecideds will eventually decide, and that the eventual result must add up to 100%, do you think it's more likely that Ricky Red overperforms his poll number in the eventual result, or underperforms it?

If this sounds like an easy question, good; you've now advanced beyond "spread" and "proportional" understanding of what poll data tells us. Given that he's polling at 46% with 10% undecided, it's far more likely he overperforms this "46%" in the eventual result than underperforms it.

We'll get into the differences between predictions and forecasts later, but for now I just want you use this poll to make your best estimate of how you think this election will end up. If granted a +/– 4% margin of error, what number would you say Ricky Red and Grace Green end at?

Even though this experiment involves mints and shovels and will soon involve other silly "pollsters" I feel obligated to remind you, as with "What's for Lunch?": statistically speaking, this is as precise of an application of polling as possible; we are taking a random sample from a larger population with the goal of knowing something about that population. No different than political polls. Collectively, these are applications of **inferential statistics**.

But in political polls, because of the large number of variables, measuring accuracy is complicated.

When inferential statistics involve humans (especially in plan polls, and *especially* political polls), it is often very hard to know *exactly* who your population is. In elections, the goal is to take a poll of actual voters – but the best pollsters can possibly do before an election is find Likely Voters. If pollsters don't get it right, this introduces the possibility of an actual source of poll error – **Frame Error**.

Beyond that, even if pollsters do an outstanding job of figuring out who is likely to vote, it can be an even bigger challenge to contact a random sample of that group and figure out what to do when some of them don't respond. This opens the door for **Nonresponse Error**.

That is to say, in a "normal" political poll, we can't quantify precisely who will eventually vote, but in Mintucky we can: all of them.

In a normal political poll, we can't quantify precisely if everyone will respond, or if groups that don't respond might have something in common; in Mintucky we can quantify this precisely: there will be 100% response.

Every poll in Mintucky has a 100% response rate, and the sample is both random and taken from the population of interest.

This experiment controls for Frame Error and Nonresponse Error: they are zero.

What else is there? Enter, the typically confounding variables for poll error calculations: changing attitudes and undecideds.

Working with mints has its advantages. Not only do we know who our exact population is, Mintuckians are transparent about their vote preference (literally, with the exception of the silvers) and never change their mind.

"Changing Attitudes" in Mintucky, considering the mints can't change color, is zero.

By eliminating the causes of actual poll error, and one major confounding variable, the Mintucky experiment has done what most people would believe to be impossible because our minds have been contaminated by political polls.

The only variable that could cause any discrepancy between each Mintucky poll and the eventual result, other than the statistical margin of error itself, are the undecideds.

Mintucky has a lot to teach us about how to read real political polls.

Let's continue.

Mintucky Special Election			
Candidate		Poll (%)	
Ricky Red (R)		46	
Grace Green (D)		45	
Undecided		9	
Nov 28, 2022	*Pollster: Twovases*	*460 mints*	*Margin of Error +/– 4%*

Here, Twovases polling has entered the mix. Next, a poll taken the same day from Bevtub Polling.

Mintucky Special Election			
Candidate		Poll (%)	
Ricky Red (R)		48	
Grace Green (D)		45	
Undecided		8	
Nov 28, 2022	*Pollster: Bevtub Polling*	*682 mints*	*Margin of Error +/– 4%*

So far, nothing too wild. Bevtub's poll has the largest sample so far, reporting Ricky Red's lead as slightly higher than other polls.

Which means it's time for the question again:

Compared to what?

You now know (and hopefully understand) that each poll's error, properly measured, must be compared to a simultaneous census, not eventual result.

What these polls tell us is that if we were to take a simultaneous census of the population by counting all the mints right now (including undecideds), each option would almost always fall within that margin of error.

I think that's a pretty valuable tool – but it requires you to read that tool correctly.

Instead of seeing Ricky Red's poll number as "48," as an individual poll may report, you should think of it as "probably somewhere between 44% and 52% right now" to account for the margin of error.

A similar calculation should be performed for Grace Green's range, given the margin of error and reported poll results, and likewise for the "undecided" range.

And remember that those poll ranges are *before* accounting for how undecideds eventually decide.

If those ranges feel unimpressive to you, it's because this poll doesn't tell us, with certainty, an exact level of support. Don't sweat individual polls. Thank them for the data, throw it in the average, and move along.

In addition to being more statistically sound, this "throw it in the average" approach allows you to not obsess or stress over individual polls. Yes, polls provide valuable data. If we want to make a prediction, we should use polls to inform them. But individual polls, even if "ideal" as in Mintucky, are limited in what they can tell us.

If you're concerned about which polls to "throw in the average" or how much each poll should be considered: fair questions. Just like different pollsters use different weights to get from raw data to reported data, different poll averagers (or aggregators) use different weights to get from simple average to reported average: aggregators weight for factors like sample size, recency, pollster quality, and more.

Just looking at Mintucky for now, and the various pollsters and their methodology, the average of a lot of polls is probably going to be better than guessing – or trying to pick which poll is best.

Time for a few more polls:

Mintucky Special Election	
Candidate	**Poll (%)**
Ricky Red (R)	47
Grace Green (D)	46
Undecided	7
Dec 1, 2022 *Pollster: Shovel* *454 mints*	*Margin of Error +/– 4%*

Mintucky Special Election	
Candidate	Poll (%)
Ricky Red (R)	46
Grace Green (D)	45
Undecided	9
Dec 5, 2022 Pollster: Box 572 mints Margin of Error +/– 4%	

Mintucky Special Election	
Candidate	Poll (%)
Ricky Red (R)	50
Grace Green (D)	43
Undecided	7
Dec 5, 2022 Pollster: Bowl Poll Co. 441 mints Margin of Error +/– 4%	

To this point, the poll data had been largely unremarkable – until Bowl Poll Co. released their results.

Remember, while I report the dates I take the polls, nothing actually has changed between each poll – at least not yet. All of these polls are coming from the exact same population of mints. Ricky Red has now polled as low as 46% and as high as 50%. Grace Green as low as 43% and as high as 46%. Is one pollster more accurate, or is the truth somewhere in between?

And of course, the ever-important question is as follows: more accurate compared to what?

Mintucky Election Approaches

With Election Day on December 23, some voters are starting to make up their minds. That is to say, in order to more closely mimic human political elections, in which undecideds decrease as election day gets closer, I decided that some undecided voters should no longer be undecided.

For Mintucky to produce data as similar as possible to real elections, some of the undecideds needed to be unwrapped. But which ones? How many? The process needed to be, for me, as "hands off" as possible. I want to observe the results just as they are without influencing them.

Enter, again, my lovely assistant. Having wrapped lots of mints of unknown-to-me quantities to make them "undecided" already, I now requested that she unwrap some of the silvers. How many? I said "more than a few, but fewer than half."

Now, like in real elections, undecideds have started to decide as the election approaches.

A few minutes later, the pollsters went back to work.

Here's a running list of poll results with some new polls added. Each poll has approximately the same 4% margin of error, with newer polls listed at the top.

Poll	Date	Ricky Red (R) (%)	Grace Green (D) (%)	Undecided
Bevtub Polling	Dec 14, 2022	49	44	6
Box	Dec 14, 2022	50	45	5
Shovel	Dec 14, 2022	47	47	6
Bowl Poll Co.	Dec 10, 2022	51	42	6
Twovases	Dec 10, 2022	54	43	4
Bowl Poll Co.	Dec 5, 2022	50	43	7
Box	Dec 5, 2022	46	45	9
Shovel	Dec 1, 2022	47	46	7
Bevtub Polling	Nov 28, 2022	48	45	8
Twovases	Nov 28, 2022	46	45	10
Shovel	Nov 25, 2022	46	44	10

If I hadn't told you that some of the undecideds were unwrapped between the polls on Dec. 5 and Dec. 10, would you have known it? Maybe, maybe not. But like I said in the introduction, I'm providing full disclosure throughout this polling experiment.

Now, these Dec. 10 polls will also mark a turning point for the election results: no more undecideds will be unwrapped until election day when all votes are counted.

There are a few more polls to be taken before then, but here's how the election day counting will work:

Before unwrapping the silver "undecideds," I will count each: red, green, and silver.

That means I will be counting what is – in every sense – a simultaneous census of each poll taken on and after Dec. 10.

After counting each red, green, and silver, I'll unwrap the silvers and count them. This added layer of analysis that we typically don't get in polls – knowing how undecideds eventually decided – will allow us to calculate "poll error" as it should be – and ultimately how we should apply this knowledge when reading political polls outside of Mintucky.

For those who want an extra challenge: we don't know (and can't know) either the number or the proportion of undecideds who were unwrapped before Dec. 10. Can we still calculate poll error for those, older polls, even just approximately? How would you approach that calculation?

Final polls from Mintucky!

Poll	Date	Ricky Red (R) (%)	Grace Green (D) (%)	Undecided (%)
Bevtub Polling	Dec 21, 2022	48	47	5
Twovases	Dec 21, 2022	49	45	6
Shovel	Dec 20, 2022	45	47	8
Box	Dec 20, 2022	48	45	6
Bowl Poll Co.	Dec 19, 2022	44	50	6
Bevtub Polling	Dec 14, 2022	49	44	6
Box	Dec 14, 2022	50	45	5
Shovel	Dec 14, 2022	47	47	6
Bowl Poll Co.	Dec 10, 2022	51	42	6
Twovases	Dec 10, 2022	54	43	4
Bowl Poll Co.	Dec 5, 2022	50	43	7
Box	Dec 5, 2022	46	45	9
Shovel	Dec 1, 2022	47	46	7
Bevtub Polling	Nov 28, 2022	48	45	8
Twovases	Nov 28, 2022	46	45	10
Shovel	Nov 25, 2022	46	44	10

Grace Green getting 50% in one poll by Bowl Poll Co. isn't a typo. Is Bowl Poll Co. biased or otherwise unreliable? In real life, with real pollsters, that's a valid question. But in Mintucky? It's just one more chance for me to say that fluctuation is normal and expected.

Before getting to the election results, instead of looking at lots of numbers and trying to figure out what they mean, let's break it down to a polling average.

First, here's an overall average going back to the first polls:

Poll	Date	Ricky Red (R) (%)	Grace Green (D) (%)	Undecided (%)
Poll Average	**Nov 25–Dec 21**	**48**	**45**	**7**
Bevtub Polling	Dec 21, 2022	48	47	5
Twovases	Dec 21, 2022	49	45	6
Shovel	Dec 20, 2022	45	47	8
Box	Dec 20, 2022	48	45	6
Bowl Poll Co.	Dec 19, 2022	44	50	6
Bevtub Polling	Dec 14, 2022	49	44	6
Box	Dec 14, 2022	50	45	5
Shovel	Dec 14, 2022	47	47	6
Bowl Poll Co.	Dec 10, 2022	51	42	6
Twovases	Dec 10, 2022	54	43	4
Bowl Poll Co.	Dec 5, 2022	50	43	7
Box	Dec 5, 2022	46	45	9
Shovel	Dec 1, 2022	47	46	7
Bevtub Polling	Nov 28, 2022	48	45	8
Twovases	Nov 28, 2022	46	45	10
Shovel	Nov 25, 2022	46	44	10

While an average of all the polls is better than no polls, or any individual poll, it's possible we can be a little more accurate in our poll average representation with a **recent poll average**. Including very old polls with many more undecideds could, possibly, skew our results.

How many polls should you include? How far back should your poll average go? Should more recent polls receive more weight than slightly older ones? All wonderful questions. There's no perfect answer. If you want to come up with your own poll averaging methodology, or look at established ones, that's up to you.

The real question is: what do those poll averages tell us? That's what Mintucky and the chapters that follow aim to explain.

In Mintucky, we know when the undecideds solidified. True, in real elections, we don't have that benefit; hence, the reality that there's no perfect answer to calculating poll averages. Nonetheless, a **simple recent average** – picking a cutoff date and taking the average of all that come after it – is a perfectly reasonable approach.

For my bona fide statistician friends, you can critique this simple recent average approach as messy if you want – in Mintucky it's not, but in real life it can be. How an aggregator decides which polls to include in their average, plus if/ how they weight for recency, sample size, and other factors, creates the possibility that a poll average *introduces errors not present in the polls themselves.* That's a topic with enough content for another book. But knowing how aggregators calculate their averages, like knowing how poll error calculators make their calculations, creates yet another opening for conflicting interests – whether partisan bias, mediafication, or both – to try and put their thumbs on the scale.

But for now, I'll just say that a poll average will almost always be preferable to trying to guess which individual polls or pollsters will be most accurate.

In Mintucky, knowing the date in which some undecideds decided (and after which none more did) gives us a huge analogue experiment benefit: we will have 10 polls which represent a random sample from the population of interest, and will have a simultaneous census to compare them to. Whether sorting by pollster, or overall average, we can analyze the data and calculate poll error.

Here is the simple recent average of polls on and after the Dec. 10 cutoff:

Poll	Date	Ricky Red (R) (%)	Grace Green (D) (%)	Undecided (%)
Poll Average	**Dec 10–21**	**48.5**	**45.5**	**6**
Bevtub Polling	Dec 21, 2022	48	47	5
Twovases	Dec 21, 2022	49	45	6
Shovel	Dec 20, 2022	45	47	8
Box	Dec 20, 2022	48	45	6
Bowl Poll Co.	Dec 19, 2022	44	50	6
Bevtub Polling	Dec 14, 2022	49	44	6
Box	Dec 14, 2022	50	45	5
Shovel	Dec 14, 2022	47	47	6
Bowl Poll Co.	Dec 10, 2022	51	42	6
Twovases	Dec 10, 2022	54	43	4

Interestingly, the "spread" between the poll average both before and after the Dec. 10 cutoff is 3. Does this point to a 3-point victory for Ricky Red?

Bowl Poll Co. reported on Dec. 10 that Ricky Red had the support of 51% of decided voters, while Grace Green had only 42%.

Not long after, they reported Red at just 44% and Green at 50%. We know, thanks to the experimental setup, that this fluctuation is probably just noise. But if these were real-life polls, would we be able to dismiss such drastic differences? Or might these polls be the subject of long debates and controversy?

Before we get to the results, I want you to make your final prediction using Figure 17.2.

Given the poll averages, what percent of the vote do you think Ricky Red will receive? Are you more or less confident in that prediction than your original prediction when you had only one poll?

Moreover, given again that you know there are only two possible options, and that those options must add up to 100%, do you think it's more likely that Red overperforms his poll average, or underperforms it?

Remember, the scale given here is based on the poll average, but you can "slide" your estimated outcome either way. This illustrates a *major* difference between poll/poll average, and prediction.

How different is your estimate for Ricky Red's vote share, compared to when you only had one poll? Did it change your level of confidence in your prediction?

Before getting into the individual polls, let's look at what each poll error method would say for this final poll average.

The Spread Method would say that this poll average says:

Ricky Red 48.5% (+3)

Grace Green 45.5%

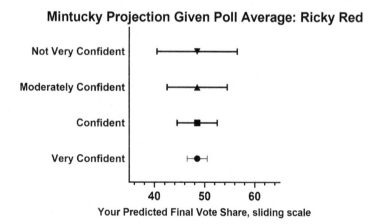

Mintucky Projection Given Poll Average: Ricky Red

FIGURE 17.2
A chart that shows sample options with varying levels of confidence for Ricky Red's eventual vote share given the poll average. Chart by the author.

Characterizing what a poll tells us by "spread" or "who is in the lead" is dangerous from a statistical literacy perspective; you understand at this point many of the problems created by such "spread analysis." The poll average doesn't tell us that "unless Red wins by 3 there was a poll error." Nor does it tell us that if Red *does* win by 3, there was zero poll error. Assuming undecideds (even a modest 6% of them) will split evenly is nothing more than an assumption. If your assumption is wrong that's an assumption error, not a poll error.

Nonetheless, this method does correctly report the two candidates' poll numbers as 48.5% and 45.5%. Despite the problems with the Spread Method's "poll error" calculation, at least it doesn't lie about what the polls say.

The same cannot be said for the Proportional Method.

Reporting Poll Results via the Proportional Method Is Fraudulent

While both the Spread Method and Proportional Method are similarly flawed when it comes to calculating poll error, only the Proportional Method reports poll results fraudulently.

The Spread Method boiling down what a poll says to a candidate's "lead" is misleading, but it at least uses the real numbers from the poll to make that claim.

The Proportional Method ignores third parties and undecideds in its reporting. In this case, with 6% undecided, the Proportional Method creates a new denominator of 94%. Since Red received an average 48.5% of the 94% "two-party" total, the Proportional Method would report the poll average as:

Ricky Red 51.6% (48.5% / 94%)

Grace Green 48.4% (45.5% / 94%)

The Proportional Method would report Red's poll average as greater than 51%, despite **the fact that only one poll out of 10 actually observed that.**

Reasonable people, even numerate ones, can be misled when they're told a lie. Based on what has been written about poll accuracy in countries where pollsters report via the Proportional Method, there is strong evidence to suggest this is what has happened.

When someone reports an "average," both the public and experts have an understanding of what that word means: the definition of an average is, I hope, not controversial. So, when a method reports an "average" that is far off what can be objectively measured, it should be done away with. The same applies to using this method to report individual polls. End it.

The Proportional Method's applications, "traditional" use, or acceptance are irrelevant. Being wrong for a long time is not an excuse to continue being wrong. As it pertains to telling us what a poll or polls say, it's time to call the Proportional Method what it is: lying.

Any attempt to characterize a poll result that has third-party and/or undecided voters that requires ignoring the existence of third parties and/or undecideds, as this method does, is lying. Want to use this method to help make your prediction or build a forecast? Go ahead. But Red's poll average is not 51.6%. That's not what the poll data says.

Both methods apply their own assumptions to what the eventual result should be as it relates to calculating poll error, and *projects their own error* onto polls, which is bad enough. But only the Proportional Method engages in deceit in its reporting of poll results.

I don't know about you, but if I spent a lot of time and money to conduct a poll, and then someone else lied about what I said when reporting it to the public, I'd be pretty upset.

While it's only the tip of the problem, the misconceptions that each of these two reporting methods create matter. Even though the "spread" between them is nearly identical, one truthfully reports that Red has received an average of 48.5% of vote, while the other lies and reports 51.6%.

Whether an informed consumer of poll data or very passive one, if you see 48.5% versus 45.5%, you might conclude that Red is "ahead" or "favored." But given the fact that 50% is required for victory, if the poll average is reported truthfully, you can at least have some information that allows you to interpret Red and Green's respective chances.

On the other hand, with the Proportional Method lying by reporting 51.6%–48.4%, it creates a false sense of certainty: that there aren't any undecideds and that Red has secured at least 50% of the vote in most polls, which we know is false.

This is without even touching on the fact that in elections where there are a lot of undecideds – thus a highly variable range of possible outcomes – the Proportional Method would not report them at all. It assumes and asserts they don't exist.

While most of my analysis to this point has only referenced US and two-party elections, be assured that all of it has applications to non-US elections as well. It is important to first explain the differences between the Spread and Proportional Methods, as many people don't realize the data is reported differently in different countries, and even by different pollsters. Second, it was important to cover what should be considered the basics of poll analysis – the standards by which poll accuracy should be measured, versus how it's currently measured. Finally, with the chapters on "What's for Lunch?" and now Mintucky, I want to explain the impact undecideds can have on the result *before* introducing another variable, third-parties.

What the Mintucky poll average tells us is that given a simultaneous census, the results should be *approximately*:

Ricky Red 48.5%

Grace Green 45.5%

Undecided 6%

The basis for any instrument error, including poll error, should be judged against this simultaneous census standard, based on what the poll observes.

In most elections, we don't have a simultaneous census – which is why I introduced Adjusted Methods for calculating poll error: this method accounts for Changing Attitudes and undecideds.

Here, since we have a simultaneous census, no adjustments are needed. Let's move forward to the results, and see how much poll error there was.

Notes

1. Tulchinsky, T. H., Varavikova, E., & Cohen, M. J. (2014). Measuring, monitoring, and evaluating the health of a population. *The new public health* (pp. 116–117). Academic Press.
2. Office on Smoking and Health (US). (2006). *The health consequences of involuntary exposure to tobacco smoke: A report of the surgeon general.* Centers for Disease Control and Prevention (US). Cancer Among Adults from Exposure to Secondhand Smoke. https://www.ncbi.nlm.nih.gov/books/NBK44330/
3. Pourhoseingholi, M. A., Baghestani, A. R., & Vahedi, M. (2012). How to control confounding effects by statistical analysis. *Gastroenterology and Hepatology from Bed to Bench, 5*(2), 79–83.
4. Skelly, A. C., Dettori, J. R., & Brodt, E. D. (2012). Assessing bias: The importance of considering confounding. *Evidence-Based Spine-Care Journal, 3*(1), 9–12. https://doi.org/10.1055/s-0031-1298595.
5. Cook, B. G., & Rumrill, P. D., Jr. (2005). Using and interpreting analogue designs. *Work, 24*(1), 93–97.
6. This experiment with mints of two colors is similar to the famous "urn" experiment conducted with marbles/pebbles by Bernoulli in the 18th century. However, by randomly obscuring the contents of some of the objects being counted (even though we know with 100% certainty their color is one of the two options) the fact that we do not – and cannot, until some later time – know their color, the calculations become different. How "different" the calculations are is an excellent topic for students and researchers.
7. Morris, G. E. (2022, July 28). *How pollsters got the 2016 election so wrong, and what they learned from their mistakes.* Literary Hub. https://lithub.com/how-pollsters-got-the-2016-election-so-wrong-and-what-they-learned-from-their-mistakes/

"After Trump's victory, the Marquette poll resumed its typical respectable record. It predicted that the Democratic candidate for governor in 2018, Tony Evers, would win the election by a one-point margin. He won by 1.1."

Excerpted from "Strength in Numbers: How Polls Work and Why We Need Them" by G. Elliott Morris.

8. Franklin, C. (2018, October 31). *New Marquette Law School Poll finds Walker, Evers tied in Wisconsin's race for governor.* Marquette Law School Poll. https://law.marquette.edu/poll/2018/10/31/mlsp50release/

9. As outlined in the "What's for Lunch?" poll, as the observed proportion moves away from 50%, the margin of error decreases. Same note as I gave there: I believe it would be unnecessarily complicated and possibly confusing ("everything you need to know … without oversimplifying") to report more precise MOEs than pollsters do.

18

Mintucky Results

Important Points Checklist

- What the "ground truth" is currently considered to be for poll accuracy, and why it's a flawed standard
- Pros and cons of analyzing "poll accuracy" by their average leading up to the election versus the single poll closest to it
- The only reason election results are relevant to our calculation of poll accuracy
- Why a Mintucky forecast is much easier than an election forecast – but still hard
- What a potential Grace Green victory would mean – if anything – regarding the accuracy of the Mintucky Simultaneous Census
- Why the argument of "which assumptions are better" in relation to undecided preference is meaningless for calculating poll error

It took me several hours – which I spread across multiple days – to count all these mints. I'm only counting the decided voters first – the undecided "silvers" will be counted last, and not unwrapped until the very end.

First batch of votes:

Ricky Red(R): 382 (50.4%)

Grace Green(D): 376 (49.6%)

A few hours later:

Ricky Red(R): 1245 (53.0%)

Grace Green(D): 1104 (47.0%)

I've made a visible dent in my counting, but the bin is still packed with thousands of mints. By my estimation, I've counted fewer than 50%. Still, all the undecideds remain uncounted and unknown.

DOI: 10.1201/9781003389903-18

A final push before taking a break for the evening, the new total is as follows:

Ricky Red(R): 1361 (52.0%)

Grace Green(D): 1254 (48.0%)

As you can imagine, counting more than 2,600 mints is quite draining. It's time to take a break. As you may have anticipated given the polls, it's too early to declare a winner in the Mintucky election, having counted around 50% of the expected vote so far.

What was 24 hours for me is only a few words for you. The counting resumed with a strong batch of votes for Green.

Ricky Red(R): 1874 (50.5%)

Grace Green(D): 1839 (49.5%)

And then Ricky Red's lead solidified, with very few decided votes left:

Ricky Red(R): 2416 (51.9%)

Grace Green(D): 2239 (48.1%)

After counting all of the *decided* votes, the results were as follows:

Ricky Red(R): 2578 (51.3%)

Grace Green(D): 2451 (48.7%)

Over 5,000 mints counted, with a few hundred "silvers" left to go. Here are the final results, *allowing for undecideds*:

Ricky Red(R): 2578 (47.7%)

Grace Green(D): 2451 (45.3%)

Undecided: 379 (7.0%)

You may not have had any major epiphanies as you have read these past two chapters, but consider what has happened, beyond the fact that I just spent a dozen hours of my life counting mints.

I took 16 random samples from a population. Ten of those random samples – after some unknown quantity and proportion of undecideds became decided – were from the *exact same population* as this result reported above.

This result is a **simultaneous census** to **10 ideal polls** taken in Mintucky. If you, like many people I discuss the concepts with, believe that the simultaneous census or ideal poll concepts are purely theoretical in application, remember: they are not. Every poll has a simultaneous census value. In this case, it is both knowable and known.

Even if the simultaneous census for some poll isn't known, it still exists. And to accurately calculate poll error, you *must* account for any non-simultaneity.

Since this census does qualify as simultaneous, since there are no changing attitudes or undecided decisions to account for, we can compare this result to our poll average directly.

Candidate	Poll Average (%)	Actual (%)	Discrepancy (%)
Ricky Red (R)	48.5	47.7	+0.8
Grace Green (D)	45.5	45.3	+0.2
Undecided (S)	6.0	7.0	−1.0

Calculate the poll error however you'd like – Most Off, mean absolute deviation (MAD), or any other – but remember, the simultaneous census, as reported above, is the only valid standard to calculate it against. There are plenty of things to debate in polls and political data, **but there's no debate or discussion to be had on this fact: a simultaneous census, not the eventual result, is the only accurate standard for calculating poll accuracy and poll error.**

This simple, true statement is one I suspect and hope most experts (and students) can understand given the example, if not a conclusion they would arrive at on their own given the opportunity, had they not been up against 100 years of misinformation.

Nonetheless, in actual political examples, even experts are so quick to subtract result numbers from poll numbers that they betray what should be a fundamental understanding of how polls work. They not only compare polls to eventual result and assume their assumptions about confounding variables can't be wrong, but say things like:

> "the election outcome can be taken to be the ground truth" for judging poll accuracy.[1]

This is not true. The ground truth for poll accuracy is the simultaneous census; this is proven by the fact that the accuracy of each poll and poll average can be calculated with 100% confidence before even knowing the election results, as shown in "What's for Lunch?" and again here for Mintucky.

The poll average has an absolute error exactly equal to the difference between the simultaneous census value and the reported poll average.

Likewise for the simultaneous census value versus each poll.

These observations lead to another valuable and interesting question.

Which *pollster* was most accurate, compared to the proper Simultaneous Census standard?

Twovases reported a result within tenths of the simultaneous census on Dec. 21, just two days before the election. Should their polls be given added weight in the future, since they were so close? Well, just 11 days prior (despite the population not having changed between those two polls), Twovases also was responsible for what happened to be the *worst poll* (by any metric) and actually reported a number for Ricky Red (54% +/– 4%) well outside the margin of error with the simultaneous census result of 47.7%.

Maybe, like with poll averages, it'd be more fair to judge a pollster's accuracy using that pollster's average.

Candidate	Twovases Poll Average (%)	Actual (%)	Discrepancy (%)
Ricky Red (R)	51.5	47.7	−3.8
Grace Green (D)	44.0	45.3	+1.3
Undecided (S)	5.0	7.0	+2.0

Numbers don't add to 100% due to rounding. Based on polls taken after Dec. 10.

Twovases has one very accurate poll and one very inaccurate one which makes them look bad overall. Now think back to the section about herding, and the "Scalester" example in which "Scalester 3" was – after seeing a trend – perhaps tempted into reporting more of a projection than observation.

If Twovases noticed that their first report, given all the polls that followed, was a major outlier – and that know they're ultimately judged as a company based on the accuracy of *the average of* their polls close to the election – they might be tempted into "overcorrecting" for their outlier poll, by reporting another outlier result in the opposite direction, which would give them a more reasonable average.

On the other hand, given the current standard I've noticed of analysts and experts judging a pollster's accuracy by only the poll they conduct closest to the election, that led me to believe they might benefit from a better understanding of "fluctuation is normal and expected" in poll data, and the problems with viewing one poll as authoritative. I'd caution against this "one poll" approach for the statistical reasons I've outlined up to now, but also apply the "mediafication" mindset to it: if a pollster knows their accuracy will only be judged on their *final* poll, their incentive to be "accurate" might not be stronger than their incentive to get clicks in earlier ones.

This is another reason why statistically valid definitions of poll accuracy are vital to the legitimacy and development of both the polling and forecasting industries. An approach that incentivizes pollsters to report results based on their *observations*, not predictions or forecasts, and whose accuracy will be compared to the simultaneous census standard, will work to "check" pollsters who release data with motives other than accuracy.

As promised, I merely conducted the polls and reported them as I collected the data, but it happens that this "two bad polls that produce an accurate result on average" case actually happened with Bowl Poll Co.

Poll	Date	Ricky Red (R) (%)	Grace Green (D) (%)	Undecided (%)
Bowl Poll Co.	Dec 19, 2022	44	50	6
Bowl Poll Co	Dec 10, 2022	51	42	6

Bowl Poll Co. had two quite inaccurate polls: both candidates on the edges of the margin of error, possibly outside it.

Yet, taking the average of those two bad polls would produce very good results:

Candidate	Bowl Poll Co Average (%)	Actual (%)	Discrepancy (%)
Ricky Red (R)	47.5	47.7	−0.2
Grace Green (D)	46.0	45.3	+0.7
Undecided (S)	6.0	7.0	−1.0

Numbers don't add to 100% due to rounding. Based on polls taken after Dec. 10.

In fact, by this pollster-average method, Bowl Poll Co. appears to have been the most accurate pollster by far – *despite producing two of the three worst polls*.

This is a tangible example of a pollster benefitting from large **compensating errors**. The only possible explanation for Bowl Poll Co.'s bad individual polls is the margin of error itself because there are no undecideds to account for; there was zero nonresponse error and zero frame error; and there are no changing attitudes. Each individual poll had a large error, yet together, they have almost none; some existing methods would say this pollster was extremely accurate, when in reality, they got lucky. Neither of their polls were accurate.

I have made the case throughout this book, with both hypothetical and real examples, that individual polls should be taken very lightly. It's not because polling isn't important, challenging, or valuable – literally the opposite. Polling is, by far, the best possible data we can have as it pertains to informing predictions or making estimations about a population, but we must understand the limitations of that tool. Polls should be expected to provide a reasonable approximation of a current state, nothing more. Individual polls are far more reliable than guesses, and poll averages are far more reliable than individual polls.

I will continue to substantiate that lesson when we get to the analysis of real election polls later. A poll that happens to be very close to the simultaneous census, or a few that happen to be far off, is not immediately indicative of poll or pollster quality.

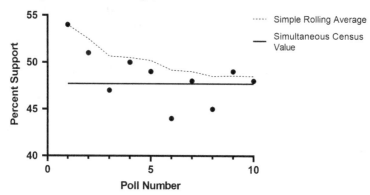

FIGURE 18.1
Chart that shows Ricky Red's individual polls, poll average, and simultaneous census value.
Chart by the author.

Given that the simultaneous census value for Mintucky is now known, have a look at Figure 18.1. The dots represent each poll's value for Ricky Red, the horizontal black line represents Red's now-known simultaneous census (true) value when each poll was taken, and the gray dotted line is the poll average, updated after each poll.

Just like the simulated "Throw it in the Average" section, most polls were far from perfect, but the overall average was very close. Fluctuation is normal and expected.

The poll average performance for Grace Green is shown in Figure 18.2, and it was even closer to the simultaneous census.

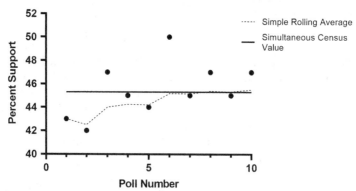

FIGURE 18.2
Chart that shows Grace Green's individual polls, poll average, and simultaneous census value.
Chart by the author.

Each poll's distance from the simultaneous census value is exactly equal to that pollster's absolute error. That statement might not sound particularly important or impactful, but when someone asserts what the "ground truth" is relative to poll data, remember, *we don't know the election results yet.*

> **The only reason election results are relevant to poll accuracy calculations is their assistance in working backwards to estimate the simultaneous census.**

The sentence above, without the context in the chapters that came before it, would have made very little sense. But it is, possibly, the most important sentence in the book.

In most real-life elections, we don't have a simultaneous census to compare to, so "eventual result" is a great starting point. But assumptions shouldn't be substituted for data when it comes to working backwards to calculate the simultaneous census.

The Mintucky Simultaneous Census provides a simple experimental proof of how and why the existing methods for calculating poll accuracy can and often do fail. And how and why they will continue to do so if they are not corrected.

Mintucky Results and the Easiest Forecast Ever

Here are the raw vote totals for decideds and undecideds for Mintucky:

Candidate	Vote Total	Percentage
Ricky Red (R)	2,578	47.7
Grace Green (D)	2,451	45.3
Undecided (S)	379	7.0

As it relates to "real" elections, 7% undecided isn't particularly notable: there have been some higher and some lower.

But what is notable is that we can calculate how many "undecideds" each candidate needs in order to win.

Given 5,408 total votes (2,578 + 2,451 + 379) and only two candidates, we can calculate with certainty how many votes will be needed to win: 50% + 1.

Since there are an even number of total votes, there's a nonzero chance of a tie. If you want to figure out that probability in your forecast, awesome. But for now, just stick to the basics: 50% of 5,408 total votes is 2,704, so the first candidate to reach 2,705 votes wins.

Ricky Red is ahead – no question – I counted. But how does his being ahead relate to his likelihood of winning? All else equal, of course it's better to be closer to winning. But does being closer to winning mean that he is most likely to win? And if he is favored, is he slightly favored or heavily favored?

Who is favored, and how you quantify their favoredness, in this situation, all depends on how you think undecideds will eventually decide. **There's no other variable to account for.** We know exactly how many votes each candidate currently has, and none can change their mind. How undecideds eventually decide will decide the election.

So with that, I'll ask you to make one final prediction about Mintucky. What percentage of the vote do you think Ricky Red will receive? Are you more or less confident than the prediction you made after all of the polls?

This starting point in Figure 18.3 is based on his simultaneous census value; you can "slide" your prediction to whatever number you think he will end up with, and whichever level of confidence you'd like.

I hope you're much more confident in your forecast than you were after the poll average, and much, *much* more confident than after one poll. Still, despite knowing with certainty how many decided votes each candidate currently has, and exactly how many votes are not counted (thus making this the "easiest forecast ever"), you must have some uncertainty around both the eventual result and spread!

Thanks to the simultaneous census, we know that there are 379 undecideds, and that in this election, 2,705 votes is a guaranteed victory.

To guarantee victory, Ricky Red needs 127 votes (2,705 to win, minus 2,578 currently). One hundred twenty-seven out of the remaining 379 undecided come to about 33%.

Grace Green, doing the same calculation, needs 254 votes, or about 67% of undecideds to win.

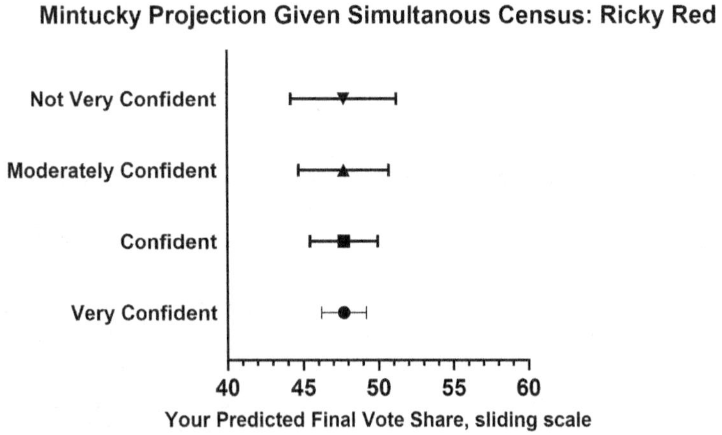

Mintucky Projection Given Simultanous Census: Ricky Red

Your Predicted Final Vote Share, sliding scale

FIGURE 18.3
Chart to estimate Ricky Red's final vote total, given varying levels of confidence, now that simultaneous census value is known. Chart by the author.

Before unwrapping the undecideds, offering the final Mintucky election results, and transitioning into real political polls, there's one final test-your-progress question that you must pass.

> *If Grace Green wins the election, in this case provably and solely because*
> *she received 67% or more of the undecided vote, does that mean*
> *my simultaneous census was wrong?*

I know this sounds like a silly question, but it's absolutely serious. If Grace Green receives so much of the undecided vote that she ends up receiving *more total votes*, is it reasonable to conclude that I, therefore, must have counted the mints incorrectly?

If you responded with anything like "no, a simultaneous census which accounted for every single vote *except* the eventual preference of undecideds can't be called 'wrong' just because it didn't account for or predict the eventual preference of undecideds."

You may continue.

If that question was easy for you, and I hope it was, remember: even in an ideal poll, there are still changing attitudes *and* undecideds who need to be accounted for before calling the difference between observation and eventual result a poll error.

With the benefit of a simultaneous census, thus no margin of error to contend with as in polls, that makes forming a prediction or making a forecast a lot easier.

On top of that, the fact that we know Mintuckians don't (and can't) change their mind, the only thing that we – if we are to act as forecasters – need to account for is how we think the undecideds will eventually decide.

For emphasis: how *we think* the undecideds will eventually decide.

Do you think they'll split 50/50? Great.

Do you think they'll split proportional to the decideds? Cool.

Maybe you're really sharp and remember that I mentioned way back in the setup portion for Mintucky that I bought an equal number of red and green mints. Even though the bin doesn't include all the mints I bought, maybe you think the most likely outcome involves Green making a comeback. Now we're getting somewhere.

The observation regarding what you believe to be the most likely composition of the undecideds may constitute a *very well-informed assumption*; you can build a forecast that incorporates this assumption if you want. You can even make the case that your assumption is better than someone else's assumption that undecideds will split 50/50. Wonderful debate to have.

But this doesn't change the fact that these are still *your assumptions.*

Whether you build a well-informed forecast or just make a guess speaks to the quality of your forecast, and its underlying assumptions: none of this speaks to the accuracy of a poll, "the polls," or simultaneous census.

If you're interested in forecasting, the calculation is now very straightforward: Grace Green's likelihood of winning is exactly equal to her probability of receiving at least 67% (at least 254/379) of the undecideds.

Even if you're not interested in forecasting, think of this is a case study into the overarching question: "how political polls should be read."

This is still a few steps away from a "real" political poll because there's no margin of error, no changing attitudes.

But in this variable-isolated experiment, in which only undecideds can impact the eventual result, there are some important lessons.

Figure 18.4 is a visual for the Mintucky election. With their respective simultaneous census numbers, we can say with certainty that Ricky Red will receive *at least* 47.7%, and that Grace Green will receive *at least* 45.3%.

The "error bars" from your estimate should not overlap impossible outcomes: Ricky Red can't receive less than 47.7% of the vote, and Grace Green can't receive less than 45.3%.

We know more than that, too.

The bars on top of each candidate's simultaneous census number represents the *maximum* number of votes they could receive; in other words, if they receive 100% of the undecided vote.

While the debate regarding the likelihood of one side receiving 100% of the undecided vote is worthwhile – whether in Mintucky, Kentucky, or elsewhere – there's no arguing with the facts: this is what this simultaneous census has told us, and the only thing we can say it tells us.

The outcome will fall within this range.

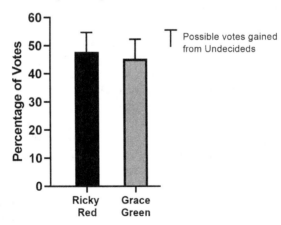

FIGURE 18.4

Range of possible outcomes given simultaneous census values for each candidate, and undecideds. Chart by the author.

And in this election, like many elections around the world, people who were undecided in the days and weeks prior to the election will play a major role in deciding its outcome. While a number such as 7% may feel insignificant, in a race to 50%, even a candidate with what is perceived as a strong lead might need a lot of them.

This is a window into the problem with characterizing what a poll tells us by its spread. "Up by 4" without context is almost meaningless. "Up by 4 with approximately 40% of decided voters" and "up by 4 with approximately 50% of decided voters" are very different; the Spread Method says they're exactly the same.

But in Mintucky, the final results are here. With 100% of the votes counted:

Candidate	Vote Total	Percentage
Ricky Red (R)	2,730	50.5
Grace Green (D)	2,678	49.5

Grace Green's voters made things close as the undecideds were unwrapped, but there were too few to overcome Ricky Red's lead.

Ricky Red wins Mintucky. I've seen enough of these mints for a while.

Working backwards, we can figure out that Grace Green received 227/379 undecided votes (59.9%), while Ricky Red received 152/379 (40.1%).

How close were your predictions?

Mintucky math is much easier than "real election" math, but again, that's by design. The beauty of an analogue experiment. Let's look at what this election can teach about poll error calculations, and how to best use polls to inform predictions.

Note

1. Shirani-Mehr, M., Rothschild, D., Goel, S., & Gelman, A. (2018). Disentangling bias and variance in election polls. *Journal of the American Statistical Association*, 113(522), 607–614. https://doi.org/10.1080/01621459.2018.1448823

19

Mintucky Poll Error and Jacob Bernoulli

Important Points Checklist

- The cause of all "error" in Mintucky according to Spread and Proportional Methods, and why it's wrong
- How this chapter further illustrates "assuming what undecideds will do" to be an invalid method
- Why Adjusted Methods should always be considered approximate
- How Jacob Bernoulli's work three centuries ago influenced the field of statistics today
- How the "urn problem" is similar to Mintucky, and the key differences

The Spread Method would characterize this 50.5%–49.5% result as "+1" for Ricky Red. That means every poll's error – by its definition – is equal to the difference between this result and the poll's spread.

Here again, each of these polls were taken after Dec. 10:

Poll	Date	Ricky Red (R) (%)	Grace Green (D) (%)	Undecided (%)
Poll Average	Dec 10–21	48.5	45.5	6
Bevtub Polling	Dec 21, 2022	48	47	5
Twovases	Dec 21, 2022	49	45	6
Shovel	Dec 20, 2022	45	47	8
Box	Dec 20, 2022	48	45	6
Bowl Poll Co.	Dec 19, 2022	44	50	6
Bevtub Polling	Dec 14, 2022	49	44	6
Box	Dec 14, 2022	50	45	5
Shovel	Dec 14, 2022	47	47	6
Bowl Poll Co.	Dec 10, 2022	51	42	6
Twovases	Dec 10, 2022	54	43	4

DOI: 10.1201/9781003389903-19

And here are the results as the Spread Method would report them, along with its error:

Poll	Date	Poll Spread	Eventual Result	Poll Error
Poll Average	Dec 10–21	+3	+1	2
Bevtub Polling	Dec 21, 2022	+1	+1	0
Twovases	Dec 21, 2022	+4	+1	3
Shovel	Dec 20, 2022	–2	+1	3
Box	Dec 20, 2022	+3	+1	2
Bowl Poll Co.	Dec 19, 2022	–6	+1	7
Bevtub Polling	Dec 14, 2022	+5	+1	4
Box	Dec 14, 2022	+5	+1	4
Shovel	Dec 14, 2022	+0	+1	1
Bowl Poll Co.	Dec 10, 2022	+9	+1	8
Twovases	Dec 10, 2022	+11	+1	10

By the Spread Method, Bevtub Polling's Dec. 21 poll grades out as perfect. Red +1 in the poll, and Red +1 in the result. No error.

One problem with that.

We *know* that Ricky Red's simultaneous census value at the time of this poll was 47.7%, very close to what Bevtub reported; but Grace Green's simultaneous census value was 45.3%, and undecideds were 7%. Here's what Bevtub, the Spread Method's "best pollster" reported, compared to the simultaneous census:

Candidate	Bevtub Polling, Dec 21 (%)	Simultaneous Census (%)	Discrepancy (%)
Ricky Red (R)	48	47.7	+0.3
Grace Green (D)	47	45.3	+1.7
Undecided (S)	5	7.0	–2.0

This poll overreported one candidate's poll by a fair amount and under-reported the undecideds by even more. Though not the worst poll on the list, it would nonetheless be characterized as the *best* by the Spread Method. Zero error, in fact. That's not correct – it is literally proven here.

Meanwhile, so far barely mentioned pollster, Box … very accurate poll on Dec. 20. But accurate compared to what?

We know that when Box took its poll, although they (and I) did not yet know the actual population of voters in Mintucky, they reported:

Candidate	Box Poll, Dec 20 (%)	Actual (%)	Discrepancy (%)
Ricky Red (R)	48	47.7	–0.3
Grace Green (D)	45	45.3	+0.3
Undecided (S)	6	7.0	–1.0

Numbers don't add to 100% due to rounding.

This Box poll was extremely close to the simultaneous census value but, according to the Spread Method, had a 2-point error. Compare that to Bevtub Polling's poll with much more error, which would be given a "perfect" rating from the Spread Method.

As for the Proportional Method:

Poll	Date	Red Poll Proportion (%)	Actual Result (%)	Poll Error (%)
Poll Average	Dec 10–21	51.6	50.5	1.1
Bevtub Polling	Dec 21, 2022	50.5	50.5	0
Twovases	Dec 21, 2022	52.1	50.5	1.6
Shovel	Dec 20, 2022	48.9	50.5	1.6
Box	Dec 20, 2022	51.6	50.5	1.1
Bowl Poll Co.	Dec 19, 2022	46.8	50.5	3.7
Bevtub Polling	Dec 14, 2022	52.1	50.5	1.6
Box	Dec 14, 2022	52.6	50.5	2.1
Shovel	Dec 14, 2022	50.0	50.5	0.5
Bowl Poll Co.	Dec 10, 2022	54.8	50.5	4.3
Twovases	Dec 10, 2022	55.7	50.5	5.2

While the Proportional Method is less dramatic about the "misses" of the first two Dec. 10 polls, which were quite bad – outside or nearly outside the margin of error in relation to the simultaneous census – it agrees with the Spread Method in characterizing Bevtub polling as the most accurate, also with 0 error.

This is a demonstrable failure of two accepted methods to accurately rate which polls were the most accurate.

If a mild 60/40 split of just 7% of undecideds can *by itself* cause these established methods to inaccurately characterize which polls are most accurate – how can they claim to be able to measure poll accuracy when the calculation gets even more complex?

Enter, the Adjusted Methods.

Now, despite the fact that we know from the simultaneous census how many undecideds there were and their eventual preference, the Adjusted Method (like other poll error calculation methods) wouldn't have access to that knowledge.

But we do have access to the poll numbers, the poll average, and the eventual results.

Adjusted Poll Value = Reported Poll Value
+ / − *Changing Attitudes + Undecided Ratio.*

In most polls, we would have to *estimate* changing attitudes and Undecided Ratio; in Mintucky, we do not have to estimate changing attitudes. That leaves the undecided ratio.

The Adjusted Poll Value would be:

$$\text{Adjusted Poll Value} \left(\text{Ricky Red}\right) = \text{Poll Value} \\ +/- \; 0 + \textit{Undecided Ratio.}$$

$$\text{Adjusted Poll Value} \left(\text{Grace Green}\right) = \text{Poll Value} \\ +/- \; 0 + \textit{Undecided Ratio.}$$

The best way to solve for Undecided Ratio would be to take all of the mints that had previously been undecided, and count them.

While I did that for the simultaneous census (approximately 60% for Green and 40% for Red), real elections would not have the benefit of knowing exactly how undecideds decided.

So, the next best way to solve (or, more honestly put, approximate) Undecided Ratio would be to take a random sample of those who had been previously undecided, and use that value in this ratio: a post-event poll.

If I had been thinking this far ahead, I'd have probably taken a large-enough sample from the 379 silvers to achieve an appropriate margin of error to estimate the Undecided Ratio *before* I went forward tallying the results. But I didn't perform that analysis this time – and that's the beauty of experimentation, do it better next time. (I promise this book will not contain any more mint counting after this chapter.)

So what's the next best way to estimate Undecided Ratio?

It's a messy assumption, but easy, to compare the poll average to the eventual result: 48.5%–50.5%. Ricky Red gained 2% above his poll average, out of a possible 6%. So we can approximate "Undecided Ratio" as 33.3% for Ricky Red.

By extension, we can approximate "Undecided Ratio" as 66.6% for Grace Green.

We know this approximation underestimates Ricky Red because of the simultaneous census, but let's see what the Adjusted Methods would have reported in the absence of poll data to provide better estimates.

Poll	Date	Ricky Red (R) (%)	Grace Green (D) (%)	Undecided (%)
Result	**Dec 23, 2022**	**50.5**	**49.5**	**Allocated**
Bevtub Polling	Dec 21, 2022	49.6	50.3	Allocated
Twovases	Dec 21, 2022	51.0	49.0	Allocated
Shovel	Dec 20, 2022	47.7	52.3	Allocated
Box	Dec 20, 2022	50.0	49.0	Allocated
Bowl Poll Co.	Dec 19, 2022	46.0	54.0	Allocated
Bevtub Polling	Dec 14, 2022	51.0	48.0	Allocated
Box	Dec 14, 2022	51.7	48.3	Allocated
Shovel	Dec 14, 2022	49.0	51.0	Allocated
Bowl Poll Co.	Dec 10, 2022	53	46	Allocated
Twovases	Dec 10, 2022	55.3	45.7	Allocated
Some numbers don't add to 100% due to rounding.				

Now, this creates a new picture. This isn't some highly advanced calculation: it's simply accounting for a variable (undecideds) that the poll itself couldn't, and doesn't try to, but *does* impact the discrepancy between poll and eventual result.

Here, we're making a big assumption that can obviously only provide a very rough approximation. Even in the case of good post-election poll data that offers insight into Changing Attitudes and Undecided Ratio, **by the nature of poll data being approximate, all Adjusted Methods should be noted as approximate.**

And if you found yourself, like me and the people reviewing this book for statistical accuracy, screaming at how arbitrary, presumptive, and statistically indefensible that method of "assume undecideds voted based on poll average versus eventual result" is, I have some promising news.

If you weren't on it already, I'd like to be the first to welcome you to the team opposed to "assume you can reliably calculate poll accuracy by making assumptions."

The "messy, but easy" assumption I made – that the poll average compared to eventual result provides an accurate measure of the Undecided Ratio – *is no more or less valid* than the methods that assume it is 50–50, or proportional to decideds. With one big distinction relating to scientific integrity:

> The existing methods assume they *know* what undecideds did, and any deviation from that assumption means there was an error *with the poll(s).*

Even the very suboptimal method of approximating undecideds by poll average is *far superior* to the current "assume you know" standard; only one of those methods admits it could be inaccurate.

To reiterate, I would not consider this "poll average versus eventual result" method to be a good way to calculate how undecideds eventually voted, it is not. It is *almost* as bad as the current Spread and Proportional standards.

Simply put, we should not place assumptions where data could go. Maybe an innovator can come up with more precise methods for estimating Changing Attitudes and Undecided Ratio, which, coupled with the Simultaneous Census standard, would give us a much better picture of poll error – and the sources of it.

The Adjusted Spread Method incorporating the admitted assumptions from above would report the following:

Poll	Date	Adj. Poll Spread	Actual Result	Adj. Poll Error
Poll Average	Dec 10–21	+1.07	+1	0.07
Bevtub Polling	Dec 21, 2022	−0.7	+1	1.7
Twovases	Dec 21, 2022	+2	+1	1
Shovel	Dec 20, 2022	−4.6	+1	5.6
Box	Dec 20, 2022	+1	+1	0

(Continued)

Poll	Date	Adj. Poll Spread	Actual Result	Adj. Poll Error
Bowl Poll Co.	Dec 19, 2022	−8	+1	7
Bevtub Polling	Dec 14, 2022	+3	+1	2
Box	Dec 14, 2022	+3.4	+1	2.4
Shovel	Dec 14, 2022	−2	+1	3
Bowl Poll Co.	Dec 10, 2022	+7	+1	6
Twovases	Dec 10, 2022	+9.6	+1	8.6

Even though we know from the simultaneous census that my suboptimal method of approximating undecided preference was "off," the Adjusted Spread Method still did a *far better job* of approximating poll error than the traditional Spread Method. It rightly reported the worst polls as worst, and the best polls (Box, Dec. 20 and Twovases/Bevtub on Dec. 21) as best.

As for the Adjusted Proportional Method:

Poll	Date	Red Adj. Proportion (%)	Red Result	Adj. Poll Error
Poll Average	**Dec 10–21**	**50.4**	**50.5**	**0.1**
Bevtub Polling	Dec 21, 2022	49.6	50.5	0.9
Twovases	Dec 21, 2022	51.0	50.5	0.5
Shovel	Dec 20, 2022	47.7	50.5	2.8
Box	Dec 20, 2022	50.0	50.5	0.5
Bowl Poll Co.	Dec 19, 2022	46.0	50.5	4.5
Bevtub Polling	Dec 14, 2022	51.0	50.5	0.5
Box	Dec 14, 2022	51.7	50.5	1.7
Shovel	Dec 14, 2022	49.0	50.5	1.5
Bowl Poll Co.	Dec 10, 2022	53.0	50.5	2.5
Twovases	Dec 10, 2022	55.3	50.5	4.8

The Adjusted Proportional Method was a huge improvement over its traditional counterpart – which wrongly identified Bevtub Polling's Dec. 21 poll as having zero error.

While this method actually underestimates the error in the poll average, as uncovered by the simultaneous census, it is – still – far superior to the traditional method.

No matter what the eventual method becomes for calculating poll error, an accurate definition will allow for improved poll quality and improved forecasts as well. This method must account for changing attitudes and undecideds, preferably by collecting post-event data, and must not make assumptions about them.

Jacob Bernoulli

If you feel a little silly learning about political data through an experiment with mints – understandable. Naming the candidates, pollsters, and their "constituency" was as much for my entertainment benefit as yours: it took me dozens of hours to conduct in total, a fact which I don't want to discourage you from replicating or modifying it.

And if you're imagining me in my basement with a colander on my head, holding a shovel, talking to myself about undecideds with a (non-voting) mint in my mouth, taking "polls" out of a giant bin, you're right on. I have it on video.

But as I did in "What's for Lunch?", I'm obligated to remind you: if you doubt the transferability of the findings of this experiment to more "realistic" applications, you should reconsider.

The mathematics that are described as underpinning "the whole of statistical inference"[1] were explained and derived in part with an example of drawing pebbles out of an urn, in a book published in 1713.[2]

The 3,000 white pebbles and 2,000 black pebbles Swiss mathematician Jacob Bernoulli spoke of in *Ars conjectandi* were used to "infer with increasing certainty the unknown probability from a series of supposedly independent trials."[3]

These findings led to what are known today as margins of error and confidence intervals.

He took one pebble at a time from the urn, replaced it, and used that data to calculate how many independent trials it would take to calculate the known composition of the urn – 60% white and 40% black.

Bernoulli's work is so well regarded that a "Bernoulli distribution" is synonymous with "binomial distribution" and his name is used to describe observations with conditions similar to his pebble experiment[4] – including coinflips and dice rolls.[5]

The Mintucky experiment bears some similarity to Bernoulli's, but instead of two options like white or black, heads or tails, yes or no, and success or failure, it considers what happens when a third, "unknown" variable is introduced, in which the future state of that unknown is one of two options, but cannot be known currently.

And this application goes even a bit further than that.

While the goal of most people who *analyze* polls is to predict some future result, the goal of the poll itself – like Bernoulli's urn – is to do nothing more than offer an approximation of the current state, which I termed a "simultaneous census."

In the terminology I've used (i.e. classifying it in a way that makes sense to me, and I hope helps others), the urn problem represents a series of ideal, present polls: the contents of the urn do not and cannot change, and all criteria of ideal polls are met. Moreover, results can be and are compared directly to a simultaneous census.

The proportion of Reds to Greens in Mintucky, if you were to ignore and never unwrap silvers, would certainly fit Bernoulli's distribution as well.

But that's not how elections work.

Incorporating Bernoulli's urn example – here's why the difference between a present poll and plan poll must be considered:

Just because *we* (the observer) know that all silvers contained within the bin will be, after some date, either Red or Green *does not mean* the poll instrument does.

The math underlying poll data is based on making accurate observations about a simultaneous census – it has nothing to do with predicting a future, changing state.

How one models or assumes the eventual future "state" of that undecided is therefore *independent* of the observation given by the poll; thus, the poll's accuracy cannot be calculated knowing only *future distribution* of the bin, urn, or voting population's proportion.[6]

To take this problem to its conclusion:

If Bernoulli's urn instead contained 3,000 white pebbles (50% of the total pebbles), 2,000 black pebbles (33%), *and* 1,000 silver pebbles (17%) – and those silver pebbles, at an unpredictable rate over two weeks, became either white or black with an unknown and unknowable proportion – this would effectively change his present poll to a plan poll.

If Bernoulli performed his experiment *immediately* to estimate the proportion of white to black in this urn, he would probably (using this term mathematically) still find that the bin contains approximately 60% white and 40% black – *if he ignores silver ones.*

More appropriately, he would find about 50% are white, 33% are black, and 17% are silver.

But how would his "immediate" data compare to a result two weeks later – when all the silvers have transitioned to either white or black?

The calculations he makes for the margin of error at a given confidence level "immediately" *cannot be said to apply* to the final result. Or even the next day, or next hour – because the silvers are known to change the composition of the population.[7] I'm not sure I could find a mathematician that would disagree, and I certainly hope not.

But poll data is not treated by this standard. And this does not even incorporate the complexities involved when five more colors of pebbles (representing more political parties) are added, and the fact that the pebbles themselves can change colors over time – often strategically, sometimes predictably, sometimes randomly.

Political polls are the hardest class of polls.

As for the way the existing methods would handle the modified urn problem:

Spread Method (US)

50% White (+17)	(3,000/6,000)	
33% Black	(2,000/6,000)	

Proportional Method (Most non-US countries)

60% White (+20)   (3,000/5,000)

40% Black   (2,000/5,000)

The Spread and Proportional methods would report conflicting "spreads" or "leads" for White. This is a problem for people who want to be informed by this data.

More troublesome, to the Proportional Method, the addition of 1,000 silvers to the urn whose color changes at an unpredictable rate over time and will become either white or black with an unknown and unknowable proportion ... didn't change anything in how the data would be reported.

> The Proportional Method would report that Bernoulli's original urn problem and this modified one observed the same thing!

I'll discuss this problem in a little more depth, and why I think it's a great problem to present on the topic of poll data – in the last chapter.

But the question becomes, how can people who want to be *informed* by data coexist with a field of experts and media who misinform?

Notes

1. Edwards, A. (2013). *Ars Conjectandi* three hundred years on. *Significance, 10*(3), 39–41. https://doi.org/10.1111/j.1740-9713.2013.00666.x
2. Polasek, W. (2000). The Bernoullis and the origin of probability theory. *Reson, 5*, 26–42. https://doi.org/10.1007/BF02837935
3. Edwards, A. (2013). *Ars Conjectandi* three hundred years on. *Significance, 10*(3), 39–41. https://doi.org/10.1111/j.1740-9713.2013.00666.x
4. Grami, A. (2023). Discrete random variables. In A. Grami (Ed.), *Discrete mathematics* (pp. 307–325). Academic Press. https://doi.org/10.1016/B978-0-12-820656-0.00017-4
   "... a random experiment with exactly two possible outcomes, in which the probability of each of the two outcomes remains the same every time the experiment is conducted."
5. Routledge, R. (2024). *Binomial distribution.* Encyclopedia Britannica. https://www.britannica.com/science/binomial-distribution
6. While I specifically mention undecided, this applies to changing attitudes as well.
7. This is the Simultaneous Census explained in another way.

20

The Point Spread Problem

Important Points Checklist

- Why "spread" is used in both Spread Method and Proportional Method applications
- The difference between calculating the error of a method and the error of a poll
- "If it disagrees with experiment, it's wrong"

I've previously outlined why the Spread Method in poll data is fallacious. While reporting the numbers this way isn't outright lying in the way the Proportional Method is, it's still problematic.

Unfortunately, and even worse, countries that use the Proportional Method seem to believe that spread is an accurate way to characterize their polls, too.

Unless my efforts achieve far more notoriety than even my most optimistic forecast would predict, it's unlikely to go away anytime soon.

For as long as we have to live with it, let's try to understand it. Starting with, why does it exist?

The Spread Method, and the use of spread in Proportional Method applications, exists to simplify what can be a difficult question: "what does this poll tell us?"

Giving a simple answer to a complex question isn't always a bad thing. The problem is, in this case, it oversimplifies to the point it's no longer useful.

"Who is ahead, and by how much?" is what spread purports to tell us. Two valuable questions to answer, if you can. But spread quickly fails this test.

If a candidate is leading 37%–35%, with 28% undecided – versus 46%–44% with 10% undecided, that's a big difference both in theory and in practice. But to the spread method, both are "up two points."

To use an analogy, if I'm watching a footrace, and I ask someone "who's winning, and by how much?" and their response is "the leader is ahead by two meters," is that all the information you need to know the state of the race? Is that even the most important piece of information?

It's only informative if you know how close they are to the finish line.

DOI: 10.1201/9781003389903-20

In a 100-meter race, two meters is a lot. In a 1000-meter race, less so. But even in a 100-meter race – being ahead two meters means much less in the 50th meter than the 99th meter.

Now we're getting somewhere.

Indeed, far away from elections, where many voters have not yet made a decision – and in many cases the candidates aren't even settled yet – potential voters are given the option of some likely or possible candidates to choose from. Fine data to have, if you're interested in which candidates might have the best chance.

But naturally, far away from an election, before campaigns and debates, a lot of voters state that they're undecided.

How does that impact our understanding of "who is ahead?" and "by how much?"

Understanding that those polls – whether there are 28% undecided or 10% undecided or 2% undecided – are only telling you about *right now* gives you an immensely better understanding of how polls work than what can be offered by the spread. Being ahead by 2 meters in the 50th meter of a race is not a prediction that they'll be ahead in the 100th meter. Nor is it a prediction that they'll "win by 2." It's just an observation about right now.

Regarding polls, even without having read about undecideds as a "confounding variable," many people intuitively understand that being "ahead" 37%–35% is not as strong as being "ahead" 46%–44%, despite the equal spread.

While people can *intuitively* understand that there's a big difference between 37%–35% with 28% undecided and 46%–44% with 10% undecided – it's a little harder to understand that there's also a huge difference between 46%–44% with 10% undecided and 48%–46% with 6% undecided. I'll show you how and why later.

But all are characterized, according to the spread, as "up by 2." Undecideds, to this method, don't matter.

While FiveThirtyEight famously, and rightly, gets a lot of credit for bringing political data to the public, they're also largely responsible for the mass propagation of the Spread Method. They ascribe so deeply to this method that many of their reports and analysis *don't even include poll numbers* – just the spread.

Their analysis, which their mission statement says is intended to "use data and evidence to advance public knowledge,"[1] is often just varying descriptions of what the poll spread tells us about a candidate's chances for winning historically, or what the election spread tells us about poll error. Their calculation of poll error relies entirely on spread.[2]

After the 2016 Presidential Election, in which Donald Trump surprisingly won, FiveThirtyEight reported that "the polls missed" Trump's election.[3] Their analysis? They subtracted the spread between their poll average in various states from the spread of the eventual result. Spread didn't match: poll error! No possibility, other than the polls were wrong.

Two years later, after the 2018 midterm elections, Nate Silver published an article entitled "The Polls Are All Right" in which he said the polls are "about as accurate as they've always been."[4] In that article, he took the obligatory jab at how the 2016 polls were "sometimes way off the mark" but characterized the 2018 poll error as "within the normal range." What normal range?

The normal range as defined by the "spread," which incorporates confounding assumptions about changing attitudes (assumed to be zero) and undecided voters (assumed to be split evenly to the major candidates). In short, if there's a discrepancy between poll and result, that means the polls were wrong, and if there's not, that means they're "all right."

I don't think these mischaracterizations of poll error are intentional, just one that requires making unsubstantiated assumptions.

Specifically, Silver writes, "If the *average* error is 6 points, that means the true, empirically derived *margin of error* (or 95 percent confidence interval) is closer to 14 or 15 percentage points!"

What he's saying is that if you *assume* the Spread Method of calculating poll error is an accurate representation of "poll error," then this is what it tells us. Great. But at no point does he (or most anyone) step back to consider the possibility that the Spread Method could be a very inaccurate way to calculate poll error and lead to very flawed conclusions about how "all right" or "off the mark" polls are.

Which it is and does.

It's the same "working backwards" problem I presented earlier: instead of assuming poll error is *unknown* and trying to figure it out, he – and as far as I know, the entirety of experts in the field – assumes poll error is *known* and that all of the factors that contribute to poll error must, somehow, add up to that number. The source of that error could be the margin of error, it could be nonresponse, it could be weighting by education! Forever a mystery. But they must add up to the number we said it does. See you next election, where we'll tell you how wrong the polls are without being able to quantify the causes and/or amount of their wrongness.

On that note, how does the Spread Method explain the fact that their "poll error" calculations often disagree with the Proportional Method by a large amount? I don't know. Neither do they. Unless I missed the lesson where maths are different depending on which country you live in, this seems to be an unspoken truce that should not exist in a field that wants (if not demands) to be taken seriously.

Both methods classify eventual undecided preference under the same umbrella of error as nonresponse error, frame error, and the margin of error. To those methods, it doesn't matter *why* the eventual result didn't match their assumptions, the simple fact *that* it didn't match their assumptions means the polls were wrong.

The Spread Method and Proportional Method's respective attempts to answer the questions "who is ahead?" and "by how much?" are not only incorrect, they mislead the public: **these methods' assumptions render**

them definitionally incapable of accounting for the eventual preferences of undecided voters.

Silver followed up 2018's "The Polls Are All Right" with 2020's "The Polls Weren't Great. But That's Pretty Normal."[5]

Why weren't the polls "great?" You guessed it. Chart number one from the article: Joe Biden's Polling Average versus Joe Biden's Eventual Result, and the difference. Spread.

This chart is titled: "There were big misses in some swing states."

The poll spread didn't agree with the election spread? Polls were wrong, big misses. The rest of the article adds no nuance. The entirety of the discrepancy in each state is attributable to poll error. Great analysis. See you next time.

For the 2022 elections, continuing with the tradition of telling us how accurate the polls were, Nathaniel Rakich wrote for FiveThirtyEight: "The Polls Were Historically Accurate."[6]

How is that "historically accurate" claim quantified? Spread.

Did pollsters improve their weighting methodologies from 2020, when they "weren't great?" How do you know?

Or, is it possible they used the same poor methodology that had some bad weights for nonresponse but benefitted from a compensating error?

They don't know. They can't know. They were owed $100 and received $100, so the transaction was historically accurate; their definitions don't allow them to go any deeper than that.

For reasons I hope I've sufficiently substantiated to this point, I don't think this spreading the "spread" fallacy advances public knowledge, and it definitely doesn't improve the field. I only specifically mention FiveThirtyEight here not because they're the "worst" offender – there are many far worse – but that I believe with their reach and brainpower they're capable of doing much better work, and have an obligation to do so, given their stated mission.

While I've done my best to provide something of a logical, and chronological, way to understand how polls work, my arrival to this point was far from a straight line.

As I mentioned earlier, my first decade of looking at forecasts and poll data was quite passive and uncritical. It wasn't until I thought "I wonder if I could build my own forecasts?" and worked backwards that I went down the path that ultimately led to me writing this book.

Naturally, my first exposure to forecasts and poll data came from FiveThirtyEight. That's what led me to understanding the Spread Method, and ultimately the statistical and logical problems it creates. Then I wondered, might these issues be limited to just FiveThirtyEight?

So I branched out. I started reading both analysis and commentary from forecasting and poll aggregation websites, both mainstream and obscure, American and non-American. Then reputable media sources, data companies and pollsters themselves – they made the same flawed characterizations about how polls work. Academic papers, those too? Yes – the same "treating polls as if they're predictions" and "assuming poll error is known" problems. Some of them used slightly different methods, but the same underlying mistakes in their reasoning, leading to the same unsubstantiated conclusions.

Being up against such a mountain of experts in an industry nearly a century old – with the full disclosure that my credentials are unimpressive compared to most of theirs – is it possible I'm just wrong?

Unlike the assumptions that underlie the various methods, I did consider that possibility. In fact, if I'm not wrong about *something* I've written, I'm probably not working hard enough; and I would welcome being corrected on anything I'm wrong about.

But when it comes to what polls tell us, I'm not wrong.

In a lecture at Cornell University, physicist Richard Feynman gave a strikingly simple-but-true characterization of how we know whether something is "wrong" in science.

> First, we guess it (audience laughter), no, don't laugh, that's really true. Then we compute the consequences of the guess, to see what, if this is right. What it would imply and then we compare the computation results ... to experiment or experience, compare it directly with observations, to see if it works. If it disagrees with experiment, it's wrong. In that simple statement is the key to science. It doesn't make any difference how beautiful your guess is, it doesn't matter how smart you are who made the guess, or what his name. If it disagrees with experiment, it's wrong. That's all there is to it.[7]
>
> **Richard Feynman, 1964**

The existing methods are wrong.

After taking the time to understand why the Spread Method was problematic, noting that "if it disagrees with experiment, it's wrong," my research led me to two different directions:

1. The Proportional Method, whose issues have been introduced previously and will be discussed more directly in non-US elections later.

2. I'm not the first to recognize the problems with these methods.

Notes

1. AllSides. (2022, December 8). *FiveThirtyEight media bias rating.* https://www.allsides.com/news-source/fivethirtyeight-0

2. Rakich, N. (2023, March 10). *How our pollster ratings work.* FiveThirtyEight. https://fivethirtyeight.com/methodology/how-our-pollster-ratings-work/
   "We compare the margin in each poll against the actual margin of the election and see how far apart they were."

3. Bialik, C., & Enten, H. (2016, November 9). *The polls missed Trump. We asked pollsters why.* FiveThirtyEight. https://fivethirtyeight.com/features/the-polls-missed-trump-we-asked-pollsters-why/

4. Silver, N. (2018, May 30). *The polls are all right.* FiveThirtyEight. https://fivethirtyeight.com/features/the-polls-are-all-right/

5. Silver, N. (2020, November 11). *The polls weren't great. But that's pretty normal.* FiveThirtyEight. https://fivethirtyeight.com/features/the-polls-werent-great-but-thats-pretty-normal/

6. Rakich, N. (2023b, March 10). *The polls were historically accurate in 2022.* FiveThirtyEight. https://fivethirtyeight.com/features/2022-election-polling-accuracy/

7. Feynman, R. (1964). *The character of physical law.* Cornell University. https://www.feynmanlectures.caltech.edu/fml.html#7

21

Remembering Nick Panagakis (1937–2018)

Important Points Checklist

- How Panagakis' research regarding the "Incumbent Rule" could have contributed to better forecasts
- Why Panagakis noted 50% as significant in his research
- What polls do not do
- The contributions made by Panagakis that seem to have been lost or forgotten

Most of this book was written based on contemporary research and findings; while polls themselves are not new, the public interest in them is certainly growing.

While it's much harder to find material that largely predates the internet, I felt that I should learn as much as I could from people who studied the topic before me. Historically, the characterizations regarding "spread" are no different from what media outlets and experts report today.

But in this research, I learned that I wasn't the only one who found these characterizations of poll data problematic.

Decades prior to FiveThirtyEight's founding, the problems with the Spread Method were known, at least to some.

Nick Panagakis, President of Market Shares Corp., and himself a pollster, wrote extensively on poll data and the problems with the Spread Method. Reading his work from decades ago gave me such a shock that I fought tears on a few occasions. One article after another in which he raises the same issues that continue today but go unheeded. For example:

> Traditional analysis says undecided voters are "... expected to split about evenly between candidates on election day. Unconditional probability is assumed, and poll point spreads are used to characterize the race."[1]

DOI: 10.1201/9781003389903-21

He continued:

> Most media still rely on this traditional assumption about undecided
> voters, and it continues to lead to incorrect characterizations about poll
> findings.

Sound familiar? He's talking about the Spread Method. Those words could
just as easily have been the introduction to this book.

He wrote them in 1997.

There's a lot more, and 1997 was far from the beginning of his work.

The problems with the Spread Method's assumptions contribute directly to
public and expert misunderstanding of what polls tell us. Panagakis' use of
the words "incorrect characterizations" struck me especially strongly; those
are the exact words I used to describe the same topic, years before finding
his work.

Panagakis noted that in most US elections to that point, "More undecideds
appear to vote for the challenger."[2]

If we know more undecideds voted for one candidate, we can – and
must, if we're being accurate – incorporate that into our calculation of poll
error.

Not just that, but *prior* to the election when we're looking at polls, if we
are trying to use them to *inform our prediction*, we must also consider the role
"uneven undecideds" could play in the eventual result.

Armed with the knowledge that undecideds tended to vote in larger numbers for the challenger, Panagakis contended on many occasions that an
incumbent with a "lead" in the polls (as characterized by spread) is, in many
cases, *unlikely* to eventually win; if a high enough ratio of people who were
previously undecided eventually vote for the candidate who is "behind" in
the polls, that candidate is likely to win.

That winning a large proportion of previously undecided voters can lead
someone who is "behind" in the polls to eventually win might seem obvious.
But in the decades Panagakis was active, he didn't just make the observation
that this could happen, he showed that it *did* happen.

Though he was a pollster by trade, he provided what should have been
the foundation for how forecasts are built, and an acceptance that polls and
forecasts are different things. To make a forecast, you start with polls, or a
poll average, and you – the forecaster – must figure out how to allocate undecideds. That polls don't predict how undecideds will eventually vote is not
an error that is induced by the poll or pollster.

The simple statement regarding how undecideds alone can lead someone
who is "behind" in the polls to eventually win might not feel like an inge-
nious observation: but when he first presented it decades ago, it was. And
it still is today because of the stagnation in the poll industry. Analysts simply, still, do not accept the fact that their assumption about what undecideds
should do is not a reflection of the poll's accuracy.

Through his experience as a pollster, and perhaps the earliest researcher to try and separate factors that cause poll error from those that don't, Panagakis had what I've come to recognize as remarkable integrity and patience.

Consider the fact that he was among the first, if not the first, to identify that undecideds tended to, eventually, overwhelmingly vote for the challenger in elections with an incumbent. As a pollster, he recognized, and lamented, that poll accuracy was judged in accordance to their closeness to eventual result.

He was active in an era where pollsters, even more so than today, were asked to make "calls" or predictions. "If your poll says someone is ahead, you must be *calling* that they'll eventually win" is the unscientific mindset that experts and the media used then, which is still frustratingly used today.

But put yourself in Panagakis' shoes for a moment, with knowledge that few others possessed regarding how undecideds were likely to decide, in an era that largely predated forecasters and one in which the media and public desperately wanted pollsters to make *calls*. How would you have handled it?

There are dozens of examples to choose from, but here's one of them:

In 1986, Republican incumbent Senator Mack Mattingly from Georgia was polling ahead of the Democratic challenger Wyche Fowler about 48%–38%.[3]

The *LA Times* had an article in late October 1986, just two weeks before the election, entitled: "Liberal Challenger Trailing in Polls: GOP Expected to Keep Its Georgia Senate Seat."[4]

In that article, experts characterized the position of Fowler, as "trailing woefully."

Another analyst said Fowler's candidacy itself was "another case of Democrats shooting themselves in the foot."

It's fair to say that public sentiment regarding Fowler's chances in that election was very slim.

As a pollster, given the high-undecided nature of that election, if you had the knowledge that a lot of those undecideds will likely, eventually, support the challenger, do you think you might be tempted to put your thumb on the scale in your reported poll results?

Think of the headlines you would garner, if you were to report that the race perceived as a blowout is actually close, or even that the challenger considered by others to be "behind by 10" is *ahead*.

Now think of what that would do for your reputation if the race ended up very close, or the challenger eventually wins. You'd be considered a genius. Your polls would become the new gold standard because you predicted the result!

Not because your poll was great, but because you made the right assumptions about undecideds.

As observed by Panagakis, the voting tendencies of undecideds suggested that this race could be very close, and even that Fowler had a chance to win, despite the 10-point spread.

But Panagakis did not put out a "blended" poll that combined poll and forecast. He instead tried to appeal to people's sense of logic, statistics, and their ability to understand the difference between polls and forecasts.

He specifically addressed "poll users and news reporters" with the following, concise messages:

- "Incumbent races should not be characterized in terms of *point spread*. If a poll shows 50%/40% and 10% undecided, a 10-point spread will occur only if undecideds split equally ..."
- "Many polls may have been improperly analyzed and reported. Many polls remembered as wrong were, in fact, right."[5]

A lot of good that did. Fast forward to today, nothing has changed.

Based on his description of how polls remembered as "wrong" were "right" Panagakis seems to be referring to a then-unnamed simultaneous census standard; I don't think there's any other way to interpret it. Where I believe he erred was assuming that the experts and media he was trying to persuade had a shared understanding of what the "true value" in a poll was.

I will not assume that because their insistence on comparing polls to elections demonstrates they do not.

If comparison alone isn't direct enough, there are plenty of examples of experts outright saying "the polls predicted ..." to justify my skepticism. In addition to the example from earlier, when G. Elliott Morris cited in his book what he believed the Marquette Poll "predicted" about the election's eventual result, he's not alone in this mischaracterization.

The American Association for Public Opinion Research (AAPOR) put together an "Ad Hoc Committee on 2016 Election Polling" comprising more than a dozen experts and researchers, whose report[6] plainly states on several occasions that polls can be interpreted to "predict" results, and this is not true. A selection of them:

> "The vast majority of primary polls predicted the right winner."
>
> "Examining the polling averages in each state, the polls correctly pointed to the winner in 86% ..."
>
> "A higher percentage of primary polls predicted the winning candidate in 2016 than ..."
>
> "Circled states indicate instances in which more than 50% of the polls predicted the wrong winner – something that happened in 9 out of the 78 contests."
>
> "the predicted margin of victory in polls was nine points different than the official margin on Election Day."
>
> "perhaps it is harder to predict the margin when more candidates are running?"

The report also furthered the falsehood that there is such a thing as a "Polls Only" forecast, which is nothing more than an attempt to place predictor/forecaster assumptions beyond criticism.

> The FiveThirtyEight Polls Only predictions are based only on data that come from the polls themselves.

This is also false. At absolute minimum, a "prediction" requires making assumptions about undecideds and Changing Attitudes. There is no such thing as a "Polls Only" forecast because polls themselves say nothing about eventual undecided preference or Changing Attitudes. The number of decided voters in an election result must equal 100%, but the number of decided voters in a poll doesn't have to – and usually doesn't.

I hope to build on the work Panagakis started by not making assumptions where we have, or can have, data. To accomplish that, it requires not assuming experts who think polls "predict" something know how to analyze poll data, or even that we have a shared understanding of what data polls are designed to give.

As for Panagakis' "Incumbent Rule," it has largely fallen out of favor in part because, as RealClearPolitics senior elections analyst Sean Trende put it, "people interpreted the rule too literally, with analysts hyperventilating every time an incumbent fell to 48%–49% in the polls."[7]

I'll take Panagakis' observations regarding incumbent races and "point spread" a step further.

No races should be characterized in terms of spread.

Panagakis provided tangible evidence of why the Spread Method is flawed and its contribution to misleading the public decades ago.

Long before I started my writing, I connected with a few researchers whose findings are regularly published on the topic. In expressing my concern with their methods and characterizations, and perhaps better ways to approach the problem, their response was not to the content of my findings, but something to the effect of: "Publish your work and then we can talk."

In fairness, not giving every message they receive consideration is understandable; I am not dismissive of that reality.

On the other hand, the basis for much of my work – as I found out recently – has *already been published*, and still goes ignored.

Panagakis was not a fringe or unknown figure to the polling community: in addition to coining the Incumbent Rule, he was the president of a market research company that conducted political polls and a regular publisher in various journals and newspapers on the topic of poll data.[8]

Regarding his explanation of the Incumbent Rule, and my disdain for making assumptions where we should use data, I thought – maybe – we would diverge at that point.

After all, just because something has been true in the past doesn't mean it will always be true in the future. If, in some election, undecideds happen to favor the incumbent (which would go against this rule) that doesn't necessarily mean the poll was wrong – just the assumption (or rule) about what undecideds should do was wrong.

In Panagakis' case, while his observation was termed as a "rule" – he at least noted it can and did have exceptions. The Spread and Proportional Methods accept no such exceptions in their calculations of error.

And just when I thought our understanding of polls and undecided voters might diverge, Panagakis brings me right back:

> This is not a system for allocating the undecided vote.[9]

He wasn't making a statement that undecideds *will* or *should* favor a challenger in future races with an incumbent, merely that they *had*. In presenting the real and tangible ways the Spread Method had failed in the past, he hoped to convince enough influential people to change the way they analyzed polls and better inform the public in the process.

In far fewer words than I, Panagakis successfully outlined why the Spread Method's characterization of poll accuracy has failed in the past, could fail in the future, and why it is a disservice to the public to continue to report them in such a way. But not only has it continued to be reported in this way, respected statisticians have adopted it as their "gold standard" measure of poll accuracy.

At that point, after the elation I felt upon finding his work, I reached a feeling of dejection: how could the problems with this method have been spelled out decades ago, only to be entirely ignored?

When I said in Chapter 1, "I have come to the conclusion that there has been a stagnation in the poll analysis industry for decades," this is a big part of what I meant.

What impact did his work have on those who came after him? Instead of building on this research to improve both expert and public understanding, it went ignored. The Spread Methods and Proportional Methods, both of which preceded Panagakis and relied on assumptions instead of data, are still accepted unconditionally today.

When I say very direct things like "experts don't know what poll data means," it's not as though I believe this seemingly confrontational (though substantiated) position will be the most persuasive. It's because the approach that should be the most persuasive – being nice, providing examples, offering improvements – is very easy to ignore. The work Panagakis did that seems to have been lost to history is good evidence of that.

A field that has herded around the idea that polls are predictions (it's better to be wrong with everyone) won't be moved by polite discord, in my opinion.

In 2006, a year before FiveThirtyEight was founded, pollster Mark Mellman issued an observation in reference to the Panagakis Incumbent Rule regarding past elections, which turned out to be a foretelling of what was to come:

> headlines heralding an incumbent's 45–35 margin as a "big lead" betray an ignorance of what the numbers mean.[10]

And yet, that "ignorance of what the numbers mean" isn't just how the public reads polls or how the media reports them, it's how the experts say we should read them, too.

The Spread Method would say a 48–38 is "+10." That candidate should win easily!

The Proportional Method would lie and say that a 48–38 lead is the same as 56–44.

Fowler overcame his perceived double-digit poll deficit and defeated Incumbent Georgia Senator Mack Mattingly 51%–49%. The candidate who polled at 38% eventually received 51%; the candidate who polled at 48% received 49%.

How much error did that poll have? The existing methods don't know; they can't know. They guess.

According to the Spread Method, that poll had a 12-point error. The Proportional Method would say it had a 7-point error. Where did that error come from? How do you account for the fact that these methods assign such drastically different "error" calculations to the same poll-to-election result?

Panagakis tried to account for it, even if only approximately. FiveThirtyEight and others, to this day, still can't and don't.

> How will poll undecideds decide? The most common answers to this question are that undecideds either break equally or break in proportion to those stating a preference for a candidate. But the answer should be: It depends on who the candidates are.[11]
>
> *Nick Panagakis, April 1989*

As you can see in that quote, Panagakis' work was not limited to discrediting the Spread Method.

Even earlier, in 1985, Panagakis wrote in *Public Opinion Quarterly*, an article entitled "On Preelection Polling." In the article, with a level of succinctness and directness that, I could only dream of, his very first words are,

> The major flaw in the arguments presented by Richard Day and Kurt M. Becker is ...

He's off and running.

He makes a few notes regarding the timing of the election that was analyzed, along the flaw inherent in applying assumptions from national polls to local ones. But just one paragraph later, he reaches this point:

> Day assumes that undecided voters distribute in proportion to decided voters on candidate preference.[12]

This is the Proportional Method.

Panagakis continued to expand that this rule for national elections does not apply to local elections and that the researchers were therefore incorrect to characterize poll accuracy in the way they did.

I dislike the characterization of what undecideds tend to do as a "rule" because while I (and you, if not before reading, certainly by now) understand rules can have exceptions, the public may view something being named a "rule" as something that can't be broken.

But terminology aside, what's important is the ability to separate the *assumptions of an analyst* from the *functions of a poll*. It doesn't matter if we can prove with certainty that undecideds broke exactly 50/50 in the next 10 elections, or exactly proportional to the decideds: it's still an assumption to say that will happen in the 11th and that any deviation from that assumption must be a poll error.

It doesn't matter if we can prove with certainty that undecideds broke exactly 50/50 in the next 100 elections, or exactly proportional to the decideds: it's still an assumption to say that will happen in the 101st and that any deviation from that assumption must be a poll error.

The existing methods don't even have *that* level of reasoning on their sides: they just guess.

As far as I can tell, Panagakis never directly offers that we must account for undecideds before calculating poll error, but it's a very short step from his work to that conclusion.

In the article in which he coined the Incumbent Rule, Panagakis noted:

> An incumbent leading with less than 50% (against one challenger) is frequently in trouble; how much depends on how much less than 50%.

Reading a poll not by how much one candidate is "ahead" but instead by the leader's poll number is another example of what seems like an unremarkable statement but is again an astute observation.

Looking at the leader's poll number is not only an accurate way to read what a poll tells us, and not only a way to simplify a poll to one number (if the media absolutely insists this must be the case), but also – as you'll see – a remarkably accurate way to inform a prediction regarding how likely a candidate is to eventually win.

For now, I'll summarize: polls are not predictions; polls inform predictions. Understanding how undecideds decide is not just *helpful* in the process of quantifying "poll accuracy" and "poll error," it is requisite. It is indispensable. There can be no discussion of poll accuracy or poll error in any reputable statistical community which does not account for how undecideds decide(d).

If their preferred method to account for how undecideds decide remains an assumption, which is an assumption whose error places blame on the instrument or pollster and not the individual or method making the assumption, let that speak to their lack of reputability on the topic.

With that, I hope to summarize the findings to this point in a way that is simplified but not oversimplified: a simple, true statement that everyone can understand. No asterisk, no fine print:

Polls do not attempt to predict the **spread** of an election result.

Polls do not attempt to predict the **proportion** of two-party vote.

Polls do not attempt to predict the result of an election.

Smoking causes cancer, and polls do not attempt to predict the result of an election.

If the fact that polls do not attempt to predict the result of an election does not make sense, if it does not resonate with you, then I regret that I have failed in my goal to explain it.

If it does make sense, and you understand that fact, then you have more knowledge regarding how polls work than most everyone in the world, including the experts.

Polls are *tools* that we can (and should, if we're interested) use to inform our prediction, but polls are not predictions.

Now that you understand what polls *don't* do, we can talk about what polls *do*.

Notes

1. Panagakis, N. (1997). *Incumbent races: A national perspective.* https://ropercenter. cornell.edu/sites/default/files/2018-07/81021.pdf
2. Panagakis, N. (1997). *Incumbent races: A national perspective.* https://ropercenter. cornell.edu/sites/default/files/2018-07/81021.pdf
3. Panagakis, N. (1989). *Incumbent races are closer than they appear.* The Pulse. https:// www.lib.niu.edu/1989/ii890428.html
4. Treadwell, D. (1986, October 23). Liberal challenger trailing in polls: GOP expected to keep its Georgia Senate seat. *Los Angeles Times.* https://www. latimes.com/archives/la-xpm-1986-10-23-mn-7095-story.html
5. Panagakis, N. (1989). *Incumbent races are closer than they appear.* The Pulse. https:// www.lib.niu.edu/1989/ii890428.html
6. Kennedy, C., Blumenthal, M., Clement, S., Clinton, J., Durand, C., Franklin, C., McGeeney, K., Miringoff, L., Olson, K., Rivers, D., Saad, L., & Wlezien, C. (2018). An evaluation of the 2016 election polls in the United States. *Public Opinion Quarterly, 82,* 1–33. https://doi.org/10.1093/poq/nfx047
7. Trende, S. (2014). *Sabato's crystal ball.* Center for Politics. https://centerfor-politics.org/crystalball/articles/what-to-expect-from-senate-polls-in-the-final-days/

8. Panagakis, N. (1989). Incumbent races: Closer than they appear. *Pollingreport. com*. https://www.pollingreport.com/incumbent.htm

9. Panagakis, N. (1997). *Incumbent races: A national perspective.* https://ropercenter. cornell.edu/sites/default/files/2018-07/81021.pdf

10. Mellman, M. (2006, February 4). *Incumbent rule broken sometimes.* The Hill. https://thehill.com/opinion/columnists/mark-mellman/6920-incumbent-rule-broken-sometimes/

11. Panagakis, N. (1989). *Incumbent races are closer than they appear.* The Pulse. https://www.lib.niu.edu/1989/ii890428.html

12. Panagakis, N. (1985). On preelection polling. *Public Opinion Quarterly, 49*(2), 261–262. https://doi.org/10.1086/268919

22

Finding the Base of Support

Important Points Checklist

- How the base of support relates to the simultaneous census
- What polls do
- The limit of what polls can, and try, to tell us
- The difference between poll error, poll average error, and forecast error – and the standards for comparison in each
- The accountability problem
- How Silver's novel techniques changed the industry for the better – and why his innovations were overlooked

Polls, as a tool, are intended to give us a reasonable approximation of a candidate's *base of support.*

Given an understanding of the simultaneous census concept, that a poll only tries to tell us about some level of support *right now*, in the context of moving from observation to forecast, I use the term "base of support."

Here's what I mean:

Given access to simultaneous census data, in which the entirety of the current population's preference is known, the only thing we can say with certainty is that each option will receive *at least* that much support.

In Mintucky, I counted all the Reds, Greens, and undecided "silvers." The eventual preference of the silvers was unknown, but all of them would be, eventually, Red or Green. What can we say that any poll, even a simultaneous census, says about the eventual ratio of those undecideds? *Nothing.* We can assume, predict, or forecast, but we cannot know.

You understood that Grace Green's position in the simultaneous census – behind by 3% – would not be properly characterized as "wrong" if she eventually won, which was proven to be possible with a strong enough Undecided Ratio alone.

What the Simultaneous Census allowed me to say, with 100% certainty, was the **base of support** for each candidate: the minimum number of votes

DOI: 10.1201/9781003389903-22

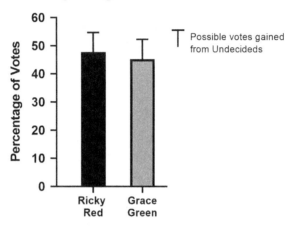

FIGURE 22.1
The Simultaneous Census results for Ricky Red and Grace Green, showing the candidate's base of support, and the maximum number of votes they could receive after undecideds were accounted for. Chart by the author.

they would receive. The vertical bar in Figure 22.1 represents votes that can be gained from undecided voters. It is the only variable in this example that needs to be accounted for between simultaneous census and eventual result.

Undecideds alone were experimentally proven to be capable of causing a candidate "behind" in a simultaneous census to eventually win: no margin of error, no nonresponse bias, no changing attitudes, and no other error. The entirety of the discrepancy between simultaneous census and result would have to be caused by the Undecided Ratio.

But in normal political elections, we don't have the benefit of simultaneous census data; in the best cases all we have are polls. That's okay, polls are great tools to inform our prediction. But to inform our prediction about what?

And here is why it's vital that you didn't skip ahead and didn't jump immediately into political polls (or even plan polls) but built your base of knowledge up to this point:

Polls are not predictions of the eventual result. They are approximations of the simultaneous census.

If it hasn't already, some more things about how polls work will start to fall into place right about now.

In the absence of a simultaneous census – as most poll applications are – poll data can only and should only be used to find an **approximate base of support**, not predict an eventual result.[1]

If you can't call a simultaneous census "wrong" just because it didn't accurately predict the eventual result, then you can't call a tool that only tries to give a reasonable approximation of a simultaneous census "wrong" for the same reasons.

But without a simultaneous census, there are more variables that must be accounted for, which can cause a poll or poll average to be wrong about a candidate's base of support.

It's possible that a poll or poll average *overestimates* a candidate's simultaneous census value, and likewise, possible that a poll or poll average *underestimates* a candidate's simultaneous census value.

And whether the simultaneous census value is known (or even knowable) is irrelevant: it exists and that's the number the poll tries to estimate.

To be clear, I'm not talking about the *eventual result* yet, which requires assigning undecideds and accounting for changing attitudes – right now, I'm only talking about the base of support: the minimum number (or percentage) of votes an option could possibly receive.

The underlying variables that can cause our base of support to be wrong are the same (real) causes of error in a poll that you already know. Primarily: nonresponse, frame, and statistical margin of error.

How can we account for this in our "base of support" approximation?

The same way the margin of error itself does: plus-or-minus.

Assuming we didn't know the results of the simultaneous census in Mintucky, the best we could do is a poll average. The methodology of that average, too, is up to the aggregator or forecaster; in this case, I had a simple average that showed:

Ricky Red(R): 48.5%

Grace Green(D): 45.5%

Undecided: 6.0%

In ideal circumstances, poll aggregation and averaging should reduce the margin of error relative to individual polls, if only for the fact that the sample size is increased when you consider lots of polls instead of just one. Mintucky is as close as it gets to "ideal circumstances." It's likely the true margin of error for the Mintucky poll average is much lower than the +/– 4% each poll had, owing to the cumulative sample size and similar methodologies of each pollster.

Perhaps if I had weighted my polls by sample size, instead of the simple average that I took, I could have produced an even more accurate poll average.

Nonetheless, in real-world applications, due to factors such as differing poll quality, poll methods, and varying poll aggregation methods, it's reasonable enough to say the margin of error for each option's base of support is at least *comparable* to each poll's margin of error. For those interested in

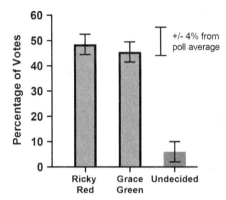

Mintucky Base of Support Approximation

FIGURE 22.2
The estimated Base of Support for each candidate, plus the undecideds, given the poll averages and approximate margin of error. Chart by the author.

building forecasts, if you can improve upon aggregation methods to better estimate a candidate's base of support, do it.

But for those of you who just want to know how you should read polls and poll averages, it starts with using those numbers as an approximate base of support. Given the margin of error in each poll was around +/– 4%, I'll use the poll average to estimate the base of support as +/– 4%.

Given the poll averages for Mintucky, *without* the benefit of the simultaneous census, that same "base of support" chart might look like Figure 22.2.

Remember, this poll average, like all poll averages, and all polls, is only estimating and can only estimate each option's simultaneous census value. This has nothing to do with predicting the eventual result yet. Different step.

This poll average methodology estimated Ricky Red's simultaneous census value – his base of support before considering changing attitudes and allocating undecideds – as somewhere between 44.5% and 52.5% (48.5% +/– 4%). It estimated Grace Green's simultaneous census value as 45.5% +/– 4%, and the undecideds as 6% +/– 4%.[2]

This is what polls try to tell us, and *this is the limit of what polls can tell us*. Given all the options, right now, this is approximately how many people would support each option.

> I don't know how many ways I can say it, and sorry if I'm repetitive here, but I'm trying to be as clear as possible: this "base of support" range of outcomes as can be visualized in Figure 22.2 is *the end* of what a poll or poll average says.

Want to talk about spread, proportion, the blended method, or anything else? Go ahead. Not what the poll says.

Want to make any prediction of the eventual result? Awesome! Do it. Nothing to do with what the poll says.

There is no such thing as a "Polls Only" forecast.

How many people who are decided today might change their mind? That's a function of an assumption, prediction, or forecast, not a poll.

How will undecided voters eventually decide? That's a function of an assumption, prediction, or forecast, not a poll.

In fact, as you'll soon learn if you didn't know already, *even poll averages and aggregators themselves can contain errors.* If your poll average includes (or excludes) certain polls, just like individual polls can have frame error, poll averages can too. If an aggregator weights certain factors in their poll average too heavily or not heavily enough – just like pollsters, that can make poll averages inaccurate, too.

And now for the question again:

Inaccurate compared to what?

A poll average attempts to measure the simultaneous census value, not the eventual result.

An individual poll, likewise, attempts to measure the simultaneous census value, not the eventual result.

A prediction or forecast is the only tool that attempts to measure the eventual result. The only thing the eventual result can be compared to without adjustment, and (rightly) have the difference called an error, is a forecast.

To summarize:

- **A poll or pollster's accuracy** is properly characterized by its closeness to simultaneous census value. The difference between the poll value and simultaneous census value is poll error. The most prevalent threats to poll accuracy include the statistical margin of error, nonresponse error, and frame error, but there are others.

- **A Poll Average or Poll Aggregator's accuracy** is also characterized by its closeness to the simultaneous census value. However, this analysis is subject to even more errors than individual polls because Poll Aggregators have the option – and ultimately, the final discretion – to include or exclude certain polls, and the extent those polls should influence its average. Threats to poll average accuracy include partisan/biased pollsters, herding, flooding, recency bias, and even poll error itself, but there are others.

- **A Forecast/prediction's accuracy** is properly characterized by its closeness to the eventual result. No adjustments or approximations are needed to calculate forecast error: what the forecast said versus what the result is equals the forecast error. Threats to forecast accuracy are too numerous to outline. As a topic of continued research for those interested: think about how much forecast error can be attributed to how the forecaster allocated undecideds, versus third-party changing attitudes, versus the model's uncertainty values, versus all of the errors inherently possible to poll aggregation, versus those inherent to poll error.

In simpler terms: if your poll was wrong, that's a poll error. If your poll average was wrong, that's a poll average error. And if your forecast was wrong, that's a forecast error. If you're going to call something an error, you must do so for valid and measurable reasons.

How undecideds decide or who might change their mind, if your forecast was wrong, that's a forecast error.

The status quo allows forecasters (or, in many cases, assumers) to deflect a considerable portion of error – error that rests on their flawed forecasts and/ or assumptions – away from themselves and onto polls.

Without accountability, the problem won't go away.

Polls Do Not Attempt to Predict Elections

It's a very natural desire, and perfectly healthy assumption given how they're discussed, to look at political polls and think they're a prediction of the eventual result. But with an understanding of how political polls (and plan polls in general) work, you can understand that this is very much not the case.

One man who Nick Panagakis debated over the years, and who deserves considerable praise for his contributions to the poll data industry himself, is Warren Mitofsky.

In their back-and-forth, from the 1990s and into the 2000s, one quote from Mitofsky – indisputably both an expert and innovator in the field – that really stuck out to me was:

> I believe the most important statistic reported by preelection polls is the one that indicates which candidate will win and by how much.[3]

Warren J. Mitofsky, 2001

Here is another expert in the field, who not only worked and contributed to it – but actively innovated and improved it with Random Digit Dialing – saying the Spread Method should be used not just to inform a prediction but to make it.

Projecting our desire to know the future onto polls doesn't change what the poll can measure. Asking a scale what you'll weigh next week doesn't change what the scale can measure. Asking people what they'll have for lunch on Friday doesn't change what they can tell you on Tuesday. And asking people who they'll eventually vote for doesn't change what they can tell you today. Unfortunately, even unquestioned experts in a field can make mistakes.

But the reputability of a scientific field isn't measured by how fervently it clings to its traditional methods, it's measured by how willing it is to be more accurate.

For decades, and largely still today, people look to polls out of a desire to know the future. In recent years, forecasting has become more mainstream – and whether out of convenience or arrogance, forecasters deflect most or all blame of "error" away from their own methods and assumptions and onto "the polls."

Members of the public, media, and even other experts, in their misplaced desire to know with certainty what poll numbers tell us about the future, are very quick and irresponsible to anoint as royalty anyone who happens to "get it right." And not at all concerned, as with poll data, to look at the underlying factors that may have contributed to their "getting it right."

Starting in more recent history, with the 2008 Presidential Election, Nate Silver and FiveThirtyEight introduced a new and innovative method that combined poll aggregation and forecasting. On his FiveThirtyEight blog, Silver "predicted the outcome" as it was erroneously reported, of 49/50 states. While veterans of US elections understand most states are not hard to "predict" the winner, the reality that the candidate he identified as the favorite eventually won all states except one is still impressive.

For those of us who are inclined to separate the "contenders" from the "pretenders" – or the people who do good work from the people who make good guesses – Silver showed his work. His blog outlined his methods, and he spoke at some length about what made his approach different from that which had been traditional in the industry. In addition to using poll averages as some others had done prior, he built a model that attempted to quantify a range of possible, and likely, outcomes.

He rightly received a lot of praise for his innovations, and his impressive accuracy catapulted him to fame for the fact that he was seemingly able to predict results in the way others couldn't. FiveThirtyEight went from relatively obscure blog to mainstream source of election commentary and forecasts.

Then, four years later – that is, one presidential term, Silver was considered dethroned as the best forecaster.[4]

Sam Wang, a neuroscientist at Princeton University, was described as the "New Nate Silver"[5] and the "election data king."[6]

On the back of one election in which he did very well, though his methods were rightly praised and credited, Nate Silver became *such* a household name, so synonymous with predicting election results, that someone could become the *new Nate Silver*.

And Wang's claim to the title "New Nate Silver"? He did really well in the 2012 General Election and then in 2014 midterms.

Silver's rise to forecasting fame came in 2008, thanks to doing well and "getting it right" in one election. Honestly, though based on previous chapters you may have perceived otherwise, I think that rise to fame was entirely justified: his innovative approach was far superior to other methods at the time. While his methods did and do make some problematic assumptions, those methods attempted to use data in areas where others used assumptions – if they attempted to quantify it at all.

But in just one election cycle, Silver fell out of favor with many, and Wang went from relatively unknown to, in the eyes of some, the best forecaster in the world.

Talk about changing attitudes.

Like analysts want polls to tell us the future, the public wants analysts to tell us the future. We are quick to crown the "best" at something based on tiny timeframes, even one election, and not at all critical of their underlying methods; we just want to know the answer.

As quickly as he was crowned, Wang's 2016 forecast proved to be an abrupt end to his reign. He said that Hillary Clinton had "more than a 99% chance" of winning the Presidency.[7]

She did not win.

That does not, in itself, mean Wang's forecast was wrong, but – I'll let his words speak for themselves.

Here's how Wang explained his forecast's failure to account for Trump's chances:

> ... there was an unusually high number of undecided and third-party voters. And usually undecided voters will break approximately equally, but this year the evidence suggests that they broke in favor of Trump ... I think the fundamental factor that made everyone wrong—and I should say, everyone wrong on *both* sides—was this polling error.[8]

"This polling error."

Not "my flawed assumptions," not "my forecast," but *this polling error*.

This is, unfortunately, the norm for this industry. Take credit when you're right, deflect blame when you're wrong.

To their credit, FiveThirtyEight was openly and loudly contrary to the prevailing public and media sentiment that Clinton was an overwhelming favorite. By their forecast, Clinton was never even a 90% favorite. In early November – just days before the election – they said her chances were under 66%, and by Election Day they calculated she had about a 70% chance to win. Favored, but far from overwhelmingly so.[9]

I'll leave analysis regarding the problems underlying Wang's model, and FiveThirtyEight's for that matter, for a separate day, and for those interested in that side of things.

But while we're on the topic of Trump-Clinton 2016, let's get into what the polls told us.

Notes

1. *Note:* Though the *"simultaneous census"* and *"base of support"* are effectively different terms for the same number, I prefer to use different terms because they have different applications. For one, speaking to people who have never heard of a *"simultaneous census,"* the term *"base of support"* is generally better understood. Moreover, the base of support is the number from which you start to build your prediction or forecast. The simultaneous census value is the necessarily specific term that should be used in place of *"true value"* to calculate poll error because *"true value"* is not properly understood in that context.
2. See Chapter 8, note 3, for explanation behind margin of error imprecision.
3. Mitofsky, W. (2001). Reply to Panagakis. *Public Opinion Quarterly, 63*(2), 282–284. https://doi.org/10.1086/297718
4. Hickey, W. (2012). *You think Nate Silver was impressive? A Princeton neuroscientist perfectly nailed the popular vote.* Business Insider. https://www.businessinsider.com/meet-sam-wang-the-neuroscientist-who-beat-even-nate-silver-in-his-election-prediction-2012-11
5. LoGiurato, B. (2014). *Meet the new Nate Silver.* Business Insider. https://www.businessinsider.com/sam-wang-nate-silver-forecasts-dem-senate-hold-2014-9
6. Nesbit, J. (2016, November 7). *Sam Wang is this year's unsung election data superhero.* Wired. https://www.wired.com/2016/11/2016s-election-data-hero-isnt-nate-silver-sam-wang/
7. Revesz, R. (2016, November 6). *The man who predicted 49 out of 50 states in 2012 has said who will win on Tuesday.* The Independent. https://www.independent.co.uk/news/world/americas/sam-wang-princeton-election-consortium-poll-hillary-clinton-donald-trump-victory-a7399671.html
8. Schulson, M. (2016, November 17). *Meet a polling analyst who got the 2016 election totally Wrong.* Pacific Standard. https://psmag.com/news/meet-a-polling-analyst-who-got-the-2016-election-totally-wrong
9. Silver, N. (2016, November 8). *2016 election forecast.* FiveThirtyEight. https://projects.fivethirtyeight.com/2016-election-forecast/

23

Trump-Clinton 2016

Important Points Checklist

- The "snapshot" analogy
- How "Temporal Error" relates to "Changing Attitudes" and their respective assumptions
- How Changing Attitudes contributed to the Trump-Clinton election, and how analysts accounted for them
- The objective standard of comparison for polls and forecasts
- Why third parties complicate the "Changing Attitude" calculation
- What happens when poor assumptions aren't tested – and then they are
- What analysts and experts mean when they say we need to "fix" polls
- The "Bill Clinton Lesson" (no not *that one*)

When asked how polls work, experts like to use a very apt comparison of a poll to a "snapshot." Many, if not all analysts, have at some point used that exact word to describe what polls try to tell us.[1] They simply, for some reason, betray that understanding when it comes time to perform analysis regarding poll accuracy. They wouldn't look at a snapshot from a footrace and expect it to predict the winner – but that's what they ask of polls. They make no adjustments for variable change between poll and eventual result.

A major contributor to that flawed analysis, as I've repeated to an extent that you're probably tired of it, is the fact that how undecided voters eventually decide is not a function of a poll, but of an assumption or forecast. However, I don't want to understate the influence (possible and actual) changing attitudes can have on the perceived accuracy of polls.

In 2008, still in the very early days of FiveThirtyEight, Silver called the fact that time is a factor between poll and result a "**Temporal Error.**"[2]

Sounding eerily like myself, he pointed out that "many things can happen" months before the election, "and (contrary to the common perception) it is not up to the pollster to predict the future."

DOI: 10.1201/9781003389903-23

He nailed it. Despite the public perception – owing largely to how the media reports polls – it's not up to the pollster to predict the future. So what happened?

He closed the definition of Temporal Error, with reasoning still applied[3] today, "For purposes of the pollster ratings, however, we can ignore Temporal Error (that is, assume it to be zero), because we are limiting our evaluation to polls taken very near to the election date."

As you know, both the Spread Method and Proportional Method assume changing attitudes as zero in their calculation of poll error. Their reasoning, at least on its face, is defensible: it's unlikely for decided voters to change their mind close to an election.

It seems defensible, but it's not.

These methods, by definition, say that mind-changing inside the three-week window they made up as the "cutoff" can be assumed as zero, or at least negligible (I have seen similar assumptions made by other sources that assume this value to be zero at two weeks, four weeks, and some that make the case that this value can be assumed to be zero for even longer).

The defense for this definition is that it's rare that this would happen, and the past data shows that it's rare.

To which the proper response is "so what?"

Panagakis' incumbent rule showed that it's rare that undecideds don't heavily favor the challenger in races with an incumbent. Can we therefore assume that 60%–100% of undecideds will favor the challenger in any such race and call any observed disagreement with that assumption a "pollster-induced" poll error?

No, it would've been invalid to do so in the election following Panagakis' research on this rule, and it would be invalid to do so today; the same can be said regarding assumptions about Changing Attitudes or "Temporal Error."

All this data about mind-changing and undecided voters can be valuable data to *inform a forecast*. If someone wants to assume that because something has generally been true in the past, that it should or will remain true in a specific future election, they're welcome to do so. Dicto Simpliciter[4] would have something to say about the logical validity of that assumption, but that doesn't mean one can't make it, nor even that it's not the best assumption one can make.

The point is, it's *the predictor's* assumption, not the polls'.

The current methods say that if their assumption about mind-changing is wrong, it's not their assumption that was wrong, nor their method, but – by some convoluted reasoning – the poll.

Of course, in the presence of a federal investigation or scandal, I think it's *indefensible* to say that at least some previously decided voters *won't* change their mind. Whether that mind-changing accounts for a candidate losing 0.1% or 1% or 10% of their support is up to a *forecaster* to try and quantify, not a poll. How is a poll conducted 21 days prior to an election supposed to be able to account for a scandal that comes out 10 days prior to an election? Too bad, poll error.

In 2016, less than two weeks prior to the Presidential Election, FBI Director James Comey sent a letter to Congress regarding an investigation into then-Secretary of State Clinton's private email server.[5] Now the Democratic candidate for President, Clinton was considered a favorite to win, both before and after the scandal broke – though how strong of a favorite depended on who you asked.

In any case, as Wang lamented after literally eating a bug as he promised he would if Trump outperformed his forecast, could the high number of undecideds have contributed to Trump's victory?[6]

Is it also possible changing attitudes stemming from the Comey Letter contributed?

Yes, and yes.

Setting aside the usual, problematic use of "spread," Silver did some great work quantifying if – and how – Comey's letter contributed to Clinton's loss.

His conclusion? "The Comey Letter Probably Cost Clinton the Election."[7]

He said, "Hillary Clinton would probably be president if FBI Director James Comey had not sent a letter to Congress on Oct. 28."

That's a bold statement. That says the letter must have contributed to undecideds overwhelmingly supporting Trump and/or caused previously decided voters to flip from Clinton to another candidate (or not vote).

Silver continued, "it's plausible that Clinton's underperformance versus the polls on Election Day had something to do with Comey."

Given this overwhelming case of a late event influencing the election, FiveThirtyEight made a one-time exception to their "poll error" calculation to try and account for this.

I'm kidding, they did not do that.

Leading up to the election (even after the Comey Letter), FiveThirtyEight continued to characterize any future discrepancy between poll and result as a poll error.

"Trump is just a normal polling error behind Clinton," they said. Their use of the term "normal" in the face of a largely unprecedented scenario is problematic.[8]

The underlying variables were different, they knew it, but still didn't bother to account for them.

Just one day after the election, once it was clear Trump had won, they wrote, "The Polls Missed Trump."[9]

Not their model, not their assumptions, "the polls." Silver and Wang hadn't agreed on much during this election cycle, but they agreed on one thing: it couldn't be their assumptions that were wrong, it's the polls.

FiveThirtyEight noted, "Trump mostly outperformed his swing state polls" and supported this claim with nothing more than the poll versus election spread difference. End of analysis.

Forecasters weren't alone in their usual mischaracterization of not predicting the eventual result as a failure of "the polls."

Patrick Murray, director of the Monmouth University Polling Institute, was quoted after the 2016 Election as saying, "The polls were largely bad, including mine."[10]

In the years leading up to 2016, by the way analysts had judged them, polls had enjoyed a high level of success in presidential elections for more than a decade: they *predicted the result* correctly, therefore were good.

"Polls performed very well" in 2008, according to Pew Research, because national polls predicted "the final margin in the presidential election within one percentage point."[11]

They continued, "Both at the national and state levels, the accuracy of the polls matched or exceeded that of 2004, which was itself a good year for the polls."

The polls were right because they had closely predicted the eventual margin, they said.

You don't have to take my word regarding the threat this flawed standard poses to the industry's credibility. Pew Research, in the same article praising 2004 and 2008's accurate polls, said:

> The performance of election polls is no mere trophy for the polling community, for the credibility of the entire survey research profession depends to a great degree on how election polls match the objective standard of election outcomes.

The election outcome *is* an objective standard, but polls *are not* a tool that attempts to predict the result or eventual spread of that standard.

The election outcome is the objective standard for *forecasts*.

The Simultaneous Census is the objective standard for *polls*.

Polls in 2012 likely had some error, even if small, considering Romney's underperformance of his national poll average, and in one crucial state, as you'll see shortly.[12]

Nonetheless, Silver assigned a very confident 91% probability to Obama's chances at re-election, and Obama won re-election.[13]

Entering 2016, public sentiment regarding the reliability of poll data was very high; Silver and FiveThirtyEight had risen to national prominence thanks to their ability to read "the polls" and predict with even more precision than other experts the all-important question: who will win?

FiveThirtyEight's rise to prominence was credited mostly to Silver and their forecast models, rarely (if ever) the accuracy of Gallup, Monmouth, Marquette, or other actual pollsters.

But when 2016 didn't go the way people expected, the fingers weren't just pointed at the forecasters, but primarily at the polls. No longer were articles written about how many states FiveThirtyEight "called correctly" (49/50 in 2008 and 50/50 in 2012)[14] but how *the polls* "missed" in 2016.[15]

And not the least of this blame on polls came from FiveThirtyEight themselves.

This pressure applied by analysts and media across the country was felt by pollsters. Forecasters struggled to make predictions using their polls, which means the polls must have been wrong. Like the public, pollsters themselves began to question the usefulness of poll data.

The pressure was accumulating and forced some out entirely.

"Polls might not be capable of predicting elections," Patrick Murray said in an interview with Business Insider.[16]

Murray and Monmouth University polling fortunately continued their "horserace" election polling beyond 2016, but unfortunately Gallup did not.

But the pressure continued to mount for pollsters, after 2016's perceived miss. After the New Jersey Governor's race in 2021, in which Incumbent Phil Murphy was considered a huge favorite but ultimately only won by a small amount, Murray went as far as to say, "Maybe it's time to get rid of election polls."[17]

I should point out, for those who aren't familiar with Murray and Monmouth, that he and they are among the best and most respected pollsters in the country – both their track record and transparency set a high standard.

So *why* Murray wrote that we, maybe, should get rid of election polls is the central point of the most telling disconnect between what good pollsters actually do, and how erroneous analysis has infected the minds of even the best in their fields.

Here was Monmouth's final poll[18] on the New Jersey Governor's race:

Candidate	Murphy (D)	Ciattarelli (R)	Undecided
Reported Poll Number	51%	40%	9%

And here was the result[19] from that election:

Candidate	Murphy (D)	Ciattarelli (R)
Final Result	51.2%	48.0%

In his article in which he said he "blew it" Murray wrote:

> The final Monmouth University Poll margin did not provide an accurate picture of the governor's race.[20]

To which, as readers hopefully understand now: so?

Is it possible that their poll was extremely accurate – by the proper simultaneous census standard – but that most of the undecided voters simply ended up voting for Ciattarelli? Mr. Panagakis' rule might have something to say about that.

None of this is to say his poll or "the polls" *couldn't have been* wrong. It's to say that to characterize "the polls" as wrong based on "spread" (or proportion) *requires valid underlying reasoning.*

To say that this poll was "off by 8" (as FiveThirtyEight did and the Spread Method would) requires a little more work than subtracting spreads and calling the difference the poll's error.

The false belief that polls should try to predict elections – or the spread of elections – is so damaging that some of the best pollsters in the world are questioning their work.

Again, analysts should take a look in the mirror.

In their paper "Disentangling Bias and Variance in Election Polls" which was published in the *Journal of the American Statistical Association*, the authors wrote:

> Nearly all pollsters declared Hillary Clinton the overwhelming favorite to win the election.[21]

Not a reader of academic papers? No problem. Natalie Jackson, then the senior polling editor at The Huffington Post, said:

> The polling data has never consistently shown anything but a Clinton win.[22]

Pew Research, by all accounts among the leading experts on polls, said:

> national and state election polling ... consistently projected Hillary Clinton as defeating Donald Trump.[23]

No, they didn't; no, it didn't; and no, they didn't, respectively.

There are thousands of quotes from experts on this subject that express this same, fundamental mischaracterization of what polls do.

With experts characterizing polls this way, it's no wonder such a misunderstanding exists in the public.

All of these characterizations are wrong. There's no nicer way to say it – and there's no other way to say it.

> Not a single poll in the history of political polls has ever "predicted" a winner. No poll has ever predicted the final "margin." Poll data does not declare "favorites." And poll data does not "show who will win."

Polls are not forecasts, and anyone who says or implies it is wrong. It doesn't matter how smart you are, or what your credentials are, those things don't preclude you from being wrong.

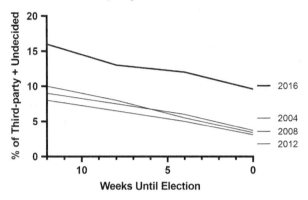

FIGURE 23.1
The National Poll Average numbers for third-party plus undecided voters showing 2016 was very different from recent history.

When statements like, "the polls predicted they would win" come from the very best researchers on the subject and from every direction in the media, it's unfair to assume the public will figure out that's not correct. When one of the best pollsters in the world apologizes for his poll not predicting the spread of an election, it's unfair to assume the public will understand that's not what polls try to do.

So, what did the polls actually tell us in 2016?

First of all, compared to the most recent presidential elections, and most US elections in general, 2016's Trump-Clinton matchup had a historic number of voters declaring that they would vote for a third-party candidate, or were undecided. Figure 23.1 shows the poll average by month leading up to each year's presidential election, among options *not* one of the two major candidates.

From the beginning, it was clear that 2016 was not a typical presidential election. I'll spare you the punditry on *why* there were a historic number of voters who were either undecided or said they would vote for a third party, and it's not particularly important for this chapter. What's important is seeing that neither Trump nor Clinton had a strong base of support much of anywhere.

Third-Party and Undecided: Very Different Calculations

Having just provided a graphic that gives an overview of why 2016 was, regardless of outcome, bound to be different from any election in recent history, it's time to go a step further in the analysis of how polls work.

I've explained at some length the impact undecideds can have as it relates to poll versus eventual election results: undecided voters can tip the scales in favor of either candidate, but how undecideds eventually decide is not something a poll tries to tell us.

Third parties are a different story. Polls *do* try to tell us something about third-party candidates: the same thing it tries to tell us about major party candidates, their simultaneous census value.

It's a helpful *starting point* to batch "third-party" and "undecided" into one category, if you're trying to demonstrate as I did above that the two "main" candidates do not have a strong base of support. But grouping third-party and undecided voters together in your calculation of poll accuracy is problematic if you, like me, are an advocate of "showing your work" when it comes to why we can call a poll "wrong."

As far as the existing methodologies for calculating poll error are concerned, the differences between what polls try to tell us about third-party and undecided voters don't matter.

To the Spread Method, a poll average that says 53%–43% with no third-party and 4% undecided is exactly the same as one that says 45%–35% with 8% third-party and 12% undecided; both are "+10."

The Proportional Method is even worse because it throws out everything except the poll numbers for the top two candidates and misinforms the public by excluding the number of undecideds. For the two polls above, since the Proportional Method is only concerned with the two main options, it would report the candidate leading with 53% of the vote as having 55% of the vote – and that the candidate leading with 45% of the vote has 56%.

> Yes, because of its flawed assumptions, the Proportional Method
> would report a candidate who is leading 45%–35% as stronger
> than one who is leading 53%–43%.

Thanks to Mintucky, "What's for Lunch?", and lots of other examples, you now understand some major reasons why those methods are flawed, if you didn't already. But up until now, the conversation has mostly centered around the impact *undecided* voters could have between the poll and eventual result. Importantly so, undecided voters can have an impact on almost every election.

But undecided voters are not the only confounding variable to account for between polls and elections.

When we think about "changing attitudes," most people, including myself prior to doing research to build forecasts and then for this book, probably only consider the two major candidates: if I originally planned to vote for one, but eventually vote for the other (or don't vote) that's mind changing. What else is there?

In Mintucky, the mints couldn't change their mind. In "What's for Lunch?", several people changed their mind, but from one option to the other.

But neither example had a "third party" option. When I said political polls were the hardest class of polls, this is on the long list of reasons why.

Here, yet another reason why better definitions to measure poll accuracy are needed, and why the existing ones which ignore third parties will inevitably continue to fail:

Individuals who say they plan to vote third-party changing their mind to vote for a major party candidate, or vice versa, *can also only be classified as changing attitudes.*

Figuring out whether that person *will* eventually vote third-party, might eventually vote for one of the major candidates, or not vote at all, *is a function of a forecast, not a poll.* Changing Attitudes is not a poll error any more than changing weight is a scale error.

Additionally, if some voters declare that they are undecided, and then eventually decide, the ratio of *how* they decide not agreeing with your assumptions is also not a poll error.

Because of the flaws in the currently used methods, they are **definitionally incapable of figuring out where the error came from.**

So, when Pew Research asks "Does polling need to be fixed again?"[24]

When Nate Silver talks about "What it would take to fix polling."[25]

The question to ask is as follows:

Fix what?

Do pollsters need to fix how they weight for nonresponse? Fix how they qualify "Likely Voters?" Fix how their poll questions are worded?

Or do pollsters need to fix their time machine so they can start including in their polls how undecided voters will eventually decide, or who might change their mind?

Analysts who ascribe to the Spread Method and Proportional Method don't know what needs to be fixed, or how much any factor contributed to poll error, their definitions prove it. They can't know.

By projecting a misplaced desire to know the future onto polls, by assuming poll error is known without first accounting for all the variables that comprise it – and understanding which ones don't – they will never be able to *fix* anything.

Here's why that mattered in 2016.

Third-party candidates in the US Presidential Elections are not rare, but for most of recent history – the 21st Century – their support had been scant. The last time a third-party candidate had received more than 1% of the Popular Vote nationally was 2000, when the Green Party's Ralph Nader received 2.7%.[26]

Certainly, 2.7% of voters is not an insignificant number, nor is 1% as it relates to closely contested elections, but the next three presidential elections came and went without a substantive third-party turnout. On top of that, presidential polls in the elections prior to 2016 also had relatively few undecideds.

Methods that don't account for third parties and undecideds in their calculations are less likely to fail when there are few third-party supporters and undecideds; even bad methods can survive as long as its flawed assumptions aren't subjected to testing. In that sense, I am an expert at building bridges as long as you don't walk or drive on it.

But 2016 had a number of undecided voters that was unprecedented for the century, and a level of support expressed for a third-party that hadn't been seen for at least 20 years. The flawed methods were subjected to testing, and the methods failed.

And to be clear, I'm not the only one who shares this belief about the poll error calculations failing – I'm just the only one who has, as far as I know, stated that fact. I previously cited Nate Silver's belief that if not for the Comey Letter, "Hillary Clinton would probably be President."

In his own words, Silver effectively said that Changing Attitudes and/or the Undecided Ratio favoring Trump – because of the Comey Letter – led to Trump dramatically outperforming his poll number, and/or Clinton underperforming hers.

In a less verbose manner, though equally as deflective, Wang stated that his forecast was wrong because of how undecideds eventually voted – and he classified his inability to provide a reasonable probability of that happening as a failure not of his model, not of his assumptions, but of the polls.

> *Both Silver and Wang, among the most highly respected and influential forecasters at the time, stated unequivocally that factors unrelated to what polls can tell us caused the polls to be wrong.*

Pollsters, media, and "the experts" agreed; whether they arrived at that conclusion independently or via influence of other "experts" is a different story. They believed the polls were wrong, and that it's a problem the pollsters need to solve.

What had been true in recent history – few undecided or third-party voters as the election approached – wasn't true in 2016. It's not just hindsight that allows us to see these differences, these factors (along with the Clinton email scandal) were well known and widely discussed. They were simply not accounted for.

Worse, even with hindsight, the analysis *still* failed.

Years later, in the aftermath of Trump losing the 2020 election, Scientific American in retrospect claimed of the 2016 election that "polls across the country predicted an easy sweep for Democratic Nominee Hillary Clinton."[27]

There's that characterization about polls "predicting" things again; it's as inescapable as it is wrong.

In their interview consulting with Wang years after his terrible Trump-Clinton forecast, Scientific American didn't consider asking him why his forecast was wrong, what wrong assumptions he made, or if and how he fixed his methods – they just asked him about how much poll error there was in 2020, in relation to 2016.

The deflection worked. If my forecasts are right then it's because I'm right, if my forecasts are wrong then it's because the polls were wrong. Despite being woefully off the mark, people who don't even understand how polls work are continually given a platform to tell people how wrong they are.

The pollsters believe it, and the public believes it.

You can't fix polls if you don't know how they work.

And to answer the question the American Association for Public Opinion Research (AAPOR) posed in their Trump-Clinton 2016 report:

> perhaps it is harder to predict the margin when more candidates are running?

Yes. This is one reason forecasting the vote share in elections with several competitive candidates is harder than when there are only two. But because the existing methods assume Undecided Ratio and Changing Attitudes is known, and make no distinction between polls and forecasts, they haven't been able to give a concise answer to what should be a very simple calculation.

Just as a large (or even mild) ratio of undecideds deciding in a way that is contrary to a method's assumptions can cause them to fail, a non-negligible third-party presence can do the same because it introduces more complex "changing attitudes" calculations.

These complex "changing attitudes" calculations – in addition to the ramifications multiple competitive third parties create for the Undecided Ratio – is why we started with two-party data and not multi-party data. Juggling multiple variables is fine when you understand they all exist – but researchers and interested members of the public who insist on immediately analyzing political polls don't properly account for those variables. Every analysis I've seen from public to expert puts these "late deciders" all in the same bin, failing to account for each, and drawing avoidably erroneous conclusions – in addition to discounting the possibility of compensating errors.

Considering "two-party" elections before "multi-party" elections is just a way to help explain poll data. It's a simplification of real-world examples that is not oversimplified.

I offered those examples to explain these concepts because there have been many *real* elections with multiple competitive parties that taught lessons that weren't understood, discarded, and/or forgotten.

The Forgotten Bill Clinton Lesson

Presidential election analyses in the United States, and any forecasts built to predict the outcome, are complicated by the Electoral College. Electoral

Votes, with a couple exceptions, are allocated in a winner-take-all manner; if you receive the most votes in a state, you receive all of that state's Electoral Votes. It doesn't matter if a candidate receives 10 more votes than their closest competitor in a state, or 10 million more votes, the winner gets all of those Electoral Votes. It also doesn't matter if the winning candidate receives 100% of the vote, 50.1% of the vote, or 40% of the vote – if the candidate receiving 40% of the vote received the *most* votes in that state, they win it.

Consequently, as it relates to the eventual result, the National Popular Vote is meaningless in the Electoral College system. The presidency is decided not by the Popular Vote of the country, but by the Popular Vote of a few "swing" states.

A **swing state** in presidential elections refers to any state that is considered likely to be competitive; one in which the number of votes the top candidates are expected to receive is close enough that either one could conceivably win.

And in case you think I was exaggerating about 40% possibly winning in a state, Bill Clinton won *four states* in 1992 with *less than 40%* of the vote in each. And that doesn't include Colorado, which he carried with just 40.1%.[28] How did he do that? A historically strong third-party candidacy by independent candidate Ross Perot, which garnered 18.9% of the vote nationwide, and even received more votes in one state (Maine) than the Republican George Bush.

Not all US elections are much different from a typical UK or non-US election, with multiple competitive parties and winners finishing with far less than 50%.

What's the lesson here? Third-party vote matters. Being able to accurately account for *every* candidate's base of support – not just the primary two – *directly leads to being able to make better predictions.* Better predictions about the winner, better predictions about the "margin."

I will always reiterate that polls aren't predictions because they aren't and anyone who tries to characterize them that way is wrong.

However, I will never say that polls aren't the best tool to *inform our predictions* because until we get that time machine analysts seem to believe pollsters should have, they are. And I'll show you why being able to account for third-party base of support – which both existing methods ignore – will allow you to make better predictions and better understand what polls told and tell us.

Notes

1. Kennedy, C., Blumenthal, M., Clement, S., Clinton, J., Durand, C., Franklin, C., McGeeney, K., Miringoff, L., Olson, K., Rivers, D., Saad, L., & Wlezien, C. (2018). An evaluation of the 2016 election polls in the United States. *Public Opinion Quarterly, 82*, 1–33. https://doi.org/1093/poq/nfx047

"...polls and forecasting models are not one and the same. As the late pollster Andrew Kohut once noted (2006), "I'm not a handicapper, I'm a measurer. There's a, difference." Pollsters and astute poll reporters are often careful to describe their findings as a snapshot in time..."

2. Silver, N. (2008, April 30). *Pollster ratings, v3.1.* FiveThirtyEight. https://fivethirtyeight.com/features/pollster-ratings-v31/

3. Rakich, N. (2023, March 10). *How our pollster ratings work.* FiveThirtyEight. https://fivethirtyeight.com/methodology/how-our-pollster-ratings-work/

4. Thompson, B. (2019). *Dicto simpliciter—Accident.* https://www.palomar.edu/users/bthompson/Accident.html
   Originally described by Aristotle in Sophistical Refutations 5, this fallacious reasoning was translated to mean "what is true in a certain respect (taken) to be true absolutely."

5. Comey, J. (2016, October 28). *Letter to Congress.* US Department of Justice. https://assets.documentcloud.org/documents/3198222/Letter.pdf

6. King, A. (2016, November 12). *Poll expert eats bug after being wrong about Trump | CNN politics.* CNN. https://www.cnn.com/2016/11/12/politics/pollster-eats-bug-after-donald-trump-win/index.html

7. Silver, N. (2017, May 3). *The Comey letter probably cost Clinton the election.* FiveThirtyEight. https://fivethirtyeight.com/features/the-comey-letter-probably-cost-clinton-the-election/

8. Enten, H. (2016, November 4). *Trump is just a normal polling error behind Clinton.* FiveThirtyEight. https://fivethirtyeight.com/features/trump-is-just-a-normal-polling-error-behind-clinton/

9. Bialik, C., & Enten, H. (2016, November 9). *The polls missed Trump. We asked pollsters why.* FiveThirtyEight. https://fivethirtyeight.com/features/the-polls-missed-trump-we-asked-pollsters-why/

10. Bialik, C., & Enten, H. (2016, November 9). *The polls missed Trump. We asked pollsters why.* FiveThirtyEight. https://fivethirtyeight.com/features/the-polls-missed-trump-we-asked-pollsters-why/

11. Rosentiel, T. (2009, June 25). *Perils of polling in election '08.* Pew Research Center. https://www.pewresearch.org/2009/06/25/perils-of-polling-in-election-08/

12. RealClearPolitics. (2012). *Election 2012—General election: Romney vs. Obama.* General Election: Romney vs. Obama. https://www.realclearpolitics.com/epolls/2012/president/us/general_election_romney_vs_obama-1171.html

13. Harding, L. (2012, November 7). Numbers nerd Nate Silver's forecasts prove all right on election night. *The Guardian.* https://www.theguardian.com/world/2012/nov/07/nate-silver-election-forecasts-right

14. Harding, L. (2012, November 7). Numbers nerd Nate Silver's forecasts prove all right on election night. *The Guardian.* https://www.theguardian.com/world/2012/nov/07/nate-silver-election-forecasts-right

15. Mercer, A., Deane, C., & McGeeney, K. (2016, November 9). *Why 2016 election polls missed their mark.* Pew Research Center. https://www.pewresearch.org/short-reads/2016/11/09/why-2016-election-polls-missed-their-mark/

16. Smith, A., & LoGiurato, B. (2016). *"Polls might not be capable of predicting elections": How everyone blew it on Trump's huge upset.* Business Insider. https://www.businessinsider.com/polls-wrong-trump-clinton-why-2016-11

17. Murray, P. (2021). *Pollster: "I blew it." Maybe it's time to get rid of election polls.* Opinion. NJ. https://www.nj.com/opinion/2021/11/pollster-i-blew-it-maybe-its-time-to-get-rid-of-election-polls-opinion.html

18. Monmouth University Polling Institute. (2021, October 27). *Murphy maintains lead.* https://www.monmouth.edu/polling-institute/reports/monmouthpoll_NJ_102721/

19. Solomon, N., & Talbot, R. (2021, November 3). *Democrat Phil Murphy is reelected in an extremely tight race for New Jersey governor.* NPR. https://www.npr.org/2021/11/02/1050183040/new-jersey-governor-election-results-murphy-ciattarelli

20. Murray, P. (2021). *Pollster: "I blew it." Maybe it's time to get rid of election polls.* Opinion. NJ. https://www.nj.com/opinion/2021/11/pollster-i-blew-it-maybe-its-time-to-get-rid-of-election-polls-opinion.html

21. Shirani-Mehr, H., Rothschild, D., Goel, S., & Gelman, A. (2018). Disentangling bias and variance in election polls. *Journal of the American Statistical Association, 113*(522), 607–614. https://doi.org/10.1080/01621459.2018.1448823

22. Nesbit, J. (2016, November 7). *Sam Wang is this year's unsung election data superhero.* Wired. https://www.wired.com/2016/11/2016s-election-data-hero-isnt-nate-silver-sam-wang/

23. Mercer, A., Deane, C., & McGeeney, K. (2016, November 9). *Why 2016 election polls missed their mark.* Pew Research Center. https://www.pewresearch.org/short-reads/2016/11/09/why-2016-election-polls-missed-their-mark/

24. DeSilver, D. (2021, April 8). *Q&A: After misses in 2016 and 2020, does polling need to be fixed again? What our survey experts say.* Pew Research Center. https://www.pewresearch.org/short-reads/2021/04/08/qa-after-misses-in-2016-and-2020-does-polling-need-to-be-fixed-again-what-our-survey-experts-say/

25. Silver, N., & Druke, G. (2021, December 10). *What would it take to "fix" polling ... and other questions from podcast listeners.* FiveThirtyEight. https://fivethirtyeight.com/videos/what-would-it-take-to-fix-polling-and-other-questions-from-podcast-listeners/

26. UC Santa Barbara. (2023). *2000: The American presidency project.* 2000. The American Presidency Project. https://www.presidency.ucsb.edu/statistics/elections/2000

27. Dickie, G. (2020, November 13). *Why polls were mostly wrong.* Scientific American. https://www.scientificamerican.com/article/why-polls-were-mostly-wrong/

28. UC Santa Barbara. (2023). *The American presidency project. 1992.* The American Presidency Project. https://www.presidency.ucsb.edu/statistics/elections/1992

24

The Law of 50% + 1

Important Points Checklist

- The percentage of votes needed to guarantee victory

In what I hope is considered a respectful nod to Nick Panagakis, this will be the most concise chapter of this book. While its conclusion may seem flippant, it is possibly – paired with an understanding of what a Simultaneous Census is – the most meaningful concept of this book.

Panagakis observed that incumbents up to 1989 fared poorly among undecided voters, which led him to name the Incumbent Rule. Partially because that rule developed too many exceptions, and partially because experts and the public alike took the rule "too literally," it fell out of use.

But one of his observations – how he accounted for a candidate's chances by incorporating what he knew could be an unfavorable Undecided Ratio – seems to have been lost to history.

Knowing that undecideds often did not split evenly or proportionally, Panagakis noted that Incumbents who receive less than 50% of the vote in a poll are "frequently in trouble."[1] More presciently, he said, "how much trouble depends on how much less than 50%."

With that in mind, I'd like to offer the following not as a rule but as a law:

If a candidate receives 50% + 1 votes in any election in which the winner is determined by who receives the most votes, they can't lose.

This is the **Law of 50% + 1.**

If this sounds overly simple, I will reassure you this is not the case. I know you know this fact. But there's a very strong chance you don't know what it means as it pertains to reading polls – and I can say with certainty that users of the Spread Method and Proportional Method don't.

Note

1. Panagakis, N. (1989). *Incumbent races: Closer than they appear.* PollingReport.com. https://www.pollingreport.com/incumbent.htm

DOI: 10.1201/9781003389903-24

25

The Clintons' Lessons

Important Points Checklist

- The "finish line" in elections
- How third parties impact the finish line
- Why it's important to consider third-party and undecided separately, if you're trying to inform a prediction

Six presidential elections and 24 years apart, a lesson that should have been heeded when Bill Clinton won multiple states with less than 40% of the vote, wasn't.

Given how few undecideds and third-party candidates there were, the elections from 2004 to 2012 were about as hard to forecast as Mintucky.

2016 was different.

True, in any election contest, we can apply the Law of 50% + 1.

But why is the Law of 50% + 1 important? If you get more than half the votes, you win, of course. This is *so* obvious. But the consequences of this law teach us more than that.

Receiving 50% + 1 votes means you can't lose. But in elections with third parties receiving votes, that number becomes smaller. Depending on the variables in a specific election, whether third-parties are on the ballot and how much support they have, *the percentage of votes that would guarantee victory can be far smaller than 50% + 1.*

Major party candidates aren't fighting over 100% of the vote. Major party candidates are fighting over a smaller number – the *available* vote.

In a race with only two candidates, 50% + 1 wins. There's no other possible outcome.

But in a race with *three or more* candidates, the math is entirely different. Treating these different calculations as *similar* would be bad enough. But the Spread and Proportional Methods treat them as *exactly the same.*

True, and importantly, 50% + 1 still guarantees a win. It's a law, after all. But with more candidates, you no longer *need* 50% to win.

DOI: 10.1201/9781003389903-25

Assume that you know that a third-party candidate will receive at least one vote. You don't need 50% + 1 to guarantee victory anymore, just 50%.

If third-party candidates account for 2% of the vote, it's no longer a race 50% + 1, fighting over 100%; 100% is not available. Now, only 98% is available. 49% + 1 guarantees victory now.

If third-party candidates account for 4% of the vote, though that number may seem small, it's no longer a race to 50% + 1, or 49% +1, but 48% +1, fighting over 96% of the vote.

All of these examples are the "finish line" moving to a number much smaller than 50% + 1. These are not hypothetical examples, either – it applies in almost every major election.

In swing states where races are often decided by 2% or less, the finish line moving from 50% to 49% or 48% isn't a small amount.

We can go on and on until we reach what happened in 1992, when third-party candidate Ross Perot was the leader in some polls.[1]

The **finish line is defined as** *50% + 1 of available votes after accounting for third parties.*

Regardless of what you might have forecasted for Perot's chances of winning the election – or any state – no one would dispute the fact that many states would be won with far less than 50% of the vote that year.

But how much less?

My approach for reading polls requires looking at polls for what they mean, not what some assumptions say they say. It's not as easy as pretending a poll is a prediction, but it's not hard, either.

The first step is understanding where the finish line is. When third parties receive votes, the finish line is reached at less than 50%. My approach accounts for this. Existing methods don't and can't.

For that reason, if your goal for looking at polls is to inform a prediction about a candidate's chances of winning, you cannot use the spread, you cannot use the Proportional Method, you must consider each candidate's base of support."

How much the finish line moves is directly related to how many third-party votes there are. Figuring out how many third-party votes there will be requires making a prediction or forecast. Polls are the best tool to inform that prediction.

After having avoided political polls for a long time – now finally tackling them head on – I'm sometimes asked the question, to which I hope you know the answer, "sure, informing a prediction about *who will win* is important, but what if you want to know *how much* that person will win by?"

A wonderful question; it requires a forecast. Polls are the best tool to help inform that, too.

There's no other way to get from plan poll to eventual result than forecast. Some forecasts – in elections with very few undecideds and very few third-party voters – are easier than others. But polls are not predictions, and polls are not forecasts. Polls can only tell you something about right now.

To add to the list of reasons the Spread Method fails, it *can't* accurately tell you how much a candidate is ahead because it can't even give you an approximate idea of where the finish line is. The Proportional Method is even worse at this task because it pretends the race is being run on a track it invented which doesn't actually exist.

To use an actual finish line as an example, consider a footrace in which you don't know whether the race's finish line is at 40 meters or 50 meters.

If I ask you "how far ahead" the leader is, and you say "about 2 meters" whether the race is to 40 or 50 makes a difference in how you might characterize the leader's chance of winning, wouldn't it? Even, perhaps, how much they could win by.

Not for the Spread Method. Is the race to 40 or 50? Doesn't matter, the leader is up 2.

How close are they to the finish line – wherever it is? Doesn't matter, the leader is up 2.

This is how the experts talk about election data and tell the public to evaluate it. This is how experts measure poll error. They were up by 2 but then they weren't! Snapshot was wrong!

But given a more complete picture, I should say, an accurate picture of what polls try to tell us, you can make much better informed predictions and analysis:

- The race is to approximately 40, and the leader is ahead approximately 39–37. Good data! What's the most either runner is likely to "win by?"

- The race is to approximately 50, and the leader is ahead approximately 39–37. Also good data! What's the most either runner could "win by?"

You'd probably be able to estimate the leader's chances of winning a little better *and* have a better idea of how much they might win by, if you knew whether they were racing to 50, 49, 45, or 40.

Where the finish line is depends directly on how many third-party voters there will be.

Polls provide that data! Better stated, polls provide a reasonable approximation of that data. But you won't be able to find it by the existing methods. The Spread Method throws undecideds and third parties into the same bin.

Normally, though certainly not always, users of the Spread Method are at least kind enough to provide the numbers from which it derives its spread, before assuming you can't figure out the difference between two numbers.

$$39\,(+2)$$

$$37$$

But worst of all, the Proportional Method reports 39–37 as 51–49. That's it.
It would also report 20–19 as 51–49.
It would also report 45–43 as 51–49.
It would also report 49–47 as 51–49.
As for the closeness to the finish line in each of these cases: what's a finish line?

You might think the chapter regarding the "Law of 50% + 1" is obvious, yet both spread and proportional analyses ignore its applications.

Now that we've covered the impact of third parties, and you know the impact of undecideds, let's take another look at the Trump versus Clinton election of 2016 – and analyze it properly.

Note

1. Dionne, E. J. (1992, June 9). Perot leads field in poll. *The Washington Post*. https://www.washingtonpost.com/archive/politics/1992/06/09/perot-leads-field-in-poll/5c0499dd-d5c5-42e4-bc63-e4c32ca083e7/

26

Trump-Clinton-Third Party 2016

Important Points Checklist

- Why US election data and analysis should focus on "swing states" more than national numbers
- Accounting for third-party changing attitudes
- The difference between poll error and poll aggregation error – and why it's invalid to assume any poll average is an objective standard for what "the polls" say
- Some potential sources of error in poll averages not present in polls themselves
- The US third-party underperformance phenomenon

Given that national polls aren't extremely relevant for presidential elections, I'll use one of the now-infamous swing states from 2016, Pennsylvania (PA).

Here are the 10 polls closest to the election[1,2] and the average of them. As I said in Mintucky, 10 isn't a magic number, nor is it a simple average the most elegant approach – but it's a fine place to start for understanding what polls tell us.

Candidate	Clinton (D) (%)	Trump (R) (%)	Third-Party (%)	Undecided (%)
Poll Average (PA)	**46.3**	**43.9**	**4.7**	**4.9**
Trafalgar	47	48	3	2
YouGov	45	43	6	6
Harper Polling	46	46	3	3
Gravis Marketing	47	45	4	4
Susquehanna	45	43	4	8
Monmouth	48	44	3	5
Remington	45	43	3	9
Quinnipiac	48	43	6	3
CNN/ORC	48	44	6	2
Morning Call	44	40	9	7

These polls were all taken within the final two weeks of the election. Some of these pollsters were funded by or otherwise affiliated with a major party, which I didn't note on the chart. This is, perhaps, a factor to consider in your poll aggregation methodology, but I opted for the simple average here.

For whatever it's worth, this is the same poll average (to the decimal) FiveThirtyEight had for Clinton, while this average for Trump is 1.5% higher than they had, and third-party is slightly lower. Not the same, but pretty close for such a simple approach. FiveThirtyEight estimated a slightly higher number of undecideds in their poll average.[3]

Just like in Mintucky, I'm going to apply a 4% margin of error to each of the poll averages because most of the individual polls are in that range. If you think that margin of error for poll averages is too high or too low, that's fine. Keep that critical energy.

Regardless of your poll aggregation method of choice, what do these numbers presented in Figure 26.1 tell us?

These polls, like all polls, intend to give us an estimate of the simultaneous census. If you want to make a prediction from that data, you can think of it as a base of support.

And this is the limit of what any plan poll tries to tell us.

Anything beyond this is a function of a prediction or forecast. No asterisks, no exceptions.

Was this poll average wrong? Were "the polls" wrong?

The only way you can possibly know is to compare them to what the tool measures: a simultaneous census. No simultaneous census available? Work backwards from the election results to make the adjustments.

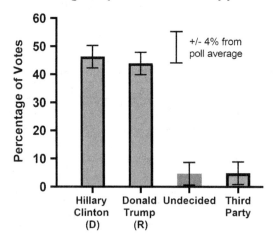

FIGURE 26.1
A chart showing a simple recent average in PA in 2016. Chart by the author.

Was Clinton's **simultaneous census value** in PA at the time of this poll average above or below 46.3%? Possible! If so, **poll average error.**

Did any individual poll give a number above or below the simultaneous census for that candidate? Possible! If so, **poll error.**

But does the eventual result disagreeing with *your assumptions, prediction, or forecast* about what undecideds decide and who might change their mind equal poll error? No. If your **forecast** was wrong, that's a **forecast error.**

As it pertains to informing a prediction, the thing that should strike you about this poll average is how low both candidates' base of support was. Neither one is safe.

It doesn't matter if one candidate is "up by" 20 or 10 or 2, as the Spread Method would report. For one, that's not what polls tell us, but as it pertains to informing a prediction:

How much trouble they're in depends how far they are from the finish line.

While that analogy talking about races in terms of distance to the finish line might sound oversimplified, given a proper understanding of what polls try to tell us, it doesn't just make sense, it's true.

Hillary Clinton's poll average was not above 47% in a single swing state. Not one. Not according to FiveThirtyEight, not according to RealClearPolitics, not according to my extremely basic calculations.

Anyone who didn't believe (better yet, *understand*) that the polls indicated Clinton had a real chance of losing doesn't understand how polls work.

To their credit, again, because it's well deserved in this instance: FiveThirtyEight was the only major forecaster (if not the only forecaster) to accurately characterize Clinton's chances as less-than-solid.

Now, back to PA specifically, it's fair to say neither candidate was close to the finish line, but how far from it were they? After all, Mr. Clinton won several states with a much smaller percentage of votes! Third parties matter. If you don't account for them in your analysis – poll, forecast, or otherwise – your analysis is bad.

The third-party base of support was estimated to be around 4.7%. That means, given the fact that polls are only intended to be reasonable approximations, it could feasibly be anywhere from about 1% to 9%.

So for now, let's just take the polls at face value and say that third-party support was about 4.7% in PA.

Subtracting 4.7% from 100% gives us 95.3%.

Given third-party base of support at 4.7%, there are about 95.3% of votes available for Trump and Clinton to fight over.

That means my poll average's estimated finish line is 95.3%/2, or 47.65%.

Is 47.65% a much shorter finish line than recent US presidential elections? Yes. Should this election being different from recent ones impact an expert's ability to understand it? No.

Unfortunately for the existing methods, and to the detriment of statistical literacy in the public, being different from recent elections did impact their ability to understand it. This theme repeats in other elections and countries.

In "What's for Lunch?", I pointed out that 100% of undecideds chose the same option, and I said something to the effect of: it doesn't matter what I (or anyone) think or believe should happen, what matters is what is true. Incorporate what is known to be true into your analysis, don't substitute your own assumptions for something that can be known. In this presidential election, higher than usual third-party vote was known, higher than usual changing attitudes was known – and yet, analysts substituted their own assumptions.

In "Mintucky," I asked you to make a prediction for the percentage of votes Ricky Red would get. Astute readers remembered I had bought an equal number of red and green mints – and thus, figured it was more likely that Grace Green would earn more of the undecideds. Great forecast work, everyone. And nothing to do with the accuracy of the polls.

Even if we have beautiful and convincing data to support our predictions and/or forecasts regarding what undecideds will eventually do, or even which voters might change their minds, that's not the poll's doing; that's the predictor or forecaster. It doesn't matter if we *assume* something to be true, or have *good evidence* to believe it will be – polls are not predictions, and polls are not forecasts. Good analysts would incorporate what is known to be true about what happened into their analysis, not substitute their own assumptions for something that can be known.

Yet, here's the consensus from Trump-Clinton 2016:

> *There were more undecided and third-party voters than usual, and it*
> *turns out they didn't split evenly between the major candidates.*
> *Therefore, there was a big poll error.*

If you can find the fault(s) in that statement, then you are more qualified to analyze polls than many who are presently considered experts.

If *your forecast* assumed that undecideds would split evenly – which is a perfectly fine assumption, by the way – but that assumption turns out to be wrong, that's not a poll error. If your forecast was wrong, that's a forecast error.

In PA, I estimated a finish line around 47.65%. Clinton's average was around 46%, Trump's around 44%. Both needed help to cross the finish line. Where might it come from?

It's unwise to assume the polls or even a poll average can't be wrong – hence the error bars on the chart. Either candidate's actual base of support could be higher or lower.

But it's perfectly reasonable, given a finish line around 48%, to say that Clinton should be considered a favorite here. But it's not reasonable to say, as proponents of the Spread and Proportional Methods would, Clinton "will almost certainly win" or "should win by about 2%" *or else there's a poll error.*

Grace Green taught us that lesson.

Poll Average Implied Base of Support: PA 2016

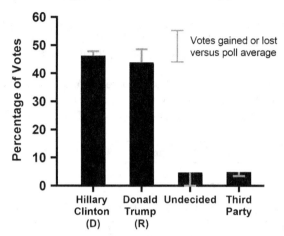

FIGURE 26.2
A chart showing each option's simple recent average in PA in 2016 along with its final result. Chart by the author.

Just like Mintucky, and this is not an oversimplification: each candidate's chances of winning depended largely on how you forecasted undecideds would eventually decide.

If you want to assume how they'll decide, fine. If you want to use data to inform your assumption, even better. If you, like legitimate forecasters, incorporate a lot of different variables and possibilities to calculate a probabilistic range of outcomes, that's awesome – great work.

But if your forecast was wrong, you don't get to deflect blame by saying that "the polls" were wrong. If your forecast was wrong, that's a forecast error.

Figure 26.2 shows the final result for PA.[4]

Spoiler alert: Trump won. He won PA and lots of other swing states in a similarly close manner on his way to winning the Electoral College and the Presidency.

What do these results tell us about the accuracy of the polls?

First of all, notice how *wrong* the polls were about how many undecideds there would be after the election! (Sorry, I had to.)

Some real analysis:

Third parties underperformed their poll average of 4.7% receiving just 3.5% of the vote.

Therefore, my poll average had a 1.2% error regarding third parties. Right? Not right – we covered this one!

If I say I don't have a dog today and then I have a dog next week, that's not a poll error.

Just like polls can't be judged on their closeness to eventual result, nor can poll averages.

Before you can calculate poll error or poll average error, you must first *account for undecideds and changing attitudes.*

The *only thing* you can compare without correction to the eventual result is something that tries to predict the eventual result: a forecast.

My *forecast* for PA in this election was:

Clinton (D)	Trump (R)	Third-Party
49%	47%	4%

Regarding Clinton's final result, my forecast was off by +1.1%. For Trump's, –1.6%. For third-party support, +0.5%. Slice it up however you want, that was my forecast error (see how easy that was, guys?).

For each of the pollsters above, comparing the poll spreads to actual results, the Spread Method would (and did) say:

Candidate	Poll Spread	Result Spread	Poll Error
Trafalgar	R + 1	R + 0.7	0.3
YouGov	D + 2	R + 0.7	2.7
Harper Polling	D + 0	R + 0.7	0.7
Gravis Marketing	D + 2	R + 0.7	2.7
Susquehanna	D + 2	R + 0.7	2.7
Monmouth	D + 4	R + 0.7	4.7
Remington	D + 2	R + 0.7	2.7
Quinnipiac	D + 5	R + 0.7	5.7
CNN/ORC	D + 4	R + 0.7	4.7
Morning Call	D + 4	R + 0.7	4.7

Of course, the Spread Method doesn't include third-party candidates in its calculations, so if your poll says 0% third-party or 20% third-party, it doesn't matter as far as the Spread Method calculation for poll error is concerned. Poll spread minus result spread for the two main candidates.

Which polls were best by this standard?

Wow, Trafalgar, least poll error by far!

Harper Polling, good job.

Monmouth, Quinnipiac, CNN, and Morning Call – ouch. You guys really need to *fix* your polls. Have you considered a time machine? Maybe call Trafalgar about it. See you in 2020.

Unless you care about accurately calculating poll error, in which case your work has just begun.

Fortunately, perhaps paradoxically given the level analysis on the subject, there is *already* some research done on these real factors that do – and don't – contribute to poll error.

In his (justified) self-congratulatory article "Why FiveThirtyEight Gave Trump A Better Chance Than Almost Anyone Else" Nate Silver wrote, "Undecideds and late deciders broke for Trump."[5]

Unlike his calculation of poll error, he didn't simply make assumptions about undecideds in this instance, he had data to back it up. He even broke it down by state!

Based on polls that were taken after votes were cast (actual voters) – 54% of undecideds in PA opted for Trump.[6]

To which you might think, 54%, that's barely more than half, that can't account for much of the discrepancy between poll and result.

But unlike the existing methods, *you* know there aren't just two parties. Third parties claimed 8% of the undecided vote for themselves. That left just 37% for Clinton (1% refused to answer).

We knew – at least had a very good idea, if we read the polls properly – that neither candidate was polling across the finish line. That pie of the 4.9% undecided was vital.

Accounting for what we now know – or, given the limitations of poll data, what we can approximate – about 54% of that 4.9% undecided eventually voted for Trump. That's 2.7%.

With 37% of them choosing Clinton, that's 1.8%. By my poll average, that would've put Clinton at 48.1%.

The **simple recent poll average** after accounting *only for undecideds*:

Clinton (D)	Trump (R)
48.1%	46.6%

Given that FiveThirtyEight had a slightly higher number of undecideds in their poll average, a 54% share adjustment would've equaled even more voters.[7]

Using their poll averages, and assigning 54% and 37% of their 6% undecideds to Trump and Clinton, respectively, their adjusted poll average would have boosted Trump by 3.2% and Clinton by 2.2%.

FiveThirtyEight's adjusted poll average, accounting only for undecideds, would be:

Clinton (D)	Trump (R)
48.5%	45.6%

In either case, accounting for undecideds – a factor we know can impact the poll-to-election discrepancy – makes Trump look more favorable. But considering the anticipated finish line (mine at 47.65%, FiveThirtyEight's around 47%), PA looks to be over after adjusting for undecideds.

So, the remainder of the discrepancy is poll error, right?

Like in Mintucky, it's a little test of how closely you're reading.

$$\text{Eventual Result} = \text{Simultaneous Census to Poll}$$
$$+ / - \textbf{\textit{Changing Attitudes}} + \textbf{\textit{Undecided Ratio.}}$$

Unlike Mintucky, we have some Changing Attitudes to account for.

If you're thinking, "Yes, the Comey Letter! Clinton might have lost some support!" Not there yet. There's another variable we're missing still. What happened to third-party support? They underperformed their poll number – and that number had to go somewhere.

Third parties only received 3.5% of the vote in PA. My simple average said it was higher, and FiveThirtyEight said it was *even higher* than that. What happened?

The most likely explanation is that many of them changed their mind.

That's not to say polls didn't or can't have *any* error with regard to third-party support, entirely possible. But the mere fact that third-parties underperformed their collective poll number is not instantly indicative of this.

And there's data to support it.

Nate Cohn, chief political analyst for the New York Times, presented findings from the Pew Research, AAPOR report on 2016 polling, in which they found only 60% of voters who previously declared they would vote for a third-party ultimately voted for a third-party.[8]

To be clear, this study is far from proof that there wasn't any poll error regarding third parties and that the discrepancy between reported poll result and eventual election result can be entirely accounted for by changing attitudes – I'm team "more data" in that sense, but it's pretty good evidence the *entire* underperformance can't be attributed to poll error. Some people just changed their mind.

Who did those "said I was voting third-party in the poll but changed my mind and voted for a major party candidate" folks ultimately support?

According to this study, 26% Trump, and 11% Clinton. Again, I'll repeat that I'm team "more data" here because applying national numbers to individual states introduces the risk of other errors.

But at least I don't just assume mind-changing has zero effect when there's data that proves this is false.

It would be useful to have state-specific data for this Changing Attitude factor – if we want to accurately calculate poll error.

Nonetheless, by this admitted approximation, it boosts another 1.3%–1.6% for Trump and just 0.6%–0.7% for Clinton. That's very different from the "zero" assumption these methods require.

The new **simple recent adjusted poll average** would be:

Clinton (D)	Trump (R)	Adjusted Spread
48.7%	47.9%	0.8%

And **FiveThirtyEight's adjusted poll average** would be:

Clinton (D)	Trump (R)	Adjusted Spread
49.2%	47.2%	2.0%

Again, Clinton appears to be across the finish line, and Trump just short. But we know that's not what happened. Both of these adjusted numbers overstate the support Clinton received in PA – though far from the "catastrophic" error it's reported as.

Which leads to the final point.

According to Silver, the Comey Letter "probably" caused her to lose the election. If some of her previously decided voters defected or didn't vote, combined with undecideds mostly supporting Trump, I think that's possible.

If you believe that the Comey letter caused Clinton to shed just 1% of support – combined with third-party defection to Trump, combined with undecided defection to Trump ... where's the error?

All of these factors, assuming a net "–1%" for Clinton due to Changing Attitudes from the late campaign controversy – suddenly the 2016 polls, properly read, *look remarkably accurate.*

To belabor the point, one last time, this adjustment for Changing Attitudes and Undecided Ratio is not magic. It's neither advanced math nor a junk assumption. The only thing the adjustment does: separate the factors that polls can tell us about from the ones it doesn't and can't.

Given the linear nature of time, I don't expect polls to give us a better sense of how undecideds decide or who might change their mind anytime soon. But being from the future is, in a way, a time machine. We have access to both the poll data and the election data. Only after you admit the confounding variables are confounding can you account for them.

FiveThirtyEight's "Poll Error" Problem

If it seems like I'm picking on FiveThirtyEight a lot, it's not because there are no other forecasters who are better or worse.

But by nature of being highly influential, they're the ones I'm most apt to cite. They have done some great work (as I hope I've given credit for), and though Silver is no longer with FiveThirtyEight, I hope they can continue their mission to advance public knowledge. I'm likewise hopeful that Silver, his successors, and the thousands of others in and adjacent to this field can contribute to its development and public understanding of how polls work.

As such, criticism and critique are sometimes necessary.

FiveThirtyEight's poll average in PA said Trump was "behind" by 3.9.

Trump ultimately won by 0.7. They don't adjust for Changing Attitudes, they don't adjust for Undecided Ratio, and they don't adjust for anything.

Their poll average said Trump –3.9, and the final result was Trump +0.7, so the polls were off by 4.6. End of story. Poll error was 4.6. That's the number reported to the media, that's the number they use to say how much the polls were "off by" to this day. That's the logic that drives the myth that the 2016 polls were catastrophically bad.

Here's how they would show their work:

Poll Average	Eventual Result	Poll Error
D + 3.9	R + 0.7	4.6

Like their definition for poll error assumes changing attitudes to be zero and that the undecided ratio is 50:50, *they also assume that their poll average can't be wrong in the calculation of overall poll error.*

Individual pollsters? Yeah, they can be wrong.

"The polls" can also be wrong.

Not their poll average, though. It's a really awkward statement, if you think about it.

FiveThirtyEight – who is, in part, a poll aggregator – by their definition of poll error, **assumes their poll average can't be wrong**, the media reports it as such, and public believes it. Their poll average is the standard by which, they say, we can detect how wrong "the polls" were.

The logic underlying this belief is no different from one pollster saying to another pollster that "your poll is off by however much it disagrees with ours" or any poll aggregator saying to another that "your poll average is off by however much it disagrees with ours."

Poll aggregation is not, whether FiveThirtyEight or RealClearPolitics or Carl's simple recent average of 10 polls, immune to error. Poll aggregation, while it is a better approach than "look at one poll," *introduces the risk of errors not present in polls.*

Which polls do you include? How much do you weight by your subjective "pollster quality" metrics? Do you include partisan polls? If you include partisan polls, do they count independently? What about pollsters who conduct polls by phone versus online, do you weight those differently? How much do you weight by recency? Like polls themselves, poll aggregators have to, and get to, decide which factors to weight for and how much. This is a fair way to judge the accuracy of a poll average.

But accuracy compared to what?

Like a pollster is responsible for weighting for several factors, a poll aggregator is responsible for weighting all of the above factors, and more. None of those potential sources of **poll average error** have *anything* to do with poll error. The current "error" definitions do not account for this.

Just as the entirety of forecast error is currently allowed to be deflected onto "poll error," poll aggregators also shirk accountability this way.

For all of the time spent debating *poll error*, the thousands of pages of academic papers committed to it, and even expert task forces reporting in 2016 that, "many believed the polls predicted the wrong winner,"[9] I can find no literature regarding potential poll averaging error.

For some reason, to date, poll aggregators, like forecasters, seem to be immune or shielded from any fault. Forecasters blame their forecast errors on "the polls," and poll aggregators blame their poll aggregation error on "the polls." The polls are what needs to be *fixed*.

Regarding the allocation of undecided voters, FiveThirtyEight says, "Empirically, an even split works better for presidential races than a proportional split."[10]

Okay. And if that assumption proves to be wrong, no matter how empirical, that's not a poll error.

My simple recent average of 10 polls in PA differed from FiveThirtyEight's in one major way: mine said Trump's poll average was 1.5% higher.

If someone wanted to argue the superiority of FiveThirtyEight's aggregation methodology which is, by all accounts, far more sophisticated than my simple recent average, they could make a very compelling case.

But better compared to what? Not compared to the result, that's for sure. Good talk, debate over.

That's how polls are treated today. Adopting the Simultaneous Census standard and accounting for confounding variables would, if not rectify this problem, be a major step in the right direction.

The existing definitions for "error" are only concerned with the eventual result, and there currently exists no basis at all for poll aggregation accuracy. Unless and until these definitions are fixed, and confounding variables are accounted for, there never will be.

Poll error is real, but not as it is currently understood. Poll error can impact poll aggregation accuracy, and poll error can impact forecast accuracy. But the entirety of a poll aggregator's error, or the entirety of a forecaster's error, cannot be attributed solely to poll error.

This is no longer just a problem with the Spread Method. Here, we have, either by accident or conceit, poll aggregators declaring themselves the gold standard to which the eventual result should be compared, and "what the polls say" should be judged.

And for the record, my simple recent average performed far better than FiveThirtyEight's advanced aggregation methods, so that means it was better. Debate over. In case my tone wasn't clear, I don't think these "spread" calculations should be the end of the debate, nor part of the debate at all.

Adjusted Methods tell a very different story about the accuracy of the polls.

Using the Adjusted Spread Method[11] to calculate poll error, even without Changing Attitudes among voters who declared they would support Clinton in the poll, even assuming FiveThirtyEight's poll average had no error:

538 Adjusted Poll Average	Eventual Result	Adjusted Poll Spread Error
D + 2.0	R + 0.7	2.7

This is nearly a full 2-point difference between stated poll error and Adjusted Poll Error. Adjusted Poll Spread Error isn't a gold standard either; if experts ever work on developing new standards that account for these factors, it might not even end up a bronze standard. But it's far better than the current "guess" standard.

The existing methods miscalculating error isn't a fluke; when you don't account for confounding variables, and those variables confound, you will end up with some very bad calculations.

You saw how this method failed catastrophically in "What's for Lunch?" and again in Mintucky – in cases where the Simultaneous Census was known. If it disagrees with experiment, it's wrong.

I would love to repeat each of these calculations for every state, but for the sake of time, I'll spare you. If you're interested, you can perform the adjustments yourself.

If you've had enough of Trump-Clinton 2016, suffice it to say, as Americans who followed that election night can attest, roughly the same thing happened in every swing state.

According to the same source cited by FiveThirtyEight, undecided voters in Wisconsin broke 59% to 30%[12] for Trump. In Minnesota: 53% to 31%;[13] in Iowa: 54% to 34%;[14] and in Michigan: 50% to 39%.[15]

Empirical.

Much was said about the polling failure in the Midwest. The Spread Method reported, and it was accepted, that there were enormous 4-, 5-, or 7-point errors in multiple swing states. To which I say: show your work.

Accounting for factors known to impact poll-versus-election spread but are not caused by any poll error (Undecided Ratio and Changing Attitudes), the actual error could easily be less than *half* what has been reported. Not poll error, but poll aggregation error and/or forecast error – was a major part of those indefensibly large numbers reported.

That is not to say that there was *no* error.

Many pollsters seem to have underweighted their polls by education, and others didn't weight them by education at all.[16] That's not good.

Of course, with the benefit of hindsight, analysts can say pollsters not weighting by education is problematic. They reached this conclusion because weighting by education would have made the "spread" appear more accurate.

But did those same pollsters weight by education in previous years? Why was it a problem in 2016, and not 2008 and 2012, when polls were "extremely accurate?" The fact that this apparently obvious flaw couldn't be corrected until after it caused polls to "miss" is another example of the "lack of vigilance" created when the result number ends up close to what you expect it to be. It's entirely possible that this error existed in previous years but was compensated for, thus not considered a problem. History shows that bad methods are allowed to continue, so long as they produce results analysts assume they should.

Nonetheless, not weighting by education can contribute to poll error, likely caused by nonresponse. Let's criticize where necessary and try to fix it going forward.

Additionally, it's now considered, well, likely, that Trump received support from a lot of voters who had not previously voted, or might not have been considered "Likely Voters."[17]

That's frame error, also a poll error. Let's criticize where necessary and try to fix it going forward. But with poll aggregators and analysts telling *pollsters* what *they* need to fix, and never the other way around, I'm reminded of the lack of accountability.

How much did frame error, and nonresponse error, and the other actual errors contribute? If you want to say "the polls" in PA were off by 4.6, and even more in other states: show your work.

Just when I think I might be being *a little* unreasonable in this chapter, I'm reminded of the fact that experts – based on this level of analysis – have the nerve to tell *pollsters* they need to fix their *polls*.

Fix what?

You know, just fix it. Fix the "off by 4.6" part.

How?

Don't worry about where the 4.6 came from, just *fix it*.

We know at least some, if not most, of that "off by" calculation was caused by factors that have nothing to do with polls or pollsters. Analysts should start by fixing their definitions.

On the note of fixing polls, remember, not all of them were bad. Trafalgar was (by far) the best pollster not just in PA but in almost every swing state. That is, according to the Spread Method.

But we know the Spread Method, comparing polls to eventual results, misses some things.

We know undecideds boosted Trump's total, and also that third-party mind-changing boosted his total. He ended up with 48.6% of the vote in PA.

Trafalgar reported Trump's poll number just a few days before the election at 48%. FiveThirtyEight at the time had them rated as a "C" pollster.[18]

Monmouth polling, on the other hand (with an A+ rating), reported *Clinton's* poll number a few days earlier at 48%.[19]

Candidate	Poll Spread	Result Spread	Poll Error
Trafalgar	R+1	R+0.7	0.3
Monmouth	D+4	R+0.7	4.7

Trafalgar was closer, therefore better. End of debate.

Since poll error definitions don't care if your poll was the day before an election or a week before, thus creating a huge incentive for cramming, there's no correction allowed for changing attitudes.

Monmouth's poll was conducted Oct. 29–Nov. 1. Trafalgar's was Nov. 3–5.

If you believe, as Silver said he did, that Changing Attitudes caused by a scandal close to an election can tilt the outcome of an election – how is it reasonable to judge pollsters who conducted polls *at different times* on the same playing field?

It's an assumption. And a really bad one.

Is it possible that Monmouth's poll was quite accurate for the time it was conducted? They reported Clinton at 48%, and over a week later, and she ended up with 47.9%.

Is it possible that Trafalgar benefitted from being last and was able to better account for those last day changing attitudes in their poll where others done earlier couldn't? I think so.

Is it also possible that Trafalgar is a partisan entity that tends to produce results more favorable to Republican candidates in a lot of elections, and in this case, with the Republican receiving favorable Changing Attitude and Undecided Ratios, they happened to get the eventual spread nearly right? Yeah. Could be all of the above! (FiveThirtyEight retroactively marked Trafalgar polls as partisan in 2020.)[20]

All of these questions, regardless of what you may believe, can never be answered until the definitions of poll error are fixed. Until then, according to the Spread Method, Trafalgar was the best pollster in 2016, by far. And so was my poll average. Whether through skill, luck, or a blend – we can't say for sure.

The (US) Third-Party Underperformance Phenomenon

I won't call this a rule even if it's almost always true because it's the poll's job to make observations, and a forecaster's to make predictions.

> It's a well-established phenomenon that third-party candidates are "likely to underperform their polls in the final vote."[21]

Dan Cassino published an analysis in *Harvard Business Review* that featured an experiment run by pollster PublicMind (a research center at Fairleigh Dickinson University). In that study, actual election polls were conducted with an interesting twist: in some of them, the names of "major" third-party candidates were replaced with the names of far lesser known ones. Specifically, minor candidates whose eventual vote total would be measured in hundredths of a percent.

They polled as high as 4%.[22]

In these real election polls, as good researchers tend to do, Cassino and PublicMind were aiming to isolate the variable of interest: how many people who declare support for "third-party" voters actually intend to support them, and how many are just responding with an answer that is "neither of the top two candidates"?

Through this process, comparing how many reported being "undecided" in each version of the poll and how many expressed support for the third-party candidates, Cassino isolated likely third-party voters from true undecideds. He concluded that some who respond "third-party" in a poll will not ultimately vote third-party; they are simply expressing a dislike of both given major party options.

In a similarly creative design, Ipsos and the *New York Times* teamed up for a poll in advance of the 2024 presidential election that included the *Times* editor, William Davis, on a list of possible third-party options. Not a candidate for any office, Davis "won the support of about 1.5 percent of respondents, putting him on par with an actual Libertarian Party candidate."[23]

Perhaps more valuable, given the potentially impactful presence of third-party candidate Robert F. Kennedy Jr., staff editor Ruth Igielnik noted that something as mundane as the order of questions matters for polled third-party support. In their experiment, they gave half of the respondents the option of many candidates (including Davis) first, and the other half the option of only presumed major party candidates Biden and Trump first, with an option for "other, specify."

Those who were asked the shorter version *first* were much more likely to select "other." Moreover, Kennedy Jr.'s polled support increased from 7% to 13%; this is the same poll with the same methodology, and the only difference here is the order in which the question was asked.

Like Panagakis decades earlier, who realized undecideds tended to vote overwhelmingly for the challenger in races with an incumbent, a sharp forecaster – or someone who wants to use polls to inform their predictions – might want to take this third-party underperformance between polls and elections into account.

But I hope all would agree that pollsters themselves should not report a lower number of third-party voters than their respondents provide, and assign those voters to who they think they'll eventually vote for.

Likewise, I hope all would agree pollsters shouldn't report fewer undecideds than their poll data suggests and assign most undecideds to the challenger in their reported results, in hopes of being "more accurate" compared to the "eventual result" standard.

Unfortunately, in both cases, that current standard gives them a substantial incentive to do so.

In contrast to the current standard, I believe pollsters and poll aggregators should be judged on their accuracy for reporting the data as it is – which is the simultaneous census standard – and not according to the eventual result, which is how they are currently judged.

That's yet another reason why accurate definitions, and understanding what polls do, matter:

2016 had more third-party voters, both polled and eventual, than any US election in recent history. While both the Spread Method and Proportional Method – and FiveThirtyEight's own characterizations – treat third-party voters the same as undecided voters, they are not the same.

It was a different state in Trump-Clinton 2016 that provided a reminder as to what happens when a bridge not meant to be driven on is forced to hold traffic. This state also introduces a direct application to how non-US elections with substantial non-major party presence should (and shouldn't) be analyzed.

Notes

1. Silver, N. (2016, November 8). *Pennsylvania: 2016 election forecast*. FiveThirtyEight. https://projects.fivethirtyeight.com/2016-election-forecast/pennsylvania/
2. RealClearPolitics. (2016). *Election 2016—Pennsylvania: Trump vs. Clinton*. https://www.realclearpolitics.com/epolls/2016/president/pa/pennsylvania_trump_vs_clinton-5633.html
3. Silver, N. (2016, November 8). *Pennsylvania: 2016 election forecast*. FiveThirtyEight. https://projects.fivethirtyeight.com/2016-election-forecast/pennsylvania/
4. Pennsylvania Department of State. (2016). *2016 Presidential election. Pennsylvania elections—summary results*. https://www.electionreturns.pa.gov/General/SummaryResults?ElectionID=54&ElectionType=G&IsActive=0
5. Silver, N. (2016c, November 11). *Why FiveThirtyEight gave Trump a better chance than almost anyone else*. FiveThirtyEight. https://fivethirtyeight.com/features/why-fivethirtyeight-gave-trump-a-better-chance-than-almost-anyone-else/
6. CNN. (2016, November 9). *Exit polls 2016 - CNN*. Exit Polls Pennsylvania. https://www.cnn.com/election/2016/results/exit-polls/pennsylvania/president
7. Silver, N. (2016, November 8). *Pennsylvania: 2016 election forecast*. FiveThirtyEight. https://projects.fivethirtyeight.com/2016-election-forecast/pennsylvania/
8. Cohn, N. (2017, May 31). A 2016 review: Why key state polls were wrong about trump. *The New York Times*. https://www.nytimes.com/2017/05/31/upshot/a-2016-review-why-key-state-polls-were-wrong-about-trump.html

9. AAPOR. (2022). *Task force on 2020 pre-election polling*. https://aapor.org/wp-content/uploads/2022/11/AAPOR-Task-Force-on-2020-Pre-Election-Polling_Report-FNL.pdf

10. Silver, N. (2016, June 29). *A user's guide to FiveThirtyEight's 2016 general election forecast*. FiveThirtyEight. https://fivethirtyeight.com/features/a-users-guide-to-fivethirtyeights-2016-general-election-forecast/

11. I use this adjusted method here not because it is good, but because it's being compared to the traditional "spread" method and thus offers the most relevant contrast.

12. CNN. (2016, November 9). *Exit polls 2016—CNN*. Exit Polls Wisconsin. https://www.cnn.com/election/2016/results/exit-polls/wisconsin/president

13. CNN. (2016, November 9). *Exit polls 2016—CNN*. Exit Polls Minnesota. https://www.cnn.com/election/2016/results/exit-polls/minnesota/president

14. CNN. (2016, November 9). *Exit polls 2016—CNN*. Exit Polls Iowa. https://www.cnn.com/election/2016/results/exit-polls/iowa/president

15. CNN. (2016, November 9). *Exit polls 2016—CNN*. Exit Polls Michigan. https://www.cnn.com/election/2016/results/exit-polls/michigan/president

16. Cohn, N. (2017, May 31). A 2016 review: Why key state polls were wrong about trump. *The New York Times*. https://www.nytimes.com/2017/05/31/upshot/a-2016-review-why-key-state-polls-were-wrong-about-trump.html

17. Cohn, N. (2017, May 31). A 2016 review: Why key state polls were wrong about trump. *The New York Times*. https://www.nytimes.com/2017/05/31/upshot/a-2016-review-why-key-state-polls-were-wrong-about-trump.html

18. Silver, N. (2016, November 8). *Pennsylvania: 2016 election forecast*. FiveThirtyEight. https://projects.fivethirtyeight.com/2016-election-forecast/pennsylvania/

19. Silver, N. (2016, November 8). *Pennsylvania: 2016 election forecast*. FiveThirtyEight. https://projects.fivethirtyeight.com/2016-election-forecast/pennsylvania/

20. Silver, N. [@NateSilver538]. (2020, October 30). *10. Adding those (and marking past Trafalgar polls as R Partisan) turned out to make no difference. Biden's chances were 89 percent before and 89 percent afterward"* [Tweet]. Twitter. https://twitter.com/NateSilver538/status/1322302679016767489

21. Brown, C. (2016). *None of the above: Polling and third-party candidates*. None of the Above: Polling and Third-Party Candidates. Roper Center for Public Opinion Research. https://ropercenter.cornell.edu/blog/none-above-polling-and-third-party-candidates

22. Cassino, D. (2017, September 21). *How polls overestimate support for third-party candidates*. Harvard Business Review. https://hbr.org/2016/11/how-polls-overestimate-support-for-third-party-candidates

23. Igielnik, R. (2024, May 14). A simple experiment reveals why it's so hard to measure R.F.K. jr.'s support. *The New York Times*. https://www.nytimes.com/interactive/2024/05/14/upshot/support-third-party-candidates.html

27

We* Don't Talk About Utah

Important Points Checklist

- How Utah relates to non-US elections
- The empirical difference between third-party voters and undecided voters
- Why the third-party underperformance is ignored in most US elections
- How third-party performance can impact poll accuracy calculations
- Why assumption errors are accepted as poll errors

There are not many similarities that could be drawn between Utah and the United Kingdom. But in 2016, Utah hosted a presidential election with underlying third-party competitiveness that will look very familiar to those who follow UK and other non-US elections.

Much attention was given to Republican-turned-Libertarian Gary Johnson and Green Party candidate Jill Stein, as the two most popular third-party candidates nationally. Indeed, most discussion of poll data and election outcomes center, rightly, on the swing states and how these candidates might impact the results. But the third-party problem that existing methods have, by pretending they don't exist, becomes even more glaring when the confounding variables become larger.

Utah is far from a conventional swing state: Republicans have carried it every presidential election since 1968, and by a lot.[1]

But something historically weird was happening there in 2016. As Nate Silver put it:

> The state with the most "undecided" voters is Utah, but most of them are actually McMullin voters.[2]

If you need to read that quote several times to make sense of it, don't bother. Never mind the problematic-enough fact that FiveThirtyEight's founder is conflating a poll average and a forecast, and undecided voters with third-party voters, let me explain why we* don't talk about Utah:

DOI: 10.1201/9781003389903-27

Evan McMullin, like Johnson, was a former Republican running as a third-party candidate for President in 2016. He placed *fifth* in the popular vote nationally, well behind Trump, Clinton, Johnson, and Stein. But in his native state of Utah, McMullin went on to receive what was, by far, the highest percentage of votes of any third-party candidate in any state since Ross Perot: 21.5%.[3]

Prior to the election, in his own words, Silver declared McMullin "has a real shot" to win Utah. Nonetheless, in his reported poll averages, Silver threw that substantial support for McMullin in the same bin as undecideds.[4]

Contradicting his own understanding of McMullin's real chance to win, he shared a chart titled, "Which states' voters have made up their minds?" in which the "3rd-party or undecided" column for Utah reported a whopping 43.2% – far more than Clinton's 25.4% or Trump's 31.4%.[5]

The two major party candidates combining to poll at just 57% of decided voters are largely unprecedented in the United States.

But by Silver's reckoning, voters who declared their support for McMullin in the polls just hadn't "made up their mind."

His reasoning for putting these different groups in the same bin:

> I'm deliberately blurring the distinction between these groups because third-party voters often have a weak commitment to their candidates and can be picked off by one of the major-party candidates.[6]

I hope I don't need to remind you at this point that you can make whatever assumptions you want as a poll aggregator or forecaster, even very justifiable and well-informed ones modified with words like "often." But you don't get to call your assumptions being wrong a "poll error." I'll come back to that in a minute.

But first, the idea that third-party candidates are the same as undecided candidates is, McMullin or otherwise, not accurate. Undecideds deciding and people changing their mind are different calculations.

A large number of people who are polled as third-party supporters *do* end up voting for a third-party, as PublicMind's experiment showed, and should be intuitive with or without experiment. Empirically, people who are polled as undecided *do not* end up casting their vote for "undecided."

More importantly, throwing a relatively large poll number for third-party support, like in 2016, was 10% or more in many states, into the same bin as a much smaller category of voters who declare themselves "undecided" *misrepresents to the public what the polls say.*

Remember when, I hope, we agreed that pollsters shouldn't try to start assigning undecideds according to how pollsters believe they'll eventually vote but report the numbers as they observe them?

Well, some poll aggregators – being immune from criticism, it seems – have decided that *they* are allowed to aggregate the pollster's numbers differently than how the pollsters report them, and also *differently than how they observe them.*

I believe this flawed methodology is the direct, possibly inevitable consequence of an industry that is shielded from criticism by being able to blame their own poor assumptions on "the polls."

As I've stated previously, poll aggregators can and should be allowed to weight and report their numbers as they see fit.

But here is one problem with the lack of standards in poll aggregation, and flawed definitions used in the industry: knowing very well that voters who declare third-party support might not eventually vote third-party and that undecideds will eventually decide, pollsters don't get to assume or forecast how they'll eventually vote, but are nonetheless judged *solely on how close their poll was to the major party candidates'* eventual result.

When the poll aggregation assumption of "third-party and undecideds are basically the same" proves to be incorrect, in one or many specific instances, does the poll aggregator say "yeah, my bad, missed on that one"?

No.

Do analysts say, "this appears to have been a poll aggregation error"? **Also no, but in bold.**

Whose error, according to current definitions, is that wrong assumption made by analysts, aggregators, and forecasters? You guessed it, that's a poll error.

In FiveThirtyEight's article "The Polls Missed Trump. We Asked Pollsters Why," they reported that Trump "mostly outperformed his swing state polls." By spread, of course.[7]

But not just any spread, but by the *major party* spread.

The largest reported error in the 2016 polls wasn't in the Midwest, as was reported and analyzed.[8] It was in Utah. How could that be?

Well, if you understand the impact of confounding variables, it's not surprising at all.

FiveThirtyEight's poll final poll average for Utah[9] was:

Clinton (D)	Trump (R)	Johnson (L)	McMullin (I)
26.5%	36.3%	5.3%	24.9%

"McMullin" is a really weird way to spell "undecided."

By the Spread Method, this result is simplified down to "Trump +9.8" because Trump was ahead of Clinton by 9.8 (their chart reported this spread as +9.9, which could be due to rounding or a typo).[10]

Regardless, this +9.9 poll spread will be compared to the result spread, and this is the gold standard by which FiveThirtyEight and the consensus of the field's experts say this state's poll error should be measured, and one of the ways the 2016 election polls will be judged for overall accuracy.

Here is the eventual result[11]:

Clinton (D)	Trump (R)	Johnson (L)	McMullin (I)
27.5%	45.5%	3.5%	21.5%

That's a +18.1-result spread. (FiveThirtyEight published their article shortly after the election and used preliminary results stating the result spread was +18.4.)

The polls in this state were, therefore, off by about 8.2 (18.1 result spread – 9.9 poll average spread).

By far the worst polls of any state in 2016, and among the worst of century. What a catastrophic miss by the pollsters, right? And that's exactly how it was reported. The public believes the polls aren't reliable. End of analysis.

Well, by what standard?

McMullin and Johnson, as third-party candidates in the United States tend to do, underperformed their poll numbers a bit. But underperforming a historically normal 2%, or even a historically outlier 10%, is very different from underperforming, as third parties had polled in Utah, *over 30%*.

This is a major reason why poll aggregators and analysts have been allowed to "get away" with throwing third parties and undecideds into one bin, and blame their flawed assumptions on "the polls": even in elections where third parties have higher-than-normal support, it's very hard for someone who doesn't dig into the numbers to detect the effect of this flawed assumption.

But I wanted to dig in.

Utah is traditionally a very safe state for Republicans, and not only was Trump the "major" conservative candidate on the Republican ticket, *both high-polling third-party candidates were former Republicans.*

The conservative third-party candidates shed about 5% of support, and the state's approximately 7% undecided eventually decided.

FiveThirtyEight reported about 41% of undecideds ultimately backed Trump in Utah, while only 19% backed Clinton.[12]

That means, adjusting for Clinton picking up a net 1.3% of voters from undecideds, compared to her poll average, she would be at 27.8%. Which is … remarkably close to her eventual number.

Adjusting for Trump picking up about 2.9% from undecideds, plus about 5% from what the conservative third-party candidates shed, gets Trump to about 44.2%. Again, very close to his eventual number.

So, where did the error come from?

Based on an understanding of third-party Changing Attitudes and the Undecided Ratio, assigning the entirety of this 8.2-point discrepancy as "poll error" is indefensible.

Based on this analysis, it is possible that the polls in Utah were, by the proper simultaneous census standard, **the most accurate polls of any highly polled state in 2016.**

And unlike their methods, I will admit as such if proven wrong. All I ask is that you show your work.

Is it the poll's fault that more than 30% of voters declared support for a third-party? Is it the poll's fault that some of them sometimes change their mind? Is it the poll's fault that many voters who said they were undecided eventually decided to vote disproportionately for one candidate?

When analysts and experts declare "the polls" wrong, they are saying, in no uncertain terms, "yes, all of those things are the poll's fault." After all, it couldn't possibly be their assumptions.

Even when their assumptions of "undecideds split evenly, and no
one changes their mind" are proven wrong, they don't correct for them.
"Unscientific" is the nicest word that can be used to describe this practice.
And they use this unscientific calculation to convince the public not to trust
polls, and cause the public to not understand polls.

In their post-2016 analysis, and still at the time of writing, FiveThirtyEight's characterization of how "wrong" the polls were in 2016 includes this 8.2-point "error" from Utah,[13] which likely had little error at all. They reported the weighted-average error of state presidential polls in 2016 as 4.9-points.[14]

FiveThirtyEight – and every analyst – by improperly defining poll error, is doing a major disservice to the public as it relates to statistical literacy. Perhaps more problematic, regarding the future of the poll industry if this incorrect definition isn't fixed, the best pollsters will be little more than the best pollcasters who are willing to put their thumbs on the scale. And it will be the analysts and experts who caused it.

Given poll data showing conservative third-party candidates as high as 30%, and knowledge of the third-party underperformance phenomenon, what's stopping a pollster from *underreporting* the number they observed for the third-party, and *overreporting* the number they observed for the conservative major party candidate?

If they're being judged by their closeness to the eventual result, I think they'd be doing themselves a disservice not to. Is that the data poll aggregators and forecasters want pollsters to provide? Does a pollster's thumb on a scale help the public?

Fast forward a few cycles, given being closer to the eventual result means, no debate, "better," this practice of underreporting-overreporting will become standard, and the "poll industry" is no longer making observations, but polldictions. Instead of weighting for nonresponse and frame errors, they're now also "weighting" for Changing Attitudes and Undecided Ratio.

If you don't think that's the path we're on, I regret to inform you that this has already started happening in non-US elections.

The whole poll industry, if these definitions aren't fixed, is going to stop reporting observations – because they have no incentive to do so. Some of the best election pollsters have already, nearly, been driven out because they weren't close enough to the eventual result. I hope the simultaneous census standard is adopted soon for the sake of public understanding and the future of the poll industry.

What happened in Utah is a rare, real-world analogue study that we can look at to answer the question "how badly might this method misrepresent poll error if there are lots of third-party voters?"

If your definition disagrees with observation, then it's wrong.

In Utah, because the third-party numbers were much larger than anywhere in recent US history, we can see why assuming third-party voters and undecided voters are "basically the same" is an error. But it's not a poll error.

What happened in many states across the country in 2016, and will happen in most elections where third-party numbers are smaller, is that these assumption errors made by analysts that have nothing to do with polls accumulate, but go undetected, such that a large portion of the reported "poll error" is the analyst's own fault.

But the public will blame the polls, because the experts tell them to. That's how they report it. And that's not even getting into the possibility of compensating errors making inaccurate polls look accurate.

Changing Attitudes is not a poll error. Stop telling pollsters to "fix" things that have nothing to do with polls, and fix your own methods.

Understanding the threat *any number* of third-party and/or undecided voters pose to the reliability of "poll accuracy" methods is a necessary first step in fixing them. Observing, as Silver did, that "historically, the more undecided and third-party voters there are … the less accurate the polling has tended to be" seems to have played a valuable role in his model's "uncertainty" calculations, offering some insight into why he gave Trump a better chance than other major forecasters.[15] Still, it stops far short of addressing the bigger question of "why?" and problematically still views "pollster accuracy" through a lens with severe aberrations: how well it predicts the result.[16]

Utah is the first example I've provided where third-party presence was substantial. This provides a necessary base for analyzing methods used outside the United States, where polls often have a lot of third-party voters *and* a lot of undecideds, but those polls are reported as having no undecideds.

*I love to talk about Utah.

Notes

1. 270toWin.com. (2023). *Utah presidential election voting history—270towin*. https://www.270towin.com/states/Utah

2. Silver, N. (2016, October 25). *Election update: Where are the undecided voters?* FiveThirtyEight. https://fivethirtyeight.com/features/election-update-where-are-the-undecided-voters/

3. 270toWin.com. (2023). *Utah presidential election voting history—270towin*. https://www.270towin.com/states/Utah

4. Silver, N. (2016, October 25). *Election update: Where are the undecided voters?* FiveThirtyEight. https://fivethirtyeight.com/features/election-update-where-are-the-undecided-voters/

5. Silver, N. (2016, October 25). *Election update: Where are the undecided voters?* FiveThirtyEight. https://fivethirtyeight.com/features/election-update-where-are-the-undecided-voters/

6. Silver, N. (2016, October 25). *Election update: Where are the undecided voters?* FiveThirtyEight. https://fivethirtyeight.com/features/election-update-where-are-the-undecided-voters/

7. Bialik, C., & Enten, H. (2016, November 9). *The polls missed Trump. We asked pollsters why.* FiveThirtyEight. https://fivethirtyeight.com/features/the-polls-missed-trump-we-asked-pollsters-why/

8. Kennedy, C., Blumenthal, M., Clement, S., Clinton, J. D., Durand, C., Franklin, C., McGeeney, K., Miringoff, L., Olson, K., Rivers, D., Saad, L., Witt, G. E., & Wlezien, C., (2018). An evaluation of the 2016 election polls in the United States. *Public Opinion Quarterly, 82*(1), 1–33. https://doi.org/10.1093/poq/nfx047

9. Silver, N. (2016, November 8). *Utah: 2016 election forecast.* FiveThirtyEight. https://projects.fivethirtyeight.com/2016-election-forecast/utah

10. Silver, N. (2016, October 25). *Election update: Where are the undecided voters?* FiveThirtyEight. https://fivethirtyeight.com/features/election-update-where-are-the-undecided-voters/

11. 270toWin.com. (2023). *Utah presidential election voting history—270towin*. https://www.270towin.com/states/Utah

12. Silver, N. (2016, November 11). *Why FiveThirtyEight gave Trump a better chance than almost anyone else.* FiveThirtyEight. https://fivethirtyeight.com/features/why-fivethirtyeight-gave-trump-a-better-chance-than-almost-anyone-else/

13. Silver, N. (2018, May 30). *The polls are all right.* FiveThirtyEight. https://fivethirtyeight.com/features/the-polls-are-all-right/

14. Rakich, N. (2023, March 10). *The polls were historically accurate in 2022.* FiveThirtyEight. https://fivethirtyeight.com/features/2022-election-polling-accuracy/

15. Silver, N. (2017, January 23). *The invisible undecided voter.* FiveThirtyEight. https://fivethirtyeight.com/features/the-invisible-undecided-voter/

16. MKS Instruments. (2024). *Optical Lens Physics.* https://www.newport.com/n/optical-lens-physics
    In optics, they define what "ideal lenses" are despite them not being "real." These deviations from "ideal" are known as aberrations.

28

Informing a Prediction

DOI: 10.1201/9781003389903-28

Important Points Checklist

- Why 2016 and 2020 US elections were very different, despite the proximity in time
- Why US elections are typically easier to forecast than non-US ones
- Using the "base of support" or "simultaneous census" standard as a starting point for a prediction
- How undecideds relate to the possibility of "underperformance" and quantity of "overperformance"
- The concept of masking as it relates to compensating (but not compensating errors)
- The lessons taught by Kentucky and why, even when poll errors probably exist, it's important to have methods that can quantify "how much"

In his book "*Learning From Loss: The Democrats 2016–2020*," Seth Masket, the director of the Center on American Politics at the University of Denver, made an observation regarding Democratic sentiment entering 2020:

> The 2016 election results caused them to question everything.[1]

By the nature of political contests, both in their relative rarity and randomness, it's hard to say what lessons you'll learn and when. 1992 featured a third-party candidate who garnered nearly 20% of votes nationally. In 2000, Ralph Nader struggled to break 2%, but that performance looked herculean considering it wasn't until 2016 that a third-party candidate received more than 1% nationally. Recent history made third parties seem mostly negligible.

In 2016, third parties nationally accounted for a combined 5.7% of votes. That's not a huge number, but it can and did move the finish line in lots of states. Utah was its own separate and interesting case study, which I recommend interested students and academics research and discuss in more depth because the competitiveness of non-major parties makes its lessons applicable to both US and non-US elections.

But 2020 saw what came to some as a whiplash return to election normalcy in a world that was very much not normal due to the COVID-19 pandemic.

Third-party voters would only muster 1.9% of the vote nationally in 2020. Donald Trump was up for a second term, but Democratic challenger Joe Biden proved to be too strong.

Given their proximity in time and that Trump was again on the ballot, comparisons between the 2016 and 2020 elections were inevitable.

Leading up to the election, FiveThirtyEight reported, "Trump Can Still Win, But The Polls Would Have To Be Off By Way More Than In 2016."[2]

And that in order to win, "Trump needs a 'bigger than normal' error."[3]

How they calculated that error, of course – was spread.

I hope, by this point, you're an expert on poll error calculations and their flaws. This chapter will focus on how to read polls, as it pertains to informing a prediction.

The underlying variables in 2020 were different from 2016. In 2016, there were third parties performing well in polls, and while it is understood (if only for the purpose of forecasting) that third parties tend to underperform their poll number, the third parties didn't have much to underperform in 2020.

Pennsylvania in 2016 had, by FiveThirtyEight's estimate, about 6% undecided and 6% third-party. In 2020, about 4.2% undecided *plus* third party; they did not list third parties on their poll average for 2020, though their final "popular vote forecast" assigned about 0.7% to third-party for 2020.[4]

By putting all of those voters into the same bin, in addition to misinforming the public, experts are not accounting for one of the most valuable pieces of data polls can give us: an estimated finish line. That lack of accounting led them to the incorrect conclusion that 2016 and 2020 were statistically comparable.

If you go back to Chapter 26 regarding the numbers in Pennsylvania, you'll see that my "finish line" calculation took the third-party poll average at face value; I subtracted the 4.7% "third-party" vote share from the 100% available, to calculate a finish line around 47.65%.

But we're smarter than that, now. My simultaneous census approximation, given poll values of around 4.7%, would be approximately 1%–9%. A simultaneous census taken to a pre-election poll should produce an outcome in that range. *But that's for calculating poll error, not for informing our prediction.*

Mr. Panagakis' incumbent rule (should have) taught decades ago that the calculations we use to measure a poll's accuracy and those that we use to inform our prediction are independent. He used the knowledge of how undecideds tended to decide to inform his predictions regarding the chances of non-incumbents.

So, let's finally heed that lesson decades later and apply that knowledge to third-party performance in the United States.

Given that we know third parties tend to underperform their poll numbers, if I'm informing a prediction, instead of 4.7%, I (the predictor, in this scenario) might have estimated the third-party would end up around 3% in 2016. That would have given a new finish line: $100\% - 3\% = 97\%$. And $97\%/2 = 48.5\%$.

Undecideds have to go somewhere, but third-party voters don't. *Only third-party voters impact the finish line.*

To the existing methods, the 6% undecided + 6% third-party means 12% of voters "haven't made up their mind."

In reality, it means that about 6% haven't made up their mind, an additional, perhaps, 2%–3% have expressed interest in the third-party candidate closest to their ideological preference but may eventually support a major party candidate, and the remaining 3%–4% are actual, eventual third-party voters that allow you to estimate the finish line.

Being able to approximate how many third-party voters there will be gives you a better idea of where the finish line will be: if 6% of voters cast a ballot for a third-party, that puts the finish line at 47%. If only 2% do, the finish line moves to 49%. In elections often decided by less than 2%, those are very different numbers.

Knowing whether 12% of likely voters expressed that they are undecided, or that 6% of them expressed support for a third-party and only 6% are undecided, is not a minor difference.

In 2016, the difficulty in estimating the finish line, along with how undecideds would decide, plus accounting for if and how the conservative-majority third-party voters might eventually change their mind, hopefully gives you an extra layer of understanding why forecasting that election was much harder, and why Trump's chances were stronger than the "spread" indicated. And also, why non-US elections are typically more complicated than US ones.

But 2016 was not 2020.

The highest polling third-party candidate, Libertarian Jo Jorgensen, wasn't polling above 2% in many (if any) swing state polls taken in the last few weeks of the election. In Pennsylvania, again among the most crucial swing states of the election, she was receiving between 1% and 2%.[5]

Underperforming 1%–2% is very different from underperforming 6% or 30%. The finish line calculation becomes pretty easy here.

At the high end of forecasted third-party turnout, around 2%, the finish line would be about 49%.

At the lower end, 1% or less, the finish line is around 49.5%. That's a very narrow range. This was true for Pennsylvania and every swing state. Unlike 2016 where 48% would put the candidate at or across the finish line in many states (and Utah, where the finish line could have easily been estimated at 35%), 2020 was very different; comparing the two was an exercise of convenience, not one that could withstand any level of scrutiny.

Just as analysts would rightly criticize a pollster for taking a poll of 1,000 registered voters in the same neighborhood for the sake of convenience, and applying that number to the whole state, the analysts' insistence on comparing 2016 to 2020 should be criticized in the same way for the same reasons.

Putting It All Together: Informing a Prediction

Given that we understand what polls tell us about the base of support, what did the polls, and what do polls, tell us?

Chances are, like most people, you want to look at a poll and have a good idea of "who is ahead?" and "by how much?"

The Spread Method fails that test, and the Proportional Method doesn't even show up for it – but short of building full-blown forecasts and calculating probabilistic ranges of outcomes, I want to give you the tools you need to do that.

Here is my proposed approach for US elections:

1. First and foremost is the simultaneous census standard. Understanding that polls are intended to be approximations of a simultaneous census – not a prediction of eventual result – is a prerequisite to understanding not just how we should judge poll accuracy but also how to use polls to inform a prediction.

2. Second, a major party candidate is roughly equally likely to have *overperformed* a poll average as they are to have *underperformed* it. Remember, this is not referring yet to the eventual result but to the simultaneous census value: the percentage of support before allocating undecideds or accounting for Changing Attitudes, which I call the "base of support." This is the same concept shown in the "Throw it in the Average" simulation, as well as Mintucky. A poll – and by extension a poll average – is intended to approximate the simultaneous census number, not the eventual result, and not the eventual spread.

3. Third, it is highly unlikely a major candidate in any election where there is substantial polling underperforms their (properly reported) poll average in the *eventual result*.

    a. Highly unlikely does not mean impossible. "Highly unlikely" in a *typical* election in my mind is somewhere between rolling a 2 and rolling a 2 or a 12. Roughly 2%–5%, though I lean toward the low end. I'd welcome more research on the subject beyond my own, and not all elections are typical.

    b. The likelihood a major party candidate underperforms their poll average in the eventual result is directly related to how many undecideds and third-party supporters there are in an inverse relationship. The *more* undecideds and third-party supporters there are, the *less likely* a candidate is to underperform their poll average, and vice versa.

    c. The amount which a candidate *can* overperform their poll average in the eventual result has a direct relationship with the number of undecideds and polled third-party supporters: the more there are, the more they *can* overperform.

d. The impact undecideds can have on the eventual result is greater than the impact of third-party supporters, though the ideological leanings of the third-party candidate(s) and their polled level of support can influence the strength of that relationship.

"3" in totality is the concept of **masking**. In the chapter on compensating errors, I co-opted a term mostly used in accounting to describe what can – and inevitably has and will – happen in poll data: two errors cancel each other out; this is the reason the direction of errors is important.

A candidate's simultaneous census value being lower or higher than a *poll's* reported result is a *poll error.*

A candidate's simultaneous census value being lower or higher than a *poll average* is rightly called a **poll average error.**

But moving from poll or poll average to eventual result, which is the closest we can typically get to a simultaneous census, introduces other variables that cannot be "assumed as known" or "negligible."

In a two-way election with a poll average of 48%, and 8% undecided, one current industry-accepted assumption would be that this candidate should eventually receive about 52% of the vote (48% from poll average + half of undecided); otherwise, the polls were wrong. And likewise, if that candidate does eventually receive 52% of the vote, then the polls were right. These assumptions create many problems.

It's possible that a simultaneous census would reveal 46% support for that candidate – thus a 2% poll average error – and that there were actually 10% undecided. But if that candidate eventually receives 6% of the 10% undecided, they would end at 52%, thus leaving analysts under the mistaken impression that there was no poll average error.

Masking is a form of compensation, but it is not a compensating error. Masking refers *primarily* to cases like above where a candidate's simultaneous census value was **lower** than their poll or poll average, but undecideds added enough votes to "mask" the poll or poll average error – partially or entirely.

While the phenomenon of masking was my way of explaining "why do candidates almost never underperform their poll averages?", there are other possible combinations in the same class.

The same candidate with a reported 48% poll average could have a simultaneous census value of 50% but receive 2% or fewer of undecided voters, finishing at 52%. If the forecaster assumed undecideds would split 50/50 from a poll average of 48%, even though the poll average had an error, *and* the forecast made a bad assumption, the low Undecided Ratio has masked a **forecast error.**

All in all, the reason it's so rare for a candidate to underperform their poll average in the eventual result is because it's rare that a candidate's simultaneous census value is substantially smaller than the poll average *and* they do not receive enough undecided support to make up that difference.

This is by no means intended to be an exhaustive list of possibilities or observations relating to poll error, poll average error, masking, and forecast error.

"3b" says that given a poll average of 50–49 with 1% undecided, it is more likely one of the candidates underperforms their poll average than one of 41–40 with 15% undecided and 4% third-party. Remember, undecided is not the same as third-party. "B" speaks to the **probability a candidate underperforms their poll average** – not the range of possible outcomes.

"3c," on the other hand, offers a more precise calculation for the range of possible outcomes. It is far more likely one candidate outperforms their poll average by 10% when there are 15% undecided, than when there are only 5% undecided.

That it's easier to get more votes when there are more votes available *should* go without saying, but because the existing methods have contaminated our thinking, this observation alone opposes the analysis of most experts on the topic.

I'm not willing to assume they understand this concept when their words and analysis demonstrate otherwise.

Yes, if you understand that given 15% undecided, it is more likely a candidate outperforms their poll number by 10% than in an election with only 5% undecided, you understand polls better than most experts.

Welcome to ... Kentucky

Michael Barone, senior political analyst at the Washington Examiner, was left without explanation (except, of course, that "polls seem to be getting it wrong")[6] when in 2015 Republican candidate for Kentucky Governor Matt Bevin trailed Democrat Jack Conway 44–41 in the RealClearPolitics poll average – but Bevin ultimately *won* by 8.7 and received 52.5% of the votes.[7]

He was behind by 3 and then won by 9. Huge poll error. End of analysis.

Similarly, in Kentucky the year prior, long-time Republican Senator Mitch McConnell seemed to be in a closer-than-normal race by his standards: he led Democrat Alison Lundergan-Grimes only 49–42 in the RCP poll average.[8]

"Mr. McConnell won, 56%–41%."[9]

Polls said up by 7 and then won by 15. Huge poll error. End of analysis.

Given Kentucky's strong Republican lean historically, the fact that a Republican might outperform their poll number by a lot more than a Democrat in a "high undecided" election contest doesn't strike me as immediate evidence of a huge polling error as the existing methods would assert. Not all undecideds are equal – and for the aspiring forecasters, not all undecideds are unknowns either.

Even without a forecast, any reasonable analysis would conclude the "hold your nose" hypothesis (eventually voting for a candidate you don't necessarily like because they are closer to the party/ideology you typically support) has at least some merit.

But just because some, even a lot of the perceived "error" can be accounted for, doesn't mean there was *none*.

Given the number of undecided voters, the fact that Democrats underperformed their poll average in both elections should be a red flag for poll error of some extent. In an environment where these terms were properly defined, reasonable people could debate whether that poll error was small or substantial, what caused it, and how it could be fixed. But that's not the environment that currently exists.

As I hope I've made clear, I'm not saying that poll and poll average errors, even large ones, don't or can't ever exist. I'm saying that if you're going to assert that there was a 12-point error (Kentucky Governor 2015) or 8-point error (Kentucky Senate 2014), you'll have to do better than subtracting numbers and concluding; therefore, the polls failed bigly. That is as deep as the present analysis can go.

Speaking of which, not to be outdone in making sure people knew what the spread said and what people should think of it, senior political writer for FiveThirtyEight, Harry Enten, wrote "What to Make of Kentucky's Polling Failure."[10]

Enten noted that many individual pollsters badly "missed the final margin" and that "Errors like Kentucky shouldn't happen very often, and they seem to be occurring more frequently these days."[11]

Whether or not there was some polling error is a wonderful debate to have. Assigning the entirety of this error to "the polls" based on what you assume the result should be is not valid. The only failure here is the deficient level of analysis allowed by the Spread and Proportional Methods.

Beyond their failures to account for what I have presented in list numbers 1–3 of the "Informing a Prediction" section, the premises of which I believe are indisputable and would strongly encourage any expert to correct or refine, sits another reason why it's important to hold aggregators to the same standards we hold pollsters: there's no reason different aggregators can't arbitrarily choose different methods to support their desired goals – which may have nothing to do with accuracy. Indeed, I believe this may have already started to happen; I'll touch on this point in Chapter 34.

But for now, you want me to just tell you how to read polls. My poll averages for the data that follows come from RealClearPolitics (RCP). It's not necessarily that I believe RCP poll averages are the best, but owing to a long history and transparent methodology, their simple recent average approach are the numbers used for the analysis that follows.

Notes

1. Masket, S. (2020). *Learning from loss the Democrats, 2016–2020.* University of Cambridge ESOL Examinations.
2. Silver, N. (2020, October 31). *Trump can still win, but the polls would have to be off by way more than in 2016.* FiveThirtyEight. https://fivethirtyeight.com/features/trump-can-still-win-but-the-polls-would-have-to-be-off-by-way-more-than-in-2016/
3. Silver, N. (2020, November 3). *2020 election forecast.* FiveThirtyEight. https://projects.fivethirtyeight.com/2020-election-forecast/
4. Silver, N. (2020, November 3). *2020 election forecast.* FiveThirtyEight. https://projects.fivethirtyeight.com/2020-election-forecast/
5. RealClearPolitics (2020). *2020 Pennsylvania: Trump vs. Biden.* https://www.realclearpolitics.com/epolls/2020/president/pa/pennsylvania_trump_vs_biden-6861.html
6. Barone, M. (2015). *Why political polls are so often wrong.* The Wall Street Journal. https://www.aei.org/articles/why-political-polls-are-so-often-wrong/
7. RealClearPolitics (2015). *2015 Kentucky Governor—Bevin vs. Conway vs. Curtis.* https://www.realclearpolitics.com/epolls/2015/governor/ky/kentucky_governor_bevin_vs_conway_vs_curtis-5692.html
8. RealClearPolitics. (2014). *2014 Kentucky Senate—McConnell vs. Grimes.* https://www.realclearpolitics.com/epolls/2014/senate/ky/kentucky_senate_mcconnell_vs_grimes-3485.html
9. Barone, M. (2015). *Why political polls are so often wrong.* The Wall Street Journal. https://www.aei.org/articles/why-political-polls-are-so-often-wrong/
10. Enten, H. (2015, November 4). *What to make of Kentucky's polling failure.* FiveThirtyEight. https://fivethirtyeight.com/features/what-to-make-of-kentuckys-polling-failure/
11. Enten, H. (2015, November 4). *What to make of Kentucky's polling failure.* FiveThirtyEight. https://fivethirtyeight.com/features/what-to-make-of-kentuckys-polling-failure/

29

One Number Can Tell You a Lot

Important Points Checklist

- What was unique about Maine as a "battleground" state
- The lesson taught by the Literary Digest, that still goes unheeded today
- What Gallup and Roper debated in the 1950s – and why it was never resolved
- The value of a 49% poll average – and why it is strong

While I love to talk about Utah, the first real forecast I built that started me on this circuitous path to poll analysis was for Maine.

I believed it should go without saying that a candidate with a poll average around 53%, especially given a not insignificant number of 5% or 6% undecided, should be considered a near certainty to maintain at least 50% (thus, win). FiveThirtyEight disagreed. This is what happened with Biden in Maine in 2020, a race which FiveThirtyEight calculated Biden's chances of losing as 1/10, and even higher in the weeks prior.[1]

As confident and assured as I may sound today, I spent the better part of several months assuming – or at least strongly favoring the possibility – that I was wrong. What was I missing?

It happened that the chance they gave Biden of losing New Hampshire – a state similar in geography and poll average – was about the same 1/10.[2]

FiveThirtyEight's forecast also said Biden's chances of losing the election itself, 1/10.[3]

I didn't consider myself an expert on conditional probability, but those numbers didn't and don't add up.

If you're a fan of conditional probability, you can try it for yourself: in an event where Biden wins the election, how many cases are there in which he doesn't win Maine? Similarly, in an event where Biden loses the election, how many cases are there in which he does win Maine?

Like I said in the beginning, my impetus for building my own forecasts wasn't to find an edge or prove anyone wrong. I still considered it the most

DOI: 10.1201/9781003389903-29

likely possibility that I was the one missing something, but by this point, I started to consider the possibility they were. And that opened the door just enough to escape the Spread idolatry.

Given that most US presidential elections have a nonzero number of third-party voters in any given state, the finish line is rarely 50%. I chose 49% as the poll average number that should represent a very strong chance a candidate wins. Moreover, I only included states that were considered "swing states" or battlegrounds. While "battleground" is a subjective term, these are the states where there is enough polling to produce a poll average with some confidence.

My first area of research leading up to the 2020 elections were past presidential races, 2004–2016, shown in Figure 29.1.

From 2004 to 2016, only 2/30 (presidential candidates (6.7%) with a final poll average of higher than 49% had underperformed their poll average in that state, and only one (Florida, 2012, Romney) lost.

As you can tell from the size of the bars, candidates tended to overperform their poll numbers substantially, and the two underperformances were very slight, less than 1%. But it's still an underperformance, and by my understanding those should be very rare – so I kept going.

Presidential Candidate Poll avg. 49%+ by Battleground State (2004-16)

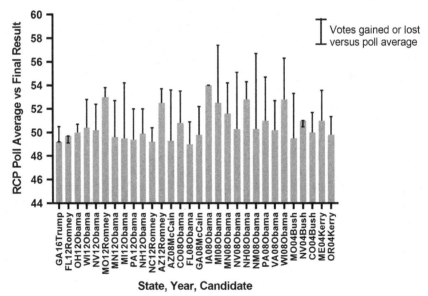

FIGURE 29.1
A chart showing presidential election results by state in which one candidate had a poll average of 49% or higher in the RealClearPolitics average, along with the final result. Chart by the author.

Romney's poll average in his underperformance was 49.7%, compared to a result of 49.1%. His final number was still strong, but the third-party vote was very low; it's the Law of 50% + 1, after all.

Notably, the poll average in that underperformance and loss featured a poll average with *fewer than 2% undecided*. Perhaps a small poll error, perhaps a small poll average error, maybe the undecideds leaned toward his competitor. In any case, that underperformance featured very few undecideds.

The other case (Nevada, 2004, Bush) was less clear. His poll average was around 51%, and he eventually received 50.5%. A slight underperformance, but he still won. This underperformance shouldn't be forgotten just because it didn't impact the eventual result; there were more undecideds in that polling average, so an underperformance is a strong indication of some poll error. But the fact that presidential candidates polling at or above 49% in a "battleground state," regardless of spread, ultimately won 29/30 times made me think, maybe, my understanding wasn't so bad.

A Note About Historical Comparisons

In my research regarding poll numbers and election results, it became evident that past elections were very different from modern ones. For one, poll aggregation is surprisingly new. As far as I can tell, RealClearPolitics was the first website to list poll averages for presidential elections in 2004; prior to that, the discussion was largely centered around individual polls, of which there were far fewer. FiveThirtyEight introduced their aggregation techniques in 2008. Given the fact that historically there have been very few pollsters, the lack of consideration given to an average makes sense.

In his book *"Lost in a Gallup: Polling Failure in U.S. Presidential Elections,"* author W. Joseph Campbell references the "Original Three" pollsters: Archibald Crossley, Elmo Roper, and George Gallup.[4] Their rivalry, which began after Gallup's rise to prominence polling the 1936 election, shares some striking – and unfortunate – similarities to the problems that persist today.

The Unlearned Literary Digest Lesson

The Literary Digest, a popular magazine of the era, had access to a huge mailing list and as part of their weekly opinion and news content, starting with the election of 1920, mailed "ballots" to their list.[5] Those returned were counted as the results of the poll.

In the elections of 1920, 1924, 1928, and 1932, they had "correctly predicted" the winner. Not just that, their polls had produced numbers within just a few percentage points of the final result in those elections.

Campbell provided a collection of contemporary quotes from newspapers across the country that offered high praise for the accuracy of Literary Digest.[6] That praise from journalists was filtered down to readers, in troublesome patterns that parallel today's landscape.

"It was so close to the actual result as to be uncanny," wrote the Wichita Beacon.

"… it should be a guide for newspapers, political leaders, and election-betting rings alike," wrote the Helena [MT] Independent.

William Allen White wrote in the Emporia Gazette, "absolute proof of political clairvoyance."

Newcomers who believe they had better methods had over a decade of public opinion working against them: no matter how good their methods, "they could claim nothing akin to the Digest's standing or prestige in soothsaying."[7]

Soothsayers, clairvoyant. People want to know the future, and The Literary Digest told them.

In the Digest's 1936 poll, which they described as a "poll of ten million voters, scattered through the forty-eight States of the Union," they received about 2.2 million responses. They reported the results as follows:

LANDON, 1,293,669; ROOSEVELT, 972,897

To their credit, in the article where they released the poll, they note:

> These figures are exactly as received from more than one in every five voters polled in our country – they are neither weighted, adjusted, nor interpreted.[8]

While their methods were unscientific, their transparency was at least commendable. Unfortunately, with their and the media's confidence – given their track record for "predicting elections" – the public was, nearly a full 100 years ago, still being misinformed. When a reader challenged the accuracy of their poll, The Digest responded in writing:

"So far, we have been right in every Poll."[9]

They had been right. Well, that settles that.

They continued, "we never make any claims before election but we respectfully refer you to the opinion of one of the most quoted citizens to-day, the Hon. James A. Farley, Chairman of the Democratic National Committee."

Farley said, prior to fellow Democrat Franklin Roosevelt winning in 1932, in reference to Roosevelt's strong performance in The Literary Digest Poll, "Any sane person can not escape the implication of such a gigantic sampling of popular opinion … I consider this conclusive evidence as to the desire of the people of this country for a change in the National Government."[10]

Conclusive evidence. Nearly 100 years ago, the same myths about what polls do were being perpetrated by the field's best and brightest. After all, if their poll had so often correctly predicted the winner, how could it be wrong?

Of course, their underlying methods would make any mildly informed reader today shudder. Their sample, while enormous, was not random. Did the 22% of people who responded to the poll have some underlying similarities not shared by the other 78%? How representative was their mailing list? Nonresponse error and frame error were among the biggest risks: no weighting, a nonrandom sample, no way to sort out who the likely voters were, among other things. The risk for error was enormous.

But, you could contend as they did, they had always been right.

Gallup's more scientific methods, though his sample size was miniscule by comparison, was more random and representative of the population. Contrary to the prevailing sentiment that Landon was a runaway favorite, Gallup reached the conclusion that Roosevelt should be considered the favorite. Roosevelt won re-election, which catapulted Gallup to a level of national prominence.

Unfortunately, to this day, Gallup's success – much like FiveThirtyEight – is credited more for who they "predicted" would win and how many "calls" they got right than the underlying data it produced, or the improved methods it utilized. And the Literary Digest poll – poorly conducted as it was – is still condemned mostly for making the wrong prediction. Only *after* catastrophic failure were its methods – in hindsight, of course – dismissed.

The Roper Center at Cornell writes:

> Gallup's breakthrough moment was the 1936 presidential election. His organization correctly predicted that Franklin Roosevelt would win the Depression Era election while The Literary Digest, an influential national magazine, incorrectly predicted Alf Landon would win.[11]

Which begs the question: if Gallup's properly done poll had produced results on the edge of the margin of error, and The Digest's poll hadn't so dramatically underestimated Roosevelt's performance, might their unscientific methods have sustained for much longer?

More directly, would the standards by which analysts judge pollster accuracy today have been able to detect The Digest's now obviously flawed methodology – regardless of closeness to eventual result?

No, they couldn't because not much has changed since then. Their definitions prove it. Polls are read, reported, and most of all ultimately judged, as predictions. Closer to result means better, end of discussion.

As mentioned earlier, as recently as 2016, pollsters who hadn't weighted by education started to be criticized for not weighting by education – not before this "error" contributed to the perceived failure of the polls, but after.[12]

And to be fair, it is *okay* to learn from mistakes and correct them going forward. That's the best we can ask of science, in many cases.

But in this field, flawed methods are allowed to catastrophically fail before they are corrected. After catastrophic failure, experts can adjust their methods and move forward; whether public trust is too far gone to be reclaimed is not considered.

The Literary Digest, in large part due to the irreparable harm done to their reputation after the 1936 election, was sold to competitor TIME. They never conducted another poll.

The same corrections after catastrophic failure cannot be said for how poll accuracy is evaluated. Poll aggregators, forecasters, and analysts have been allowed to continue judging poll accuracy solely on closeness to eventual result without contention; Utah and the rest of 2016 should have offered a Literary Digest-sized reckoning for that.

But it's easier to just blame the polls.

The Original Three pollsters, with varying techniques but a more scientific approach, shaped the way polling would be done and reported since.

At the American Association for Public Opinion Research (AAPOR) conference in 1953, Roper and Gallup debated how polls should be relayed to the public.

Roper said:

> We learned that the press is not ready to handle the survey tool properly ...
> our problem is to get the press to accept polls as analytical and speculative,
> not as predictive tools.[13]

Decades ago, Panagakis lamented the media's inability to relay to the public how polls work.

Decades before that, Roper rightly observed polls were being wrongly portrayed as "predictive tools."

And today, not only do the media still make those *same wrong portrayals*, but the consensus of the field's experts does the same exact thing – even the Center that bears Roper's name says Gallup's "breakthrough moment" was based on what his polls "correctly predicted."

But Gallup contended that the media's portrayal of polls as predictions held them accountable:

> You pay a terrible penalty for mistakes if you're wrong.

The biggest names in polling at the time were competitive, though cordial enough to hold debates. In fairness to Gallup, it was common at the time – though I think it would be and is bad practice today – for pollsters to opine eventual results. Given that forecasting was not a formal industry then, and there were very few pollsters, pollsters were effectively using their own observations to inform their own predictions. Today, that would be like FiveThirtyEight conducting their own polls to make their forecasts instead of aggregating others'.

The problem with that, beyond the hopefully obvious flaw in trying to forecast from one poll when there are many available, is that the quality of the prediction or forecast is independent of the quality of the polls. My Mintucky polls produced a poll average very close to the simultaneous census – that means it's a good poll average.

If I wanted to play forecaster and predicted that Ricky Red would win by 4, that doesn't mean my polls or poll average were wrong, it means my forecast was wrong; separate subjects.

After the Roper-Gallup debate at the AAPOR conference, it was reported in *Public Opinion Quarterly* that, "polltakers themselves disagree on these points, the public may be pardoned for being somewhat confused about the true role of political polling."[14]

This was in 1953.

For perspective, it was in 1954 that George Gallup conducted his first poll on whether the public believed smoking was *one of* the causes of lung cancer, and only 41% answered "yes."[15]

Every election before, and every election since, poll accuracy has been characterized by the same method, closeness to eventual result, treating polls as predictions or forecasts.

Kennedy-Nixon, Reagan-Carter, Bush-Gore, Clinton-Trump, Trump-Biden: what's the spread? That's all we need to know. Not much has changed except for the names on the ballot.

The Roper-Gallup debate frames a valuable historical lesson. While pollsters who work independently can and should *compete* for who can do the best work, as impartial observers we – the public – can view their work for what it is: different ways to observe the same data.

With dozens of pollsters active today, they should compete as well. But how we judge who does the "best" work should not be limited just to pollsters. Poll aggregators and forecasters should also be rated with a proper understanding of what each of those respective tools try to tell us, not with Literary Digest reasoning the anoints "The Closest" as "The Best," nor one that lays all blame on polls and ignores the existence of forecast error. The public and the field would be better served for it.

With a disclaimer that elections of the recent past are not a crystal ball to the future, but a way to analyze data in a way that is consistent with what polls truly tell us, I'll continue.

A 49% Poll Average Is Really Good

From 2004 to 2016, presidential candidates who had a poll average in a state at or above 49% ultimately won 29/30 (96.7%) of those instances. In 28/30 (93.3%) of those instances, the candidate outperformed their poll average. I dread the possibility of oversimplification, but reducing what are considered to be competitive elections to one number and predicting the winner at a 96.7% rate from it certainly merits consideration.

(Note, that in this characterization of "predicting the winner," I am the one making the predictions, not the polls, and not a poll average. If that prediction is wrong that doesn't instantly mean "the polls" are.)

However, as my understanding of how polls work improved, it became clearer that this characterization of a poll average of 49% being very hard to beat is not an oversimplification at all. Here's why:

> In elections with a nonzero number of votes going to a third-party, the finish line is short of 50%. Given the fact that masking makes it unlikely a candidate underperforms their poll average, the 49% poll average threshold is not saying much more than "candidates who receive more than 49% of the vote in elections with a third party usually win."

As promised, this is not an advanced concept, but it is a statistically valid one. And it's a concept the Spread Method and Proportional Method are unable to capture or account for.

Analyzing data from 30 presidential elections was a nice step, and the win percentage was strong, but given the statewide nature of presidential contests, why stop there? Senate and Governor's races are also statewide, some competitive and highly polled. I applied the same reasoning as before to all statewide races.

From 2004 to 2016, there were 88 elections in "battleground" statewide contests in which the leading candidate's poll average was 49% or greater. There were only three such instances of that candidate losing the election. That's 85/88, or 96.6%.

You can find Mr. McConnell's win at (49,7.2) in Figure 29.2. The race had approximately 5% expressing support for the Libertarian third-party candidate, who finished at 3%, and 4% undecided. McConnell outperforming his poll average by 7% is considered evidence of a massive "polling failure."

For those interested, in addition to the 2012 instance of Romney slightly underperforming his 49.7% poll average and losing, the other two instances (through 2016) of eventual losers carrying a poll average of 49% or higher were Senate races in 2010 and 2012.

In 2010, Ken Buck (R) in his bid to join the Senate from Colorado underperformed his 49.3% poll average by 2.5% and lost to Michel Bennet (D).[16]

In 2012, Rick Berg (R) in the North Dakota Senate race carried a poll average of 49%, and ultimately lost to Heidi Heitkamp (D). In this instance, despite losing with a poll average of 49%, Berg did not underperform that average: he ultimately received 49.3% of the vote.[17]

Only 5/88 times from 2004 to 2016 did a leading candidate whose poll average was 49%+ *underperform* their poll number (5.7%) – and of those underperformances, only 2/88 times (2.3%) did that candidate lose.

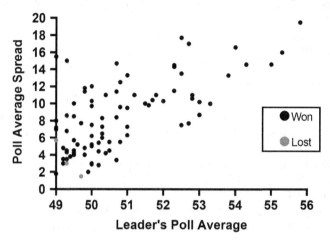

FIGURE 29.2

Chart plotting poll averages versus poll average leads in "battleground" statewide Governor, Senate, and presidential races from 2004 to 2016, and whether that candidate eventually won. Chart by the author.

Dissect it however you'd like, this lends enormous reinforcement to what is a statistically valid way to interpret poll data: poll averages provide a remarkably good estimation of a candidate's base of support, and from that, can help you estimate chance of winning. Not the spread, not the proportion, not an individual poll.

If you're thinking to yourself, as I did, that the top right of Figure 29.2 features a very generous interpretation of the word "competitive," don't forget about what FiveThirtyEight said about Maine and New Hampshire in 2020. While they didn't characterize them as tossups, they certainly believed they could be competitive. Having data at and beyond the point of interest – which was around (53, 12) for Maine and New Hampshire, is valuable to informing predictions.

As close as the elections were in time, the underlying variables between 2016 and 2020 were very different. In 2016, Hillary Clinton – who was considered a strong favorite to win – did not have a polling average over 47% in a single swing state. Given a stronger third-party presence in 2016, the "49%" finish line might be a little high, but even for a finish line closer to 48%, her poll average was not at or near it anywhere.

How much trouble a candidate is in depends how far they are from the finish line.

Joe Biden, meanwhile, had a poll average *at or over 50%* in two swing states (four if you include New Hampshire and Maine).[18]

His poll average was sitting beyond the finish line in multiple states. That's not to say the polls or poll average *can't* be wrong – or that attitudes can't change – but a poll average beyond 49% or 50% is very strong.

You wouldn't know it by Spread or Proportion Analysis though.

RealClearPolitics published a chart showing that Biden was only 1.2 points ahead of Clinton's poll average in battleground states four years earlier. Not by poll number, but by spread.[19]

Nate Cohn, the chief political analyst at *The New York Times*, shared a chart performing the same analysis, but with *two columns*: applying what the eventual spread would be if the poll spread was "off" by as much as 2012 *and* 2016.[20]

FiveThirtyEight similarly concluded, "Trump Can Still Win, But The Polls Would Have To Be Off By Way More Than In 2016" in which they showed the amount Trump outperformed his poll numbers in 2016, and applied that to their averages for 2020 to show how much Biden would still "win by." Silver concluded, "a 2016-style polling error *wouldn't* be enough for Trump to win."[21]

Which is to say, he and most others tried to compare 2016 and 2020 apples-to-apples when the underlying variables – where the finish line was and each candidate's base of support – were very different. Another "apples-to-apples" analysis problem.

Also problematic, in that article describing Trump's 2020 chances by spread, Silver said that Clinton's 2016 poll average spread in Maine was +6.9 and that the eventual result was −3.0, which would mean that Clinton lost Maine.[22]

That's not correct – the final spread in Maine in 2016 was +3.0 because Clinton won Maine.[23]

Input errors aside, they said that even with a 9.9-point "error" that they attributed to the pollsters of Maine, that Biden – with his nearly 14-point advantage in the poll average at the time – would still win the state by 3.9. I can't say for certain that this 6-point overestimation of "error" contributed to their 10% probability that Biden could lose in Maine, but it seems possible.

At both RCP and FiveThirtyEight, Biden's poll average in Maine was around 53%, with Trump around 40%. There were even a relatively high number of undecideds in Maine compared to other states – but Biden was polling well across the finish line.

In 2016, Trump overcame his "spread" deficit in large part because he won over a large percentage of undecided voters and benefitted from Changing Attitudes. Both candidates were racing to the finish line.

But there weren't any number of undecided voters that could get Trump to the finish line this time. It would require a huge number of Changing Attitudes, or a large poll error (and/or poll average error) just to put Maine in the realm of "possibly, maybe, a little competitive." And *even then*, Trump would need to win nearly all of the undecideds.

Looking only at the "spread" does a major disservice to informing a prediction.

A candidate underperforming their poll average – which is what would have to happen for a candidate polling over 50% to lose – is rare. A candidate with a poll average over 50% underperforming their poll average by *more than 3%*, which is what would have been required for Biden *just to have a chance to maybe lose* Maine or New Hampshire, might have never happened in modern history of polling. And if it has, I have a hard time believing its chances are in the universe of 1/10.

The reason "spread" does such a disservice here is that the odds a candidate overcomes a certain spread deficit and the odds that candidate underperforms a poll number are different calculations. They're different calculations because the underlying variables that must be accounted for are different.

There are lots of reasons polls can be "off." If your understanding of polls is made up largely or entirely of what the spread or proportion says, and you don't or can't account for the underlying variables in that calculation, you do not understand polls well enough to analyze them.

If you're informing a prediction about a candidate's likelihood of winning, and their poll average the day before the election is 53%, your calculation should start with: "is it possible this candidate underperforms their poll average by over 3%? How likely is it?"

This is not a difficult calculation. And it has nothing to do with spread.

Having examples that challenge the edges of your understanding is both valuable and necessary in science and statistics. Whether Utah, the United Kingdom, Maine, Kentucky, or Mintucky: if it disagrees with observation, then it's wrong.

This is not to say that I believed Biden couldn't lose *the election*. While he was a much stronger favorite than Clinton was – owing to his stronger poll averages across the board – he still needed a little help pushing past the 47–48.9 range in a few, critical swing states.

Biden's position relative to Clinton's was characterized based on spread. By that metric, Biden was only slightly ahead of Clinton's position (if at all).

Here's how Biden's position was reported:

State	Clinton's Poll Avg. Spread 2016	Biden's Poll Avg. Spread 2020
Pennsylvania	+1.9	+1.2
Wisconsin	+6.5	+6.7
North Carolina	−1.0	+0.2
Michigan	+3.4	+4.2
Arizona	−4.0	+0.9
Nevada	−0.8	+2.4

If you only looked at spread, it seemed close! But for these same states, if you want to compare a candidate's relative strength across years, here's how I think it should be viewed:

State	Clinton's Poll Avg. 2016 (%)	Biden's Poll Avg. 2020 (%)
Pennsylvania	46.2	48.7
Wisconsin	46.8	51.0
North Carolina	45.5	47.8
Michigan	45.4	50.0
Arizona	42.3	47.9
Nevada	45.0	48.7

Very important to note here: Wisconsin and Michigan. Two midwestern states with allegedly huge poll errors in 2016, with Biden's "spread" eerily similar for Democrats who remembered 2016.

But Biden's base of support was much higher in the most important swing states, and he was much closer to the finish line. You wouldn't know it by listening to Wang or Huffington Post, who wrongly forecasted Clinton as a near-certainty to win, but Biden was in a much stronger position than Clinton.

I'm probably not the first to note that candidates with a polling average at or above 50% (or 49%) are in a very strong position, though I couldn't find any other research on it. But given the obsession with spread by the most active researchers in this field, I'm almost certainly the first to attempt to quantify how strong.

Notes

1. Silver, N. (2020, November 3). *2020 election forecast*. FiveThirtyEight. https://projects.fivethirtyeight.com/2020-election-forecast/maine/
2. Silver, N. (2020, November 3). *2020 election forecast*. FiveThirtyEight. https://projects.fivethirtyeight.com/2020-election-forecast/new-hampshire/
3. Silver, N. (2020, November 3). *2020 election forecast*. FiveThirtyEight. https://projects.fivethirtyeight.com/2020-election-forecast/
4. Campbell, W. J. (2020). A tie "Would Suit Them Fine." *Lost in a Gallup: Polling failure in U.S. presidential elections* (pp. 85–107). University of California Press.
5. TIME Magazine. (1938, May 23). *Press: Digest digested*. Press: Digest Digested—TIME. https://web.archive.org/web/20100826092219/http://www.time.com/time/magazine/article/0,9171,882981,00.html
6. Campbell, W. J. (2020). A time of polls gone mad. *Lost in a Gallup: Polling failure in U.S. presidential elections* (pp. 40–61). University of California Press.
7. Campbell, W. J. (2020). A time of polls gone mad. *Lost in a Gallup: Polling failure in U.S. presidential elections* (pp. 40–61). University of California Press.

8. Campbell, W. J. (2020). A time of polls gone mad. *Lost in a Gallup: Polling failure in U.S. presidential elections* (pp. 40–61). University of California Press.

9. Campbell, W. J. (2020). A time of polls gone mad. *Lost in a Gallup: Polling failure in U.S. presidential elections* (pp. 40–61). University of California Press.

10. Campbell, W. J. (2020). A time of polls gone mad. *Lost in a Gallup: Polling failure in U.S. presidential elections* (pp. 40–61). University of California Press.

11. Roper Center for Public Opinion Research. (2023). *George Gallup.* https://roper-center.cornell.edu/pioneers-polling/george-gallup

12. Hatley, N. (2020, August 18). *A resource for state preelection polling.* Pew Research Center Methods. https://www.pewresearch.org/methods/2020/08/18/a-resource-for-state-preelection-polling/

13. Campbell, W. J. (2020). A tie "Would Suit Them Fine." *Lost in a Gallup: Polling failure in U.S. presidential elections* (pp. 85–107). University of California Press.

14. Campbell, W. J. (2020). A tie "Would Suit Them Fine." *Lost in a Gallup: Polling failure in U.S. presidential elections* (pp. 85–107). University of California Press.

15. Gallup, G. H. (1972). *The Gallup poll: Public opinion 1935–71* (Vol. 2). Random House. *http://legacy.library.ucsf.edu/tid/qpb42c00*

16. RealClearPolitics. (2010). *2010—Colorado Senate—Buck vs. Bennet.* https://www.realclearpolitics.com/epolls/2010/senate/co/colorado_senate_buck_vs_bennet-1106.html

17. RealClearPolitics. (2012). *2012—North Dakota Senate—Berg vs. Heitkamp.* https://www.realclearpolitics.com/epolls/2012/senate/nd/north_dakota_senate_berg_vs_heitkamp-3212.html

18. RealClearPolitics. (2020). *Top battlegrounds: Wisconsin, Michigan, Pennsylvania, North Carolina, Florida, Arizona.* https://www.realclearpolitics.com/elections/trump-vs-biden-top-battleground-states/

19. RealClearPolitics. (2020). *Trump vs Biden battleground states 2020 vs 2016.* https://www.realclearpolitics.com/epolls/2020/president/us/trump-vs-biden-top-battleground-states-2020-vs-2016/

20. Cohn, N. (2020, November 3). The upshot on today's polls (published 2020). *The New York Times.* https://www.nytimes.com/live/2020/presidential-polls-trump-biden

21. Silver, N. (2020, October 31). *Trump can still win, but the polls would have to be off by way more than in 2016.* FiveThirtyEight. https://fivethirtyeight.com/features/trump-can-still-win-but-the-polls-would-have-to-be-off-by-way-more-than-in-2016/

22. Silver, N. (2020, October 31). *Trump can still win, but the polls would have to be off by way more than in 2016.* FiveThirtyEight. https://fivethirtyeight.com/features/trump-can-still-win-but-the-polls-would-have-to-be-off-by-way-more-than-in-2016/

    Their 2016 "Final Margin" chart indicates that Hillary Clinton lost Maine by 3.0, but she won it by 3.0.

23. Bureau of Corporations, Elections & Commissions. (2017). *Tabulations for elections held in 2016.* https://www.maine.gov/sos/cec/elec/results/results16-17.html#tally

30

Don't Call It a "Rule" and How to Report Polls

Important Points Checklist

- Why a candidate's poll average is a better number to inform a prediction than "spread"
- Polls aren't predictions, probabilities, calls, or naming favorites – stop it
- What the "leader reporting standard" is
- My proposal for ending the Proportional Method (even if pollsters continue to use it)
- What "flooding" is and why it matters
- Why standards are needed for poll aggregators

The "49%+" threshold *is not a rule*. It is an observation that, given an understanding of what polls do, should make mathematical sense. It is intended to help inform a prediction, not take its place. What applied in 2004–2022 may not apply by 2028, or even 2024; when variables change, you must account for them. Every election is unique, and some are more unique than others. No matter how historically compelling, no matter how strong the correlation may appear, don't allow confounding variables to cloud your reasoning or judgement.

Figure 30.1 shows statewide battleground election data through 2022 for candidates with a poll average over 49%.

2022 gave another reminder that "49%" is not a rule: Arizona Republican candidate for Governor Kari Lake's polling average was north of 50%, but she ultimately underperformed that number and lost.[1] It was a true two-way election in which Lake's average according to RCP was 50.8%, but she ultimately lost with 49.7% of the votes. Another example of a candidate underperforming a strong poll average and losing – but also another example where this happened with only about 2% of voters undecided close to the election.

DOI: 10.1201/9781003389903-30

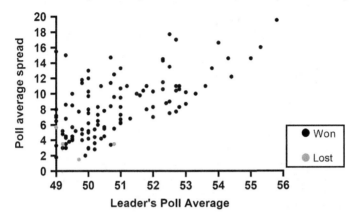

FIGURE 30.1
A chart showing statewide battleground results for governor, senate, and president from 2004 to 2022, in which one candidate had a poll average of 49% or higher in the RealClearPolitics (RCP) average, along with the final result. Chart by the author.

Even still, from 2004 to 2022, using the simple recent poll averages from RCP, candidates leading with a poll average of 49% or higher, are 112/116 (96.6%).

It's Not the Spread

Presenting the numbers as I have in the past two chapters, without the context that preceded it, doesn't give the full significance of their meaning. The data is merely an observation of past events. The chart could just as well be the simulation from "Throw it in the Average" that demonstrates the normal fluctuation of ideal polls for a given margin of error.

The reason I'm presenting this data in this chapter and not Chapter 2 is not because the data can't speak for itself, but because the *underlying reasoning* that explains this trend requires the background of what polls do and don't try to tell us. And neither "spread" nor "proportion" can account for it. The fact that candidates with a poll average of 49% or higher almost always win, while simple, is also valid – unlike the contradictory ideas of what a "spread" of a certain amount means, which I'll share shortly.

If "49%+" is all you take away from this, I have done you a disservice. The context with which I presented what 49% *means* (though, I'll admit, I got a little chatty at times) was necessary for understanding the "*why*" behind the observation.

Predicting the winner with great precision is incidental to the underlying methodology. The Literary Digest can pick the winner. A coin flip

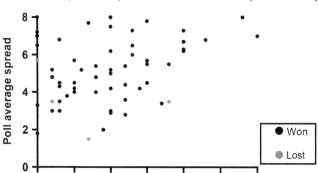

US President, Senate, Governor Elections (2004-2022)

FIGURE 30.2
A chart showing statewide battleground results for governor, senate, and president from 2004 to 2022, in which one candidate had a poll average of 49%–52% in the RealClearPolitics average and a "spread" lead of 8 or less, along with the final result. Chart by the author.

can predict the winner. I can predict the winner, FiveThirtyEight can predict the winner, you can predict the winner. All of us, sometimes, will be wrong.

But when evaluating a tool – whether a scale, a poll, a poll average, or a forecast – understanding the reasons why we can and can't call them wrong matters.

While it's important to have data to "test the edges" of an observation, I think a more rigorous test is worth conducting in the middle: exclude the candidates that polled with a large "lead." Have a look at Figure 30.2.

All four eventual losses fall in this range, but still, out of 68 observations, 64/68 (94.1%) is great, as should be expected.

I excluded the candidates with a poll average of 49%+ who also had a double-digit "spread" lead in this figure, and you may have deduced that candidates meeting these criteria won 48/48 times (100%) since 2004. Some who insist on clinging to the Spread Method, after presented with this, have contended with me that even high *single-digit* "spreads" are too dissimilar from the rest to fairly form a "win-loss" record.

For one, this is why I find it important to present FiveThirtyEight's "10% win probability" forecasts from 2020 for Maine and New Hampshire, for which the "spread" was approximately +13 *and* the leader's poll average was approximately 53%.

If one were to assume the validity of the Spread Method and grant that "10%" was a reasonable win probability for those states, it should follow that spreads *smaller* than that should be, at least, worth including in evaluating it.

But the proponents of the Spread Method don't stop there – and, frustratingly for those who care about accountability – they don't even have a consistent explanation of what they believe the spread says about win probability.

In an article published in 2023, Nathaniel Rakich at FiveThirtyEight wrote, "races within 3 points in the polls are little better than toss-ups – something we've been shouting from the rooftops for years."[2]

That article is a significant evolution from founder Nate Silver's findings in 2012 that, "There are no precedents in the database for a candidate losing with a 2- or 3-point lead in a state when the polling volume was that rich."[3]

The fact that a pseudoscientific metric would draw conflicting conclusions which are both unreliable is not surprising.

In the 2012 article, Silver characterized a 2- to 3-point lead as an "inflection point." He provided a probit regression for state poll averages showing that somewhere between a 2- and 3-point "spread" candidates cleared an 80% chance to win.[4]

Silver used this data to inform his forecast that Obama was about a 76% favorite to win in Ohio, and that it was "misinformed" given a 2.4-point advantage in the poll average, to refer to Ohio as a toss-up. That article contained four charts, lots of explanation, and I'm not exaggerating, there was not a single poll number or poll average mentioned in the entire article, or any chart therein – only spread.

In 2012, FiveThirtyEight said a 2.4-point lead means it's misinformed to call that race a tossup because a candidate with a lead that strong is much more likely to win. In 2023, they said leads smaller than three points are little better than toss-ups, and they've been shouting it for years. To borrow from the 1953 report after Roper and Gallup offered conflicting views on the role of polling, I'll say the public may be pardoned for being somewhat confused about what a particular spread means.

Silver was right, though 2024 FiveThirtyEight would disagree, it was misinformed to call Ohio a toss-up given his poll average. But it had little to do with the spread. You can find former President Obama's 2012 Ohio position in the poll average at (50, 2.9) in Figures 30.1–30.5.

I wanted to put the "spread" question to the ultimate test, but also account for the leading candidate's actual poll average – which I use to approximate base of support. So, I squeezed the numbers even tighter.

In Figure 30.3, I've reduced the threshold for the leader's poll average to 48%+, and the "spread" threshold to 3. We are looking at what would be – by FiveThirtyEight's modern spread definition – considered "little better than tossups."

The range of poll averages on this chart is 48.0–47.7 (Virginia Electoral College, 2012) to 50.0–47.0 (Massachusetts Senate, 2012). A 50.2–47.4 average (Nevada Electoral College, 2012) managed to sneak its way on here too. All within a "spread" of three points.

From 2004 to 2022, there were 40 statewide battleground races that fit the "48% leader poll average, up by 3 or less" criteria. The poll average leader won 34 times, 85.0%. (Races with the same poll average and margin are plotted on top of each other, so all 40 races don't show. All losses are shown.)

Of the six losses in this range, five occurred when the candidate was polling under 49%.

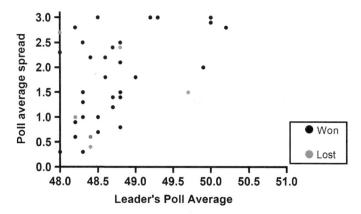

FIGURE 30.3
A chart showing statewide battleground election results since 2004 in which one candidate had a poll average of 48% or higher in the RealClearPolitics average and a "spread" lead of 3% or less, along with the final result. Chart by the author.

For such a tight range of "spread" advantage, 85% is exceptionally strong. So where does this "little better than toss-up" mentality come from?

The vast majority of instances where a candidate loses – regardless of spread – come when their base of support is not strong. As Figure 30.4 shows, the farther you are from the finish line, the more susceptible you are to being overtaken.

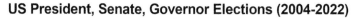

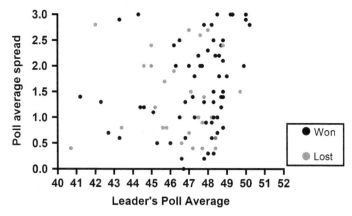

FIGURE 30.4
A chart showing statewide battleground election results since 2004 in which the leading candidate had a "spread" lead of 3% or less, along with the final result. Chart by the author.

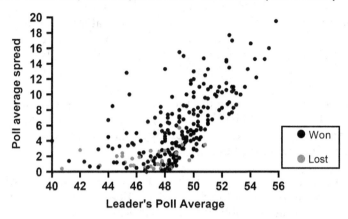

FIGURE 30.5
A chart showing statewide battleground election results since 2004, along with the final result.
Chart by the author.

The belief that "the polls were wrong" usually lives here, when the leader's poll numbers were under 48%. Far short of the finish line, leads of any size aren't safe. I don't believe the proportion of gray-to-black becoming much more "black" as the finish line is approached is a coincidence, though anyone who still insists on using the Spread or Proportional methods to measure poll accuracy *must* believe this. If these were valid metrics, then there would be no significant difference for any given spread or proportion regardless of closeness to the finish line – which is untrue.

It should be theoretically obvious that being very close to the finish line with a "small lead" is preferable to being farther from it with a larger lead in terms of win probability. Moreover, that the result "spread", or "margin," is more unpredictable in elections with a lot of undecideds compared to when there are very few. Indeed, these are hypotheses a reasonable person would form before analyzing political polls, if not for the established traditions in the industry.

And if being theoretically obvious weren't enough, these hypotheses are also supported by observation.

The first, natural question that arises when presented with the "spread vs. finish line" analysis is something like "how much more preferable is it to be closer to the finish line, versus farther from it with a larger lead?"

For a given election, all else equal, is it better to be "ahead" 45–40 in a poll average, or 49–46? Spread says 45–40, because +5 > +3. Proportion says "45–40" because it believes this poll says "leader should get 53% of two-party vote," whereas in 49–46, the leader should only get about 51.5%.

And observation says 49–46 has been far better. Which is correct?

Given a proper understanding of how polls work, these are reasonable questions to ask, and I believe they're important questions to answer – for anyone who cares to use poll data to inform a prediction. But Spread and Proportion analysts can't. The limitations imposed by their flawed definitions which say, "bigger spread = better chance to win" is a betrayal of what the data mean.

The idea that races "within 3 points in the polls are little better than toss-ups" stems from an incorrect understanding of what polls – and poll averages – tell us. So is a probit analysis that concludes there's an "inflection point" around a 3-point "lead" in which the leader has an 80%+ chance to win. Both are nothing more than a statistically unjustified obsession with spread. It didn't start at FiveThirtyEight, it has existed as long as polls themselves, and its survival seems to be predicated upon tradition and status quo, instead of improving methods.

By looking only at spread, analysts have done themselves a disservice when it comes time to inform their predictions. More importantly in my opinion, by telling the public *they need only look at spread*, the public believes that polls are predictions, and that if the candidate with a higher poll number (or poll average) loses, there's no other possibility except that the polls were wrong. Panagakis noted the error of this mindset decades ago, Roper decades before that, and nothing has changed.

Note that 50%, 49%, and 48% are not magic numbers. I will not characterize any poll average as an "inflection point" though it is plenty possible to find some. What has been true historically is not necessarily going to be true in the future, or for any given election. My goal is simply to demonstrate that, properly read, polls *can* tell you a great deal about a candidate's chances to win.

As it relates to the spread versus win probability calculation still employed by nearly every forecaster, the candidate's poll average and closeness to the finish line represent an enormous confounding variable.

When I said that a poll average of 48–46 compared to one of 46–44 was more different than they seemed – despite the Spread Method's protest that they are the same – I hope you would now agree.

> Rules of analysis are necessary, rules that are not as shortsighted as "8 points is comfortable" and "2 points is a close race."[5]
>
> *Nick Panagakis, The Pulse, 1987*

He's right. He wrote this in 1987, substantiated it, and the fact that he was right changed nothing. That's not how a proper scientific field should operate. So here I am, decades later, doing the same thing. That the public would cling to antiquated and far inferior forms of analysis is forgivable, that experts would do so is not. If experts do better, the public could too.

I won't be offering any rules or postulating inflection points with regards to Figure 30.5, but the fact that there's a lot more gray to the left of 48% than there is to the right of it is not a coincidence.

How to Report Polls

Whether you at this point perceive me as an expert or not, that's okay. Nothing I say or recommend should be taken as authoritative or unquestioned. But I hope my findings and presentation stand or fall on their merits, not dismissed because "it's always been done this way." Yes, it has always been done this way, and that's not a good thing. The fact that a lot of people who have worked in the field for a long time have always done it this way is not an excuse either – though it's the most common one I hear. Being wrong doesn't make you not an expert; being wrong and not trying to fix it does. I ask that my proposals be adopted, modified, criticized, and built upon as needed.

I have no expectation to what extent, if any, these findings, terms, analogies, or bad puns will be adopted or ever read – regardless of how much I believe they should be. I'll feel better having done what I can for statistical literacy in my limited expertise by writing a book than shouting on the internet (which I will also still do).

But I've written this book with the primary goal of educating the public, and providing some tools the public needs to be informed consumers of data. That goal would be diminished if I didn't recommend changes for how polls are reported by experts and the media.

1. Stop calling polls predictions, any such near-synonym, and characterizing them as such. This should be the easy one. Not predictions, not probabilities, not "calls," and not favorites. Stop it. Get it out of your vocabulary. They're just polls. Snapshots, observations, not predictions.

The best way to read and report polls and poll averages, by far, is the honest and complete way: all of the numbers, with the statistical margin of error listed next to each. No spread, no proportion, no lumping undecideds in with third-party voters.

Ricky Red: 46% +/– 4%

Grace Green: 44% +/– 4%

Eddie Euler: 5% +/– 4%

Undecided: 5% +/– 4%[6]

For print publications and social media, this standard should be easy enough. The real challenge is a standard for live media (TV, podcast, radio, streaming, etc.).

In his column "How undecideds decide," Panagakis lamented: "in on-air discussions of poll results reporters and anchors slipped into a point spread characterization. One number is easier than three."[7]

The idea that we need to be able to characterize a race with just one number is frustrating, but I do understand it. The fact that the number currently used is the spread, however, makes the problem much worse.

In addition to being a bad metric, "spread" brings with it a risk of overconfidence or apathy. If someone believes a candidate is "up by" a large amount, whether they intended to vote for or against that person, it's reasonable to see why some might be deterred from voting.

I believe there is a solution, for media and social media alike:

2. Adopt the **leader reporting standard**. When reporting poll data, if you don't report the complete results – report only the leader's number. This goes equally for on-air media and written headlines. *Do not use spread even secondarily.*[8] No "up by" or "ahead" characterization. The only numbers that should be reported secondarily are other poll numbers, the margin of error, or the number of undecided voters.

2a. If a poll reports its data with the Proportional Method and says that it observed zero undecideds even though it didn't, **blacklist them until they stop**; do not participate in promoting misinformation. I understand this proposal is probably controversial because of the reverence for convention and tradition, and because lots of reputable pollsters currently do this, but in the absence of tradition I believe this practice would be rightly considered scientific misconduct. If the goal is to end misinformation – and I think it should be – then this is not an unreasonable proposal.

If you report that Ricky Red has the support of approximately 46% of decided voters, that's both a true and accurate reflection of what the poll says (and what it means). If you wish to say more, that this poll reported Grace Green has the support of approximately 44% of decided voters, that's great too.

If someone wants to do the math themselves and conclude "Ricky Red is up 2," while that's not an accurate representation of what the poll reported, that's their own mistake. A consumer reaching that conclusion on their own, *not* because a perceived or actual expert erroneously believed "up 2" is the most important piece of information, makes a difference. Note that 46–44 is very different from 48–46; it's important the media and experts do not participate in misinforming consumers.

There's no one number that better or more accurately represents what a poll or poll average says, other than the leader's number. If you insist on reporting just one, use that one.

> "Ricky Red has the support of about 46% of likely voters."

For one, this is a valid way to report poll data: it relays what a poll actually says to the public. No projecting winners, no misinformation about "who is ahead" and "by how much," and no mischaracterizations about polls as predictions – just the observation. It also doesn't lie and say Ricky Red has about 51% of voters and none are undecided, as the Proportional Method would.

The use of "about" or "approximately" or any other appropriate modifier I think is simple enough to fit in reports and headlines, without giving the unjustified precision that current ones do, citing a specific number – usually spread.

This practice should be adopted for both individual polls and poll averages.

Perhaps equally as important as being more accurate, this standard prevents mischaracterizations about "lead" that can lead to overconfidence or apathy. Whether a candidate is far from the finish line or, reportedly, slightly across it, hearing someone has "about 51% of likely voters" should not inspire undue confidence or apathy in voters. If just 2% of voters don't turn out, if people change their minds, or there is in fact a poll or poll average error, that candidate can still lose. The same accurate representation of data cannot be said for reporting someone is "up by 8."

Threats to Inferential Statistics

As I've covered in some depth how the conflict between what pollsters observe and how analysts' judge their accuracy jeopardizes the field's legitimacy and trust in the public eye, I'll let those sections speak for themselves. How the media reports on polls, experts or otherwise, plays a role in this as well.

But I would like to briefly address some concerning trends that appeared to gain even more traction in 2022. Trends that should be addressed *before* the industry is embarrassed again, not after.

In the aftermath of the 1936 Literary Digest debacle, James Farley – the same James Farley who said in 1932 that the Literary Digest poll results favoring the candidate of his choice offered "conclusive evidence" – now claimed that he "always contended that The Literary Digest only covered those in the upper income brackets" and that people "who supported President Roosevelt didn't receive ballots in the same proportion."[9]

An individual accepting what a pollster reports can be largely related to whether it confirms their desires, and this has been around since the beginning of political polling. Confirmation bias – the tendency to accept new information that you agree with and reject information that you don't – is not unique to politics.[10]

For most of history, political polls have been conducted by a few pollsters with a singular, at least primary, incentive of being *right*. Now, with increasingly polarized media and consumers, there's a larger than ever financial incentive to tell people what they want to hear, and show them what they want to see. If most of the polls say a certain candidate is "behind" but one, bold pollster confidently asserts they're actually "ahead" (and with data!), guess which poll media outlets and social media influencers favoring that candidate will report?

And for the consumers who prefer to look at a poll average instead of individual polls, while more shielded, those aren't immune to the threat posed by biased pollsters either. With substantial weight given to whichever polls are most recent, FiveThirtyEight, RCP, and others offer a means for potentially unscrupulous pollsters to attempt to manipulate poll averages.

In 2022, there were a historically low number of polls conducted by independent and nonpartisan companies, and a historically high number conducted by partisan ones. This is another threat to the industry's accuracy and directly influences public perception.[11]

Flooding, and the "partisan poll flood" as it was called, had a tangible impact on poll averages.

Fortunately, though the effect on the poll average was probably small in most cases, it wasn't zero; flooding is one of the many factors that can cause **poll average error** that is attributable to only to individual polls and poll aggregators, not "the polls."

Finding poll average error is not (usually) a simple calculation in real-world cases. The distinction between poll error and poll average error might seem narrow, but the effects are important to distinguish. If you're having trouble understanding how this concept applies – and why it's important to hold poll aggregators accountable – test the extremes.

If I use *only* partisan polls funded by one candidate's party in my "poll average" calculation, and someone else excludes them entirely, whose poll average is better, and how do you know? The answer is *related* to poll accuracy, but it is not entirely attributable to poll accuracy: it's related to the methodology chosen by the poll aggregator.

Silver posted on social media in response to the partisan poll flood that:

> I think generally the complaint that GOP-leaning pollsters are "flooding the zone" is not sharp … it's a free market.[12]

In Silver's eyes, like Gallup's 70 years earlier, the "reputational risk" pollsters accept when releasing poll results is deterrence enough from any ulterior motives they may have to influence poll averages, public perception, or get attention.

He didn't stop there, however.

> D-leaning pollsters could release polls too if they wanted; that they don't says something.[13]

I'll reiterate: how poll aggregators determine the weight to give which polls is up to them. But if their poll averages are wrong as a result – who is to blame?

The dismissal of the possibility that partisan pollsters could be trying to influence the public or generate attention for themselves is bad enough – but he said the fact that Democrat-leaning pollsters weren't doing the same thing "says something."

The hypothesis that pollsters are primarily interested in being "right" (by whatever metric) is a perfectly reasonable one. But just as Silver previously observed "herding," I think the possibility that some pollsters would hedge in the direction that gets them more exposure, or "put their thumb on the scale" in a way that their preferred media outlets might prefer, is not unreasonable either.

The aggregators, forecasters, and analysts not accounting for known and prominent threats to the reliability of data, and the public's trust in it, are symptomatic of an industry in danger.

At one point, Politico reported that the majority of polls comprising FiveThirtyEight's average in each of Pennsylvania, Georgia, and New Hampshire were partisan polls.[14]

Incidentally, FiveThirtyEight's absolute forecast error (their projected result spread vs. the actual result spread) was 5.8 points, 2.1 points, and 5.7 points in each of those states, respectively.

How much of that forecast error is attributable to poll error or poll average error is debatable, but their forecast error isn't just the fault of "the polls" as is traditionally asserted, and as they insist.

And to be clear about reported poll results, partisan or otherwise, I'm not saying the weights pollsters use are entirely subjective. But I am saying it's nearly impossible to distinguish the subtle differences between partisan bias and different weighting methods just by looking at the numbers, which may be as little as one or two percentage points in either direction.

In an interesting case study that demonstrates this point effectively, Nate Cohn of *The New York Times* gave "Four Good Pollsters" the same raw data and asked them to weight it; they produced four different results.[15]

While the magnitude of those differences is dramatically overstated by the "Spread Method," there are undeniable differences. The uncomfortable fact this industry must contend with is that there is no gold standard or perfect weighting method, and that is why transparency is so important *for both pollsters and poll aggregators.*

If a pollster or poll aggregator consistently and conveniently adjusts their techniques in a way that favors one candidate or party, it's fair to conclude they're not acting with accuracy in mind.

> Which is to say, the world of political polling may be a free market, but it might not be *accuracy* driving demand and paying the bills.

As for poll aggregators, while I would rate FiveThirtyEight's poll average accuracy above those of RealClearPolitics if I had to choose, I used RCP's for my examples in this chapter for the simple reason that they have a longer history and more consistent and transparent methodology. Unfortunately, though still transparent, their possible bias became increasingly evident in 2022. RCP has always used a simple recent average methodology, which is good for both simplicity and transparency. In an election with a higher number of partisan (Republican) pollsters, a Republican bias could be expected.

But their simple recent average methodology was not applied consistently across elections, or even in the same state.

Pennsylvania had two major statewide races: Senate and Governor. Naturally, pollsters took polls on both races, and these polls helped form RCP's average. For the Governor's race, they took the *three* most recent polls to form that average,[16] but for the Senate race, *five*.[17]

The same polls in the same state, conducted at the same time – but some were excluded from their average.

I cannot know for certain whether this was purposeful or an oversight, and in this case doesn't matter; the point is that pollsters know that their polls can influence poll averages. Poll aggregators know their poll averages can influence the public and media cycles.[18]

In the case of just three or five most recent polls forming a poll average, flooding is an easy game to play. If poll aggregators aren't held accountable, this problem will get worse.

While the Proportional Method thankfully hasn't made its way to American pollsters, at least one major poll aggregator, The Economist, started reporting polls this way in 2020.

In their coverage, they showed the results of hundreds of polls not as they were reported, but with the top two candidates adding to 100%.[19] The pollsters didn't report it that way, there were plenty of undecideds and some third parties, but The Economist took it upon themselves to change what the poll said, with no clarification or differentiation between what the pollster said and what their analysis assumed. One pollster, Quinnipiac, reported Biden's nationwide support at 50% just days before the election[20]; The Economist reported that Quinnipiac said Biden's support was 56%.

This was the case for every poll: The Economist applied their own "Proportional" assumptions and reported that the poll result said something different than what the poll said. If I were a pollster, I'd have a big problem with my data being lied about.

The Economist used this "Proportional" data to inform their "modeled popular vote," that's fine. They can use whatever forecasting method they want. But the results they attributed to each pollster are not correct, and the poll average they derived from it was not taken from what the pollsters reported. The possibility of a poll average error in this instance is even more highly disconnected from poll error than usual cases. In the absence of standards, aggregators can apparently report whatever they want. I suspect they wouldn't be so forgiving if people similarly lied about what their poll averages or forecast said.

A Closing Thought (for This Chapter)

When I started researching it, I expected the winning percentage for candidates polling above 49% to be high because it should be. I did not expect it to be nearly as high as it was, and I suspect no one would. The same can be said for 48%.

Having provided empirical evidence that candidates who have a strong poll average tend to win, and the underlying theory as to why, considering the fact this statement should be obvious, I suspect many analysts will shadow James Farley and invent a world in which they've "always contended" this is the case – despite their reverence of spread proving otherwise.

I'm not particularly concerned with that because it makes no difference to me how they arrive at better methods. Long before publishing, or even considering writing a book, I openly shared my findings with the hope (if not expectation) that the industry's more analytical researchers would incorporate them. Because I had less experience than those researchers, I figured they'd be far better equipped to explain the findings. That was obviously very naive of me.

Starting with this chapter, I effectively wrote this book backwards – the upcoming chapters on UK elections came later. I had the data long before I could explain why it showed what it did. Naively, I believed that given the content of my findings, experts in the field would consider the possibility that spread isn't a good metric for either poll accuracy or election win probability. And because I'm not that smart or experienced, I figured they would be better equipped to do so.

That's when I learned more about what experts considered the "ground truth" for measuring poll accuracy and how they interpreted margins of error, the assumptions they cling to even in the face of evidence that those assumptions don't hold true – and still more flaws of consequence, like the Proportional Method. I found Mr. Panagakis' work only shortly after he passed away, deeply regretting that I had spent the final years of his life contacting other experts in the field, having not yet come across his.

I understand, like Panagakis did, that polls which may seem wrong on the surface can be "in fact, right."

But what does that mean? How did he know that, and how did I? Knowing (or thinking you know) something means very little if you can't explain it.

And assuming other people in the field understand something as basic as "polls aren't predictions" was my biggest blunder. So, I set out to try and explain things.

That internal, "how do you know?" is what led me to terms like "Simultaneous Census," and categorizing the differences between "plan polls" and "present polls," and trying to quantify in my limited ability some concepts like "ideal polls," "masking," and "compensating error."

Having since used this book's concepts and analogies to successfully explain to non-experts the data provided by polls, another internal voice asks, "why does this matter?"

If all I can accomplish is offering some new ways to think about or teach poll data, that's great. But it's clear by their words, analysis, and methods that experts don't understand the concept I've termed the Simultaneous Census, the impact (or existence) of confounding variables in their current calculations, nor do many of them understand how a poll's margin of error works – ideal or otherwise.

What I'm more concerned with, which unfortunately holds a strong correlation to how much impact my work has, is the possibility that some pollsters or poll aggregators report a higher number than they observe, even for the same "spread," in an effort to get their favored candidate closer to 49%. To a pollster, reporting a poll as 49–47 or 48–46 might be a matter of squinting, rounding, or putting their thumb in the right spot. But as it relates to the finish line in an election, and using that data to inform a prediction of the outcome, it makes a huge difference. Especially if your poll is released closest to the election, and one of only a few used to form the poll average.

The solution, if it can be called that, is to adopt a new, valid set of standards by which pollster *and poll aggregator* accuracy is measured, promote competitive collaboration that allows for innovation and demands transparency, encourage more independent pollsters to be active, and perhaps most of all – something everyone everywhere can do – remember polls are not predictions.

Notes

1. RealClearPolitics. (2022). *2022 Arizona governor: Lake vs. Hobbs*. https://www. realclearpolitics.com/epolls/2022/governor/az/arizona_governor_lake_ vs_hobbs-7842.html?eType=EmailBlastContent&eId=dddd0202-e0d5-49f5- a199-a800b4c1f6fb
2. Rakich, N. (2023, March 10). *The polls were historically accurate in 2022*. FiveThirtyEight. https://fivethirtyeight.com/features/2022-election-polling- accuracy/
3. Silver, N. (2012, October 27). *Oct. 26: State poll averages usually call election right*. FiveThirtyEight. https://fivethirtyeight.com/features/oct-26-state-poll-averages- usually-call-election-right/
4. Silver, N. (2012, October 27). *Oct. 26: State poll averages usually call election right*. FiveThirtyEight. https://fivethirtyeight.com/features/oct-26-state-poll-averages- usually-call-election-right/
5. Panagakis, N. (1987). *Making sense out of poll stories*. The Pulse. https://www.lib. niu.edu/1987/ii870874.html
6. Yes, I'm still using the same MOE for numbers very far from 50% as those close to it, for the same reasons I explained in previous chapter notes. If that imprecision bothers you, then you can call it an error. It is an error. Can I call it a poll error? If not, and you contend that this error is mine (and it is), please hold the field's experts to the same standard.
7. Panagakis, N. (1987). *How undecideds decide*. The Pulse. https://www.lib.niu. edu/1987/ii871136.html

8. The only reason I used "spread" on my charts, the only reason it mattered, was because it offered a consistent way to calculate, given the leader's poll average, the "second place" candidate's poll average from a single point on a chart. When you're comparing dozens or hundreds of elections, this is just a way to plot both candidates' averages.

   For a poll whose topline shows two candidates at 49% and 46% respectively, the only use "+3" serves is that it allows you to calculate both numbers from "49%, +3." When reporting individual polls or poll averages, it's just as easy to report "49% to 46%" as it is "49%, +3." So, given the misconceptions around what "spread" means, it should be deliberately eliminated from discourse and analysis.

9. Campbell, W. J. (2020). *Lost in a Gallup: Polling failure in U.S. presidential elections.* University of California Press.

10. Ethics Unwrapped. (2023). *Confirmation bias.* McCombs School of Business—The University of Texas at Austin. https://ethicsunwrapped.utexas.edu/glossary/confirmation-bias

11. Shepard, S. (2022). *The Biden gap and the partisan poll flood: Breaking down the latest Senate surveys.* POLITICO. https://www.politico.com/news/2022/11/01/biden-gap-senate-surveys-00064362

12. Silver, N. [@NateSilver538]. (2022, November 5). *I think generally the complaint that GOP-leaning pollsters are 'Flooding the zone' is not sharp* [Tweet]. Twitter. https://twitter.com/NateSilver538/status/1588916015869878272?t=b4udis6G20gQjXAoTcDt1A&s=19

13. Silver, N. [@NateSilver538]. (2022, November 5). *I think generally the complaint that GOP-leaning pollsters are 'Flooding the zone' is not sharp* [Tweet]. Twitter. https://twitter.com/NateSilver538/status/1588916015869878272?t=b4udis6G20gQjXAoTcDt1A&s=19

14. Shepard, S. (2022). *The Biden gap and the partisan poll flood: Breaking down the latest Senate surveys.* POLITICO. https://www.politico.com/news/2022/11/01/biden-gap-senate-surveys-00064362

15. Cohn, N. (2016, September 20). We gave four good pollsters the same raw data. They had four different results. *The New York Times.* https://www.nytimes.com/interactive/2016/09/20/upshot/the-error-the-polling-world-rarely-talks-about.html

16. RealClearPolitics. (2022). *2022 Pennsylvania Governor: Mastriano vs. Shapiro.* https://www.realclearpolitics.com/epolls/2022/governor/pa/pennsylvania_governor_mastriano_vs_shapiro-7696.html#polls

17. RealClearPolitics. (2022). *2022 Pennsylvania: Oz vs. Fetterman.* https://www.realclearpolitics.com/epolls/2022/senate/pa/pennsylvania_senate_oz_vs_fetterman-7695.html#polls

18. Snyder, B., Rothschild, D., & Malhotra, N. (2012, October 30). *How polls influence behavior.* Stanford Graduate School of Business. https://www.gsb.stanford.edu/insights/how-polls-influence-behavior

19. The Economist Newspaper. (2020, November 8). *President-forecasting the US 2020 elections.* The Economist. https://projects.economist.com/us-2020-forecast/president

20. Quinnipiac University. (2020, November 2). *Florida and Ohio: Biden has the edge over Trump, Quinnipiac University poll finds; Nationally, Biden maintains a strong lead.* Quinnipiac University Poll. https://poll.qu.edu/images/polling/fl/fl11022020_bgbw76.pdf

31

UK Elections and Brexit

Important Points Checklist

- Why non-US polls often report 0% undecided, even if they observe undecideds
- Why 2015 and 2017 UK general elections were so different
- The options for voters in Brexit
- How the Proportional Method treats "undecided" versus people who "will not vote"

My first exposure to UK Politics was when Ali G ran for MP of Staines. Ali G is a fictional character played by Sacha Baron Cohen. Hilarious as it was to teenaged me, and might help if I'm ever in a tough debate and need a good comeback, it wasn't an ideal starting point.

Plus, elections in the United Kingdom are far more complex than the United States' (primarily) two-party elections because at least 10 parties currently hold seats in Parliament.[1]

However, since the winner of each constituency is determined by who gets the most votes, whether two parties or 10, the applications offered to this point can be thought of as anywhere from Mintucky, to Barack Obama 2008, to Mitch McConnell 2014, to Bill Clinton 1992, to Utah 2016. Accounting for all the underlying variables isn't always easy, but it's a lot easier once you admit they exist.

So away I went, "In for a penny, in for a pound."

While there are two major parties in the United Kingdom (Labour and Conservative), many constituencies have at least three competitive parties, and Labour and Conservative are not always the top 2. Nor is it always clear who the top two are. If you thought Utah was fun, there are dozens of constituencies with party parity like that in the United Kingdom.

And if you thought the poll error calculations were bad for Utah, you haven't seen anything yet.

Since most general elections contain three or more competitive parties, for most UK constituencies figuring out where the finish line is a good first step. In many cases, it will be well short of 50%.

 DOI: 10.1201/9781003389903-31

In 2019, in 65 of the 650 elections, the winner won with less than 45% of the vote, and in 23 of them 40% or less.[2] Interesting, but no problem as long as you don't try to characterize things by spread or proportion.

I went back to look at polls leading up to past elections, putting them in a spreadsheet, advanced stuff. I started with the 2017 general election. It took me way too long to notice: why are all of these polls adding up to 100%?

As I alluded to previously – they're using the Proportional Method. The results pollsters report are after they allocate "undecideds" and "refused to answer."

When I say "allocate," I'm being generous. A more accurate characterization is to say they are lying about what their poll says.

Each pollster has their own methodology for allocating undecideds. Some use a direct proportional method as described in earlier chapters, others use data from past elections to allocate undecideds how they believe they'll vote. But they're all doing what should be considered a threat to the industry's credibility: building a forecast into their observations and reporting their forecast as if it were the poll.

It doesn't matter how defensible or beautiful your assumptions are, those are still your assumptions, not the poll's.

Once you try to predict how undecideds will eventually decide *anywhere*, you are introducing the potential for assumption/forecast errors. That risk is greater when:

1. There are more undecideds
2. There are more candidates

In 2017, multiple polls taken in the days leading up to the general election showed a double-digit percentage of undecided nationally. But you can only find those numbers if you read the full reports: they're not the numbers reported to the media.

ICM Unlimited had 11% "don't know" in their poll taken in the two days just before the election, and 2% "refused."[3]

Among their "Likely to Vote" base, Survation's poll also had 11% "undecided" in the same timeframe, with 6% "refused."[4]

Ipsos MORI, 6% "undecided" and 6% "refused."[5]

YouGov surveyed 9% who responded "don't know."[6]

Those are not the results they reported though. They reported 0% undecided. How is this an accepted practice?

They didn't just weight for turnout or who the likely voters were. They went well beyond that, and *tried to predict how undecideds would eventually decide*. That prediction is largely or directly proportional to the already-decided.

If analysts and pollsters who utilize these methods dislike undecided voters so much that they'd rather pretend they don't exist, then they should find a different profession. The most generous explanation for why they do this is that they want their poll to be a forecast, or they believe that since this is what the public wants, they must try to give it to them. It's a bad practice that starts with misunderstanding the value of poll data, and a practice accepted by the field's analysts as a measure of poll – not forecast – accuracy.[7]

An estimate of the percentage of undecided voters is one of the most valuable pieces of data given by a poll. Throwing them out or assuming what they will do to make your data appear more meaningful than it is a textbook example of "lying with statistics."

As discussed in earlier chapters, if someone wants to assume undecideds will split a certain way, or even that they all go to one candidate, that's fine! If they have very good data to support it, even better. But those are assumptions of a *forecast*, not a *poll*.

By blending their polls and forecasts, many pollsters in the United Kingdom (and pollsters in the many countries who ascribe to variations of this method) are introducing *forecast error* to their *polls*.

Every pollster who reports this way has introduced a source of error to their polls that has nothing to do with the accuracy of the poll, which goes unaccounted for when it comes time to analyze how accurate "the polls" were.

Bad pollsters with a good forecast can be rated as good, and good pollsters with a bad forecast are rated as bad. The Proportional Method makes no distinction between poll and forecast.

Like US elections, forecast errors being projected onto polls are magnified when confounding variables confound. In this case, the primary confounding variable is high undecideds which many pollsters pretend don't exist.

There is a known impact that high undecideds can have on an election: it increases the range of possible outcomes. The possibility that one party far outperforms their poll number is higher when there are more undecideds.

A party outperforming its poll number, or poll average, is not instantly indicative of a poll error.

The Proportional Method, by ignoring undecideds, will drastically underestimate the amount by which one party can overperform its poll number or poll average – especially in elections with a lot of undecideds.

With three or more parties likely to have significant support in each constituency, the calculations become increasingly complex as it relates to approximating the finish line, as you saw in Utah. Moreover, as the election approaches, some strategic voters may change their minds from what they truthfully told pollsters and vote for the competitive party they more closely align with, to maximize their vote's impact.

These are all things that can be tested and accounted for, but only if researchers understand the fact that Changing Attitudes is not a poll error.

The post-Brexit 2017 election in which Conservatives (seemingly) underperformed their expectations came as a shock to many, when in reality, like Trump-Clinton 2016, it's unlikely the polls had major errors. There were two factors that made most of the UK general election polls in 2017 *seem* poor:

1. The 2017 election had the highest percentage of two-party vote (Conservative and Labour) in decades, over 82%.[8] Just two years before, that number was around 69%. That huge change proved difficult to account for in methodologies that often assumed (or forecasted) that undecideds would vote similarly to how they did in 2015.

   The assumption that people who are undecided are most likely to vote how they did in the previous election is a perfectly reasonable one; it is arguably *the most reasonable* one. Doesn't matter. If you make that assumption and that assumption proves to be wrong, that's an error – but it isn't a poll error.

2. The same "undecideds will probably split evenly and changing attitudes are negligible" assumptions that perplexed US analysts when they didn't hold true in 2016 did the same to UK analysts in 2017. Researchers from the University of Manchester reported that in the final two months of polling, third-party "mind-changers" had split evenly in 2015 (about 25% each to the two major parties), but in 2017 a massive 54% of mind-changers went for Labour, compared to only 19% Conservatives.[9] Combined with the fact that Labour won "more than half" of undecided voters (in elections where there were three or more competitive parties) suddenly, an alleged Conservative underperformance quickly points to very little poll error and a huge forecast error.

There are a lot of moving parts in polls and elections done at the constituency level with three or more competitive parties, and there's plenty more research that needs to be done there. To borrow from a quote in the introduction, my findings are not conclusive, but they all point in the same direction.

I will go into more depth on how those general election polls should have been reported, understood, and analyzed, but it requires a few steps.

Fortunately for educational purposes, though unfortunately for how it was wrongly reported on and viewed, there is a much simpler and very real example to introduce the inevitable failure of the Proportional Method.

And it doesn't require mints or pebbles!

The Brexit Poll Analysis Catastrophe

The sentiment among experts and the public is that the polls around Brexit were bad. I don't think that's the case because that's not what the data shows.

Brexit poll reporting was a worse catastrophe than The Literary Digest one and it's not even close; at least The Literary Digest didn't pretend to be scientific. This catastrophe should have broken the Proportional Method and poll analysis that accompanies it in the same way it broke The Literary Digest; it shouldn't have been taken seriously in the first place, but now that it *really* messed up, it's time for some changes.

That did not happen, and still has not happened.

As usual, "the polls" were blamed. Not assumptions of forecasters, not predictors, not flawed methods, "the polls."

CNBC wondered how a majority of polls could "predict a wrong outcome."[10]

The Guardian reported that it was a "bad night" for pollsters because "fewer than a third" of polls predicted a "leave" vote.[11]

BBC asked, "Why were the polls wrong again in 2016?"[12] invoking both Brexit and the Trump election.

This leads me to why I had to save Brexit and the United Kingdom for my final "poll analysis" chapters. Brexit combines the worst of all types of errors and mischaracterizations, a Frankenstein of statistical invalidity that would go unappreciated by those who hadn't read up to this point. The cascade of erroneous assumptions led to a consensus belief that "the polls" were wrong when if **properly read, they were extremely accurate**.

Finding out where the finish line is in a typical UK election, though important, is a hard forecasting job, and one of approximation. Even understanding what polls tell us, and having very good data to inform a prediction, reasonable people could disagree on a finish line forecast for any competitive constituency in the United Kingdom. No such approximations are necessary for Brexit.

The "EU referendum," which became better known as Brexit, was a referendum in which eligible voters could vote on Britain's continued membership in the European Union.[13]

Brexit was a rare non-US election where there were only two options: Remain or Leave. No third-party, no write-in candidates, Remain or Leave.

With just two options, 50% + 1 wins, anything less than that does not. No exceptions, no other possible outcomes.

As you know, however, when taking a plan poll, it's not that simple. Despite the *election* only having two options, the options *before this election* are not binary. The options are Remain, Leave, or Undecided.

The trouble with that: when pollsters lie about their observations and report no undecideds – and poll aggregators choose to report their forecasts as averages – it becomes really hard to understand what the polls actually say.

Even well-informed and numerate people can be deceived when they're lied to.

There were some actual *forecasters* who made clear that they were *forecasting* based on different data and tried to predict the eventual percentage (and win probability) for each option. This is not about them. This is about poll aggregators and pollsters.

What UK thinks: EU describes itself as "Britain's leading independent social research agency" and "the place to come to for all of the key poll and survey data on what the UK public thinks about Britain's future relationship with the EU."[14]

They reported a "poll of polls" that said:

"Remain" on 52%, and "Leave" on 48%.[15]

They did not report any undecideds.

The Telegraph similarly on their "poll tracker" reported no undecideds:

"Remain" at 51%, and "Leave" at 49%.[16]

Different methods, different numbers – that's not a problem. Aggregators can aggregate however they want. Most recent five polls, 10 polls, weighting by data collection method, weighting by sample size – all fine. But *aggregators* do not get to *forecast* and simultaneously claim "this is what the polls say." At least, they shouldn't without some major reprimanding.

The only way to get from some undecided to no undecided before an election is to forecast.

Poll aggregators aren't the only ones at fault here, however.

There are also – by the Proportional Method's insistence on lying – *pollsters* reporting forecasts.

Many pollsters said, as you'll see: "this is what our poll says" but reported a forecast not allowing for undecideds.

Many poll aggregators said: "this is what "the polls" say" but reported a forecast not allowing for undecideds.

Analysts and experts not understanding the differences between polls, poll averages, and forecasts have consequences for how the public is informed.

To sum up why I referred to Brexit as a "Frankenstein of statistical invalidity": some pollsters were forecasting; some poll aggregators were averaging those forecasts; and reasonable people, experts included, took those reports to reach their conclusions about what "the polls" were saying.

To further complicate things, *some* pollsters *truthfully* reported their poll results as they found them: Remain, Leave, *and* Undecided. Others didn't.

Financial Times, on their "Brexit poll tracker" page, was the first poll aggregator I found that reported any undecideds: 6%.[17]

That's a pretty reasonable number of undecideds to have just before an election. At first glance, I figured this was likely much closer to a reasonable poll average.

Their "poll of polls" page reported:

Remain 48% and Leave 46%.[18]

No problem. But then I scrolled down to view the polls they used to form that average. It said this:

Remain (%)	Leave (%)	Undecided (%)	Date	Pollster
55	45	0	Jun 22, 2016	Populus
48	42	11	Jun 22, 2016	ComRes
41	43	11	Jun 22, 2016	TNS
44	45	11	Jun 22, 2016	Opinium
51	49	0	Jun 22, 2016	YouGov
52	48	0	Jun 22, 2016	Ipsos MORI
45	44	11	Jun 20, 2016	Survation
42	44	13	Jun 19, 2016	YouGov
53	46	2	Jun 19, 2016	ORB
45	42	13	Jun 18, 2016	Survation

Keeping with the "ten most recent polls" theme.

What on Earth.

Financial Times described their poll average methodology as follows: "We take the seven most recent polls from seven different pollsters, remove outliers by dropping the poll with the highest share for Remain and the one with the lowest, and adjust for how recent the data were."[19]

This is a great methodology for taking a poll average: it is, in effect, a simple recent average that limits the influence of outliers.

However, *Financial Times* introduced a major source of poll average error that I hadn't before considered: taking an average of polls that are reported using different methods.

Some of the polls in their average included undecideds, and some didn't. Even excluding the 55% Remain poll (Populus) and the 41% Remain poll (TNS), you're left with a poll average that consists of three pollsters (ComRes, Opinium, and Survation) who reported undecideds, and two (YouGov and Ipsos MORI) who did not report undecideds.

This gives substantially more weight to the pollsters who used the Proportional Method compared to those who reported their data honestly.

This "poll average" was effectively taking an average of three forecasts and two polls because the polls that *reported* 0% undecided didn't actually *observe* 0% undecided.

As for the poll aggregators, all of them reported "Remain" as ahead: one reported 52%–48%, another reported 51%–49%, and a third reported 48%–46% with 6% undecided. Whose average was best? And best compared to what?

When reported this way, it's easy to see why so many people wrongly believed one side was heavily favored.

In fact, it wasn't just the general public and media perceiving one side as heavily favored. Betting on election outcomes is legal in the United Kingdom, and leading up to the election oddsmakers implied the chances of "Leave" around just 25%.[20]

Bettors – "punters" if you'd prefer – risked upwards of £20 million in total on the outcome of the EU referendum, including a single wager of £100,000 on Remain.[21]

With "Remain" reported as high as 55 and as low as 41 – with polls taken on the same day no less – other analysts perceived what should have been properly read as normal fluctuation as a "sudden shift in the polls."[22]

It just so happens that the perception of this "sudden shift" required comparing a poll which reported no undecideds (and "allocated" them) to a poll that reported double-digit undecideds.

To bring things all together in this Frankenstein of invalidity – yes, pollsters and the media reported on these polls that used the Proportional Method on the basis of their spread.

A few weeks before the vote, *The Independent* reported a "10-point swing towards Brexit."[23]

Regarding the polls the day before the vote, *Huffington Post* reported, "YouGov had the status quo up by 4 points, and Ipsos-MORI had it up by 6 points."[24]

YouGov reported in their "Eve of Poll" survey on the referendum, that 51% of respondents said "Remain" and 49% said "Leave" as listed in the previous table. The headline reported "Remain leads by two."[25]

In the methodology section of their report, YouGov said this number is derived from "turnout weighted and including squeeze question."

Weighting for turnout, great; figuring out who's likely to vote is part of conducting a poll.

As for the "squeeze question" – this refers to the methodology choice in which pollsters ask "undecideds" a follow-up question regarding which way they "lean." Those "leans" are typically reported as part of the "decideds." This is the gray area where pollsters can choose *if* to ask a squeeze question, and also how to report them. Still, a perfectly reasonable approach.

But this is where things get even more frustrating.

YouGov *reported* of their poll that, "we asked people who said they didn't know how they would vote which way they were leaning and re-allocated them on that basis."

On its face, this is fine: they chose to ask the squeeze question and put "decideds" and "leans" together. But after the lean question, were they left with 0% undecided? No.

Their unweighted report was Remain at 45% and Leave at 45%. Note that 2% said they "would not vote" and 8% said "don't know."[26]

Of that 8% "don't know," which YouGov said they re-allocated based on lean, *fewer than half expressed a lean.*

What happened to the rest of the undecideds? They were ignored.

Twenty-seven percent of the undecideds, after the squeeze question, said "will probably end up voting to remain" and 13% said "will probably end up voting to leave." Seven percent said "will probably end up not voting" which left them with, as it states in their report but was not reported as such: *53% of "don't knows" also did not express a lean.*

Assigning these "leans" from the "undecided" subset can be done as follows: 27% "lean remain" given 8% undecided is about 2.2%. That adds about 2.2% to "Remain" (now ~47.2% total).

Thirteen percent "lean leave" given 8% undecided is about 1%. That adds about 1% to "Leave" (now ~46% total).

Seven percent "lean not vote" given 8% undecided is about 0.6%. That adds about 0.6% to "will not vote" (now ~2.6% total).

That leaves 53% "don't know" of 8% originally undecided – who express no lean after the squeeze question – or about 4.2%. A total of 4.2% were undecided.

Accounting for the fact that 2.6% of people polled said they will not/probably won't vote (thus throwing them out of a Likely Voter report) that gives a new denominator of 97.4 (100% − 2.6% = 97.4%).[27]

While weighting for turnout is a sensible and valid thing to weight for – and one thing a poll does try to tell us – the Proportional Method does not stop there.

The Proportional Method *treats "undecided" voters who are likely to vote the same as people who state they "will not vote."* It removes *both* groups from the denominator.

The denominator reduction for this calculation appears to be 2.6% (would not vote) *plus* 4.2% undecided: 2.6 + 4.2 = 6.8.

The new denominator used for this report looks to be 100–6.8, or 93.2.

The new calculations, following these calculations, match what was reported:

- Remain: 47.2/93.2 = 50.6% (reported as 51%)
- Leave: 46/93.2 = 49.4% (reported as 49%)

Does counting "undecideds" and "will not vote" as the same make a difference?

If they had simply reported what they observed – and not ignored the undecideds – this poll would not have reported "51–49."

It would have reported something like:

- Remain: 48.5%
- Leave: 47.2%
- Don't Know: 4.2%

And given the known finish line of 50% + 1, this looks very different.

With "leave" winning the election, this creates another very important question that remains, to my knowledge, unanswered: did late deciders/don't knows eventually break disproportionately toward leave?

This creates another very important question that remains, to my knowledge, unanswered: did late deciders/don't knows eventually break disproportionately toward leave?

If so, this answers a huge portion of how "Remain" polled mostly "ahead" near the election, but "Leave" eventually won. If not, this raises the question of a legitimate poll error, even if it may have been small. Serious researchers should want to know.

But due to their flawed definitions, they can't know. By the accepted definitions, the polls were off by a lot because Proportional Method overestimated Remain, or because it didn't accurately predict the final spread. No other possibility.

"How Brexit Polls Missed The 'Leave' Victory," read a *Huffington Post* headline.[28]

They reported that "internet polls seem to have performed better than telephone polls" because, of course – internet poll spreads were closer to the result spread.

Ironically, and unfortunately, *Huffington Post* was one of two companies (the other being *The Economist*, not yet reporting via the Proportional Method) in a prime position to "scoop" all the other media outlets, and even the betting markets – and report long before the election, up to the day of it, that neither side was clearly favored.

While most poll aggregators used polls that included the Proportional Method, causing them to dramatically underestimate the possible range of outcomes because of undecideds, both *Huffington Post* and *The Economist* had their own poll averages. Those poll averages told a very different story.

The *Huffington Post* poll average said

Remain: 45.8% and Leave: 45.3%, with 9% undecided.[29]

The Economist's poll average said

Remain: 44% and Leave: 44%, with 9% undecided.[30]

Reasonable people could debate the accuracy of the very different poll averages, some which reported 9% undecided and some with 0%, and such a debate would contribute to the progress of the industry, but that analysis is always limited to "the polls."

Poll aggregators can and should be allowed to *average* polls however they please – up to and including obviously invalid ways. Similarly, a pollster should be able to *take* (and report) polls however they please. But only once "error" is assigned properly, and standards are set, can progress be made.

The accuracy of a poll, and a poll average, should be judged not on spread or proportion, but on the Simultaneous Census standard: how close it is to each possible result, including third-party and undecided, at the time the poll was taken. If your poll or poll average reports 0% undecided, and there are more than 0% undecided, you have a huge error. And it's not a poll error.

How "error" is defined plays a direct role in the legitimacy and credibility of this industry; pollsters can make errors, poll aggregators can make errors, and forecasters can also make errors. Until these terms are better defined, and the difference between poll, poll average, and forecaster is understood, it can never be fixed.

Let me explain why and how I can say with such confidence that polls weren't wrong about Brexit.

Notes

1. UK Parliament. (2023). *State of the parties*. UK Parliament. https://members.parliament.uk/parties/Commons
2. House of Commons Library. (2023). *General election 2019: The results*. General Election 2019: https://commonslibrary.parliament.uk/general-election-2019-the-results-so-far/
3. The Guardian. (2017). *Campaign poll 10 prediction poll*. ICM Unlimited. https://www.icmunlimited.com/wp-content/uploads/2017/06/2017_guardian_prediction_PRELIM_1500.pdf
4. Survation. (2017). *GE 2017 telephone voting intention final poll*. Survation. https://survation.com/wp-content/uploads/2017/06/Survation-GE2017-Final-Poll-2d71918.pdf
5. Ipsos MORI. (2017). *Ipsos Mori Political Monitor—2017 election final poll tables*. Ipsos. https://www.ipsos.com/sites/default/files/2017-06/pm-election-2017-final-tables.pdf
6. YouGov. (2017). *YouGov/The Times Survey Results*. YouGov/The Times. https://d25d2506sfb94s.cloudfront.net/cumulus_uploads/document/xalfiwu0ed/TimesResults_170118_VI_Trackers_MaySpeech_W.pdf
7. Sturgis, P., Kuha, J., Baker, N., Callegaro, M., Fisher, S., Green, J., Jennings, W., Lauderdale, B. E., & Smith, P. (2018). An assessment of the causes of the errors in the 2015 UK General Election Opinion Polls. *Journal of the Royal Statistical Society Series A: Statistics in Society, 181*(3), 757–781. https://doi.org/10.1111/rssa.12329
8. Uberoi, E., Loft, P., & Watson, C. (2020). *General election results from 1918 to 2019*. House of Commons Library. https://commonslibrary.parliament.uk/research-briefings/cbp-8647/
9. Fieldhouse, E., & Prosser, C. (2017, August 1). General election 2017: Brexit dominated voters' thoughts. *BBC News*. https://www.bbc.com/news/uk-politics-40630242
10. Saiidi, U. (2016, July 4). *Here's why the majority of Brexit polls were wrong*. CNBC. https://www.cnbc.com/2016/07/04/why-the-majority-of-brexit-polls-were-wrong.html
11. Duncan, P. (2016, June 24). How the pollsters got it wrong on the EU referendum. *The Guardian*. https://www.theguardian.com/politics/2016/jun/24/how-eu-referendum-pollsters-wrong-opinion-predict-close
12. Cowling, D. (2016, December 24). Why were the polls wrong again in 2016? *BBC News*. https://www.bbc.com/news/uk-politics-38402133

13. National Library of Scotland. (2023). *EU referendum*. https://www.nls.uk/collections/topics/eu-referendum/

14. What UK Thinks: EU. (2023). *About the site*. https://www.whatukthinks.org/eu/about-the-site/

15. What UK Thinks: EU. (2016). *EU referendum poll of Polls*. https://www.whatukthinks.org/eu/opinion-polls/poll-of-polls/

16. Kirk, A., Coles, M., & Krol, C. (2018, January 10). *EU referendum results and maps: Full breakdown and find out how your area voted*. The Telegraph. https://www.telegraph.co.uk/politics/0/leave-or-remain-eu-referendum-results-and-live-maps/

17. Financial Times. (2016). *Brexit poll tracker*. EU referendum poll of polls. https://ig.ft.com/sites/brexit-polling/

18. Financial Times. (2016). *Brexit poll tracker*. EU referendum poll of polls. https://ig.ft.com/sites/brexit-polling/

19. Burn-Murdoch, J. (2016, June 9). *How accurate are the Brexit polls?* Financial Times. https://www.ft.com/content/6a63c2ca-2d80-11e6-bf8d-26294ad519fc

20. Smith, M. N. (2016). *Brexit betting: The biggest single bet on remain is 10 times larger than the biggest bet on a Brexit*. Business Insider. https://www.businessinsider.com/brexit-betting-william-hill-says-the-biggest-single-bet-on-remain-was-in-the-five-figures-2016-6

21. Smith, M. N. (2016). *Brexit betting: The biggest single bet on remain is 10 times larger than the biggest bet on a Brexit*. Business Insider. https://www.businessinsider.com/brexit-betting-william-hill-says-the-biggest-single-bet-on-remain-was-in-the-five-figures-2016-6

22. Casselman, B. (2016, June 17). *What a "Brexit" could mean for the economy*. FiveThirtyEight. https://fivethirtyeight.com/features/what-a-brexit-could-mean-for-the-economy/

23. Grice, A. (2016, June 14). Leave opens up massive lead in exclusive Brexit poll for the Independent. *The Independent*. https://www.independent.co.uk/news/uk/politics/eu-referendum-poll-brexit-leave-campaign-10point-lead-remain-boris-johnson-nigel-farage-david-cameron-a7075131.html

24. Jackson, N. (2016, June 24). *How Brexit polls missed the "leave" victory*. HuffPost. https://www.huffpost.com/entry/brexit-polls-missed_n_576cb63fe4b017b379f58610

25. Wells, A. (2016, June 22). *YouGov's eve-of-vote poll: Remain leads by two*. YouGov. https://yougov.co.uk/topics/politics/articles-reports/2016/06/22/final-eve-poll-poll

26. YouGov/Times Survey Results. (2016). https://d25d2506sfb94s.cloudfront.net/cumulus_uploads/document/atmwrgevvj/TimesResults_160622_EVEOFPOLL.pdf

27. Unlike the invalid practice of throwing out third-party and undecideds who are likely to eventually vote, this is just weighting for Likely Voters. Imagine that if in Mintucky, instead of just "Red" and "Green" and "Silver (undecided)," there was a fourth group: Blue. And in this example, just as we assigned silver "undecided (will eventually be red or green)," we could assign Blue as "will not vote."

    If I take a sample of 100 that includes 45 Red, 45 Green, 8 Silver, and 2 Blue – and I'm only trying to estimate proportions of Likely Voters – while my sample was 100, I would not report Red as 45% (out of 100) I would reduce the denominator by 2 to account for nonvoters – thus report the numbers with a denominator out of 98 instead of 100.

28. Jackson, N. (2016, June 24). *How Brexit polls missed the "leave" victory*. HuffPost. https://www.huffpost.com/entry/brexit-polls-missed_n_576cb63fe4b017b3 79f58610

29. Huffpost Pollster. (2016). *UK European Union Referendum*. https://elections.huff-ingtonpost.com/pollster/uk-european-union-referendum

30. The Economist Newspaper. (2016). The Economist's "Brexit" poll-tracker. *The Economist*. https://www.economist.com/graphic-detail/2016/06/23/the-economists-brexit-poll-tracker

32

The Polls Weren't Wrong About Brexit

Important Points Checklist

- How the quality of polls versus forecasts is viewed by the Proportional Method
- Why declaring methods that were "closest the result" as "best" based on one or several elections is shortsighted and unscientific
- Different poll aggregation methods
- The number of undecideds in Brexit, and how that related to the finish line

Pollsters who report zero undecided when they observe undecideds are engaging in deceit. If I reported zero undecideds in "What's for Lunch?" or Mintucky, there could be no other explanation except that I lied.

The most generous explanation, which I think is the most common case, is that this deceit is unintentional, or simply because "this is how it has always been done." Given that polls are judged on closeness to eventual result, I believe some pollsters have caved to the pressure, and started reporting forecasts.

A term probably familiar to the statistics community, multilevel regression with poststratification (MRP) is one such method of forecasting. This method is used by YouGov for their reports.[1]

This is a perfectly fine method for building a forecast; you can forecast however you want. But you can't call your forecast a poll. The simple fix to this problem would be to separate "poll" and "forecast" entities and report polls as polls and forecasts and forecasts. Through 2024, the use of the term "vote intention" is becoming more prevalent, but media outlets still report this vote intention forecast as a "poll lead."[2] This isn't simply an informational gap however, as poll companies themselves are still describing and reporting these forecasts as polls.[3]

Instead of reporting their observations, each pollster who ascribes to these variations of the Proportional Method is deceiving the public, taking what they observed and reporting a forecast instead. And unlike a legitimate

DOI: 10.1201/9781003389903-32

forecast that reports a range of outcomes with a given probability, they typically report the number as if it were their poll.

As more pollsters do this, instead of dozens of independent polls, you get a dozen independent forecasts based on individual polls.

Now, the valuable data about where the finish line may be and how many undecideds there are doesn't exist. The range of possible outcomes, which can largely be determined by how many undecideds there are, is also unknown.

What has happened in the United Kingdom is the inevitable, unfortunate consequence of analysts who blame the polls for every "error" even if it's not their error, and pollsters saying "if we're going to be judged like we're a forecast, we're at least going to try and forecast."

Here's why it's a problem for the poll data community and public as a whole:

If a reported poll result – which happens to be a forecast – ends up being "wrong," is it because the poll was wrong or because the forecast was bad? **You can't know if you view them as the same thing.**

It's entirely possible that the poll was bad and the forecast was good, thus giving a false sense of security regarding the quality of the polls.

On the other hand, if the poll was good but forecast bad, owing to the misplaced sense of blame in the industry, it's likely an otherwise good poll would be combed over for errors – and blamed for the entirety of it regardless of the findings.

Moreover, perhaps largest of all, is that applying MRP (or any method of forecasting) to an individual poll ignores one of the most vital and central rules of poll data: **fluctuation is normal is expected.**

A poll conducted very well, **even an ideal poll**, has a risk of being unrepresentative of the population. That was shown with the "Throw it in the Average" simulation, "What's for Lunch?", Mintucky, and other previous examples. And that's okay! An approximation is better than a guess. That's why it's better to take an average.

But by trying to create a forecast from only one poll, forecasters are introducing an additional source of error to their forecasts: artificially restricted sample size. I suspect most qualified forecasters and even readers of this book would consider me quite foolish if I said "I'm going to build a forecast using only Monmouth polls" or "only Trafalgar polls." The foolishness of that approach is a fraction as ill-considered as trying to build a forecast from one poll, as these methods do.

Instead of just *reporting* what the poll said, they're trying to derive from one, small sample what the poll would predict. You learned why that was a bad approach way back in "Throw it in the Average" section, if you didn't understand already. Yet, highly reputable statisticians and analysts employ this "look at one poll" as their preferred method.

This forecasting from one poll introduces a third possibility in addition to "poll bad, forecast good" and "poll good, forecast bad." It's possible that "poll good, forecast good" also produces inaccurate results. If a poll produces results on the edge of the margin of error, which even a great poll could, how is a forecast supposed to correct for that?

The fourth and final possibility "poll bad, forecast bad" is the most troublesome. In the mind of most people, bad + bad = really bad. That could be the case, but understanding compensating errors, you know this is not always the case. Yes, it is entirely possible for a bad poll and a bad forecast to cancel each other out in such a way that the final result looks very accurate. The current definitions do nothing to correct for this, they can't.

If it's close to the result, it must have been right. It's been nearly a century since the *Literary Digest* poll fiasco, but smart people who mock it don't realize they haven't moved past that same level of reasoning.

YouGov confidently declared of Brexit polls, after the result was tallied and with the full benefit of hindsight, "The Online Polls Were Right."[4]

How do they know that? Because online polls (in contrast with telephone polls) showed the race was "very close." The same conclusion was drawn by *Huffington Post* for the same reasons.

The Literary Digest Polls Were Right, too, using the exact same reasoning. Until they weren't, after which all of its flaws seemed obvious.

In the words of YouGov regarding the debate over online versus telephone poll accuracy, "finally, this controversy can now be settled once and for all."[5]

That's right. Because the reported results of online polls were closer to the eventual result than telephone polls, in this one election, they are more accurate. Settled. End of analysis.

As for the long-term value of that analysis, wanting to know the future, people might want those same sibylline, online pollsters who showed "Leave" ahead to try their hand in the later US elections.

Only survey that correctly predicted Brexit says Clinton wins U.S. election.[6]

That is what a USA Today headline shared.

Settled, once and for all.

In Brexit, like other UK elections, and inevitably most election analysis in the future – because of varying methods and a misunderstanding of how to quantify polls versus poll averages versus forecasts, analysts will remain entirely incapable and unqualified to say how much error there was and who or what caused it. It's a guessing game of who was "better" solely on closeness to eventual result, and little to no concern for methodology or validity.

Statistical literacy has suffered, and I don't think even the experts in this field realize how big of a problem they've created. This domino effect of misinformation starts with not understanding what polls do at a fundamental level, with polls simple as "What's for Lunch?" and "Mintucky."

Thanks to that understanding, instead of accepting at face value what the polls forecasted (and understanding that anyone who reported 0% undecided was not being forthright), I asked, "what did these polls actually observe?"

I went back to each poll that didn't report undecideds to figure out how many undecideds they would have reported if they didn't use the Proportional Method. I used the same method as outlined in the previous chapter for the June 22nd YouGov poll.

Here is the first round of recalculation for the reports of 0% and 2% undecided

Remain (%)	Leave (%)	Undecided (%)	Date	Pollster
48.4	44.1	7.5	Jun 22, 2016	Populus*[7]
48	42	11	Jun 22, 2016	ComRes
41	43	11	Jun 22, 2016	TNS
44	45	11	Jun 22, 2016	Opinium
48.5	47.2	4.3	Jun 22, 2016	YouGov*[8]
48	46	6	Jun 22, 2016	Ipsos MORI*[9]
45	44	11	Jun 20, 2016	Survation
42	44	13	Jun 19, 2016	YouGov
49.4	46.4	4.2	Jun 19, 2016	ORB*[10]
45	42	13	Jun 18, 2016	Survation

* *Result adjusted to allow for undecideds and leans as indicated in the poll report.*

After reading through the YouGov report from June 19th, I stumbled upon another problematic fact regarding how undecideds were reported by the *Financial Times* (FT) average. YouGov's raw data (unadjusted for turnout) reported 9% "don't know" and 4% "would not vote." This was put in the FT polling average as "13% undecided."

Just when you thought frame error only applied to pollsters – surprise! Here, it is applying to a poll average. A poll average counting people who say they don't plan to vote as "undecided" is an error, but it isn't a poll error. This wasn't a one-off mistake, I had to fix this for several "undecided" numbers reported in the *Financial Times* average, as you can see below.

To recap so far, we have some *pollsters* reporting undecideds with others ignoring them and some *poll aggregators* reporting undecideds and others ignoring them.

To complete my recalculation of poll averages in a way that allowed for undecideds, I now needed a third step to generate a poll average for these 10 polls: separate actual undecideds from people who said they don't plan to vote.

Remain (%)	Leave (%)	Undecided (%)	Date	Pollster
48.4	44.1	7.5	Jun 22, 2016	Populus*
47***	42	11	Jun 22, 2016	ComRes**[11]
42	49	8	Jun 22, 2016	TNS**[12]
44	45	9	Jun 22, 2016	Opinium**[13]
48.5	47.2	4.3	Jun 22, 2016	YouGov*
48	46	6	Jun 22, 2016	Ipsos MORI*
45	44	11	Jun 20, 2016	Survation[14]
46.4	47.4	6.1	Jun 19, 2016	YouGov**[15]
49.4	46.4	4.2	Jun 19, 2016	ORB*
45	42	13	Jun 18, 2016	Survation[16]

* *Result adjusted to allow for number of undecideds and leans as indicated in the poll report.*
** *Result adjusted to remove "would not vote" from undecideds.*
*** *Financial Times poll average reported 48%, but ComRes report said 47%.*

None of these recalculations were done arbitrarily. The only thing this recalculation does is apply the fundamental understanding that polls are observations, not predictions, and therefore a plan poll must allow for the possibility of undecideds.

I strongly encourage other researchers to corroborate or contest the findings as I have reported them – having shown the work as to how I reached those conclusions – because what it tells us about Brexit, and how polls should be viewed and reported in the United Kingdom and non-US countries in future elections is consequential.

Say "the polls" were right or wrong on the merits, not assumptions. Report an appropriate level of uncertainty in a range of outcomes, too.

While the discrepancy in the number of undecideds reported in each poll may seem indicative of a level of variation not explainable by mere fluctuation, I found the resolution was in the pollsters' respective methodologies. Some pollsters used "squeeze questions" and others didn't.

As with how they qualify and weight for turnout and "Likely Voters," that is a matter of preference. With my "forecaster" hat on, I tend to lean toward a diversity of methodology in this area because how many undecided voters there are (and how many only express a lean) is valuable data. But how I inform my predictions versus how others do so is not a matter of science, it's a matter of preference. If my forecast is wrong, that's not "the polls'" fault. If my poll average is wrong, that's not entirely attributable to "the polls'" either.

If another poll aggregator wants to use only polls with or without "squeeze questions," they can. They can also use only "online" or "telephone" polls, if they think the accuracy of those methods is "settled, once and for all." And they can weight any of those factors, among others, however they choose – **as**

long as they admit their methodology introduces the risk of error not entirely attributable to the polls.

That level of accountability does not exist at this point.

The purpose of transparent methodology is not to twist arms into adopting the same standards – that approach undermines the value of independent research and novel methods. The purpose of transparent methodology is simply to confirm the methodology is valid and potential sources of error. In the cases of these polls and poll averages reporting 0% undecided, no matter how strong the analysis is, or how beautiful the assumptions, the methodology is not valid.

In my case, I have again opted for the simple recent average approach in calculating my poll averages, using 10 polls, not because I believe that it is the best approach, but because as a researcher transparency and consistency are among the most valuable assets. My reported poll average below is not intended to be a gold standard about "what the polls said" but to give a better idea of what the poll data actually observed about the United Kingdom's opinion on Brexit, allowing for undecideds.

Remain (%)	Leave (%)	Undecided (%)	Date	Pollster
46.4[a]	45.3[a]	8.0[a]	Jun 18–22, 2016	Poll Average
48.4	44.1	7.5	Jun 22, 2016	Populus*
47***	42	11	Jun 22, 2016	ComRes**
42	49	8	Jun 22, 2016	TNS**
44	45	9	Jun 22, 2016	Opinium**
48.5	47.2	4.3	Jun 22, 2016	YouGov*
48	46	6	Jun 22, 2016	Ipsos MORI*
45	44	11	Jun 20, 2016	Survation
46.4	47.4	6.1	Jun 19, 2016	YouGov**
49.4	46.4	4.2	Jun 19, 2016	ORB*
45	42	13	Jun 18, 2016	Survation

* *Result adjusted to allow for number of undecideds and leans as indicated in poll report.*
** *Result adjusted to remove "would not vote" from undecideds.*
*** *Financial Times poll average reported 48%, but ComRes report said 47%.*
[a] *Numbers don't add to 100% due to differences in poll reporting & rounding.*

Leading up to and very close to the election, voters on the UK Referendum to remain in or leave the European Union were nearly evenly split. For such a politically heavy topic, a surprising number of voters reported being undecided. Most pollsters, poll aggregators, and members of the media failed in their duties to properly report the uncertainty in this election.

Not a single poll observed 50% support for either side in the 10 polls closest to the election, but because of the Proportional Method, they said many did.

In Mintucky, I warned how the Proportional Method would wrongly report each poll and poll average – and how that method could report a poll average far greater than 50% despite there being very few observations of that. That's almost exactly what actually happened with Brexit.

Whether you believe the true number of undecided voters in the days before the Brexit Referendum vote was closer 4% or 13%, a debate reasonable people could have, it's indisputable that there were more than zero. Given the fact that the finish line would be – with no approximation necessary – 50% + 1, even a low-end estimate of undecided voters could swing the result one way or the other.

But reading and reporting polls by "spread" or "proportion" blind people to the reality of what polls tell us.

Even poll aggregators who found themselves in the best position to properly characterize the race for what it was – high undecided and a close election that could go either way – missed that opportunity due to this blindness.

Huffington Post, citing their 45.8%–45.3% poll average, reported "The average of all surveys conducted prior to the polls opening on Thursday showed that 'remain' was up by 0.5 percentage point over 'leave.'"[17]

The Economist with their 44%–44%–9% poll average said that "a tenth" of voters remained undecided with the decided voters evenly split.[18]

Reporting how many undecideds there are, while not quite going as far to make the analysis that this vote could go either way, was at least in the right vicinity.

But after voting had started, what *The Economist* had to say about their model that tried to predict referendum results based on early voting reports was eye-opening.

They said that their model "suggested that 'leave' would win the referendum comfortably—precisely the opposite conclusion from the one reached by betting markets, which see 'remain' as the overwhelming favorite. To bring our estimates in line with the wisdom of crowds—if you think you're smarter than a prediction market, think again—we simply reduced the projected 'leave' share by the same amount in every counting area."[19]

Remember, the betting markets implied "Remain" as around a 75% favorite prior to the election. While this reporting was not related to a poll or poll average (and I believe *The Economist* should be praised for the transparency of their methodology), it raises a very serious question about whether pollsters or poll aggregators would put their thumb on the scale, using observations a mere suggestion, and report what they – or "the market" – believe to be true instead of what their data says. If nothing else, it shows the keen sense of awareness these entities have for outside influences – and the threat posed by analysts who judge the accuracy of polls by their closeness to eventual result.

It would be unfair of me to characterize "all" pollsters and media falling victim to the polls-as-predictions mindset with Brexit, even if the vast majority did. An Opinium spokesperson accurately characterized their final poll as, "Everything rests on 9 per cent still undecided."[20]

Time reported it as "too close to call."[21]

But for every quality report, there were hundreds that relied solely on "spread" and "who is ahead." The public remained misinformed.

Brexit, like Trump-Clinton, are perfect case studies of what can happen in tightly contested elections in which both candidates are far from the finish line. In the case of Trump, it is known that a highly favorable Undecided Ratio is largely responsible for the election going in his favor; with Brexit, less is known. Yet, even with undecideds voting largely for Trump, polls *still* take the blame for not predicting the future well enough.

The Referendum resulted in a vote to leave the European Union, with "leave" carrying about 52% of the vote.[22]

Not just because of their proximity in time or surprising result, but in how they were poorly reported on and analyzed, the combination of Brexit and Trump-Clinton 2016 should have dealt a serious wake-up call to poll analysts, if not a total overhaul of the field, but it didn't. The conclusion that "polls were wrong" seems satisfactory enough, and the public believes them. With Brexit being more similar to US elections than a typical UK election, direct comparisons may be imperfect but are certainly worth considering.

One way to look at what the polls said about Brexit is through this lens: in high-undecided elections in which there are only two major options, any perceived "lead" is not safe, especially not a small one far from the finish line. But with Brexit, there weren't just two *major* options, there were only two options; it's a very rare instance of being able to know with certainty where the finish line is.

With that knowledge, as you can see in Figure 32.1, we can look at how far an option has to go before crossing it; in the Brexit referendum of 2016, it was a long way.

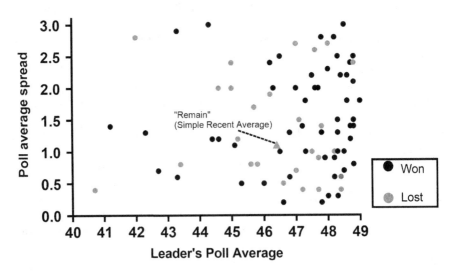

FIGURE 32.1
A chart showing a historical statewide averages from US elections through 2022, plus where "Remain" would have fit on it. Chart by the author.

But I don't want you to think this is purely a hindsight project of "if we knew then what we know now." It's much more than that.

Anyone who understands what polls tell us – as a tool – would predict exactly what you see, past, present, or future: highly unpredictable results that become more predictable as the leader's poll average approaches the finish line. Unfortunately, blinded by spread, proportion, and other metrics made up to judge "poll accuracy," they miss it.

Figure 32.2 shows what the Brexit poll aggregators **who allowed for unde-cideds** would have looked like, plotted only against results through Brexit.

Whose averages were better? Compared to what? Does disagreeing with the eventual result mean, with no other explanation, "the polls" were wrong?

And finally, to hopefully illustrate the chaos created by the lack of account-ability around poll averages, Figure 32.3 shows poll averages and "poll of polls" for all poll aggregators.

I would encourage anyone with an interest in this area to perform the same analysis I have for US battleground states – plotting poll averages for highly polled elections – with the eventual result relative to the finish line. Unlike the United States, where the finish line is almost always around 49%, non-US elections should be handled on a "similarity" basis. Constituencies with similar third-party support levels can be better compared because the third-party support offers an estimate of the finish line. Elections with 30% third-party support (finish line around 35%) are quite different from those with

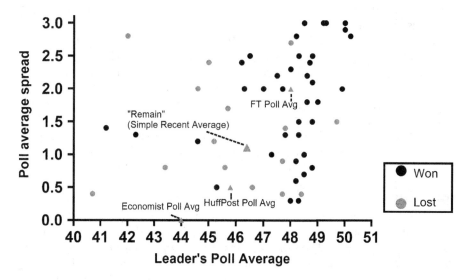

FIGURE 32.2
A chart showing a historical statewide averages from US elections through 2014, plus where the poll aggregators who allowed for "undecideds" would have fit on it. Chart by the author.

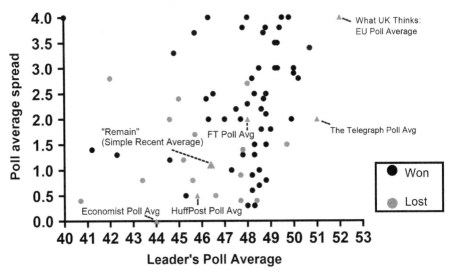

FIGURE 32.3

A chart showing a historical statewide averages from US elections through 2014, plus where the all poll aggregators would have fit on it. Chart by the author.

18% third-party support (finish line around 41%). Comparing Utah 2016 to Kentucky 2016 (or even Utah 2020) is not a wise exercise.

In countries like the United Kingdom where some constituencies have three or more competitive parties, but others only have two, these rules of analysis are necessary.

As with FiveThirtyEight's assertion that their poll average is the gold standard by which we can judge "the polls" – which seems to be accepted with no critical energy by experts and the media – the poll averages that showed "Remain" ahead seemed to form the basis of that judgement for Brexit. But the amount Remain was "ahead" – if at all – was simply too small and too far from the finish line to draw such strong conclusions. Yet, because of the Proportional Method, overly confident numbers were reported, and the public again is left with the feeling "the polls" failed – when in reality, it's the assumptions, predictions, and forecasts of analysts that did so.

The polls weren't wrong about Brexit. The polls weren't wrong about Trump-Clinton.

Unless and until the industry's accepted definitions are fixed, the experts within it will continue to misinform the public, and there will be no progress; no progress in polling, no progress in poll aggregation, no progress in forecasting. Public confidence in polls and inferential statistics will suffer as a result, and it may push out the truly good and honest pollsters in favor of pollcasters, if it hasn't already.

None of the proposed changes I've offered, if accepted on their merits, would take more than one or two election cycles to truly implement. Like the "smoking and cancer" link, the understanding that polls aren't predictions – and the more complex "what that means" – may be received among experts more quickly than the public at large, but right now we're still in the 1930s; there aren't even many experts on board.

I fear the appeal of "the way we've always done it" will cause slower than necessary adoption among experts, leading to more decades of stagnation. Some conversations I've had with those experts give me the impression they're very set in their ways. But I'm hopeful some influential folks in the media and innovators in this field will adopt them. If it takes more "Gallup vs. Roper" style debates about how polls should be judged, so be it; I just hope it accomplishes something this time.

If nothing else, as members of the public, we can use these tools to sort through the rubbish to be better informed consumers of poll data.

And you now understand polls better than the experts.

Polls do not attempt to predict the spread of an election result.

Polls do not attempt to predict the proportion of two-party vote.

Polls do not attempt to predict the result of an election.

Notes

1. English, P., & Thornton, L. (2023, April 28). *Local elections 2023: YouGov MRP predicts conservative losses in key Battleground Councils.* YouGov. https://you-gov.co.uk/topics/politics/articles-reports/2023/04/28/local-elections-2023-yougov-mrp-predicts-conservat

2. Reuters. (2024, May 9). *UK's Labour holds 30-point poll lead over Sunak's conservatives.* https://www.reuters.com/world/uk/uks-labour-holds-30-point-poll-lead-over-sunaks-conservatives-2024-05-09/

3. Smith, M. (2024, April 30). *London 2024 mayoral Race: Khan 47%, hall 25%.* YouGov. https://yougov.co.uk/politics/articles/49286-london-2024-mayoral-race-khan-47-hall-25

4. Sayers, F. (2016, June 28). *The online polls were right, and other lessons from the referendum.* YouGov. https://yougov.co.uk/topics/politics/articles-reports/2016/06/28/online-polls-were-right

5. Sayers, F. (2016, June 28). *The online polls were right, and other lessons from the referendum.* YouGov. https://yougov.co.uk/topics/politics/articles-reports/2016/06/28/online-polls-were-right

6. Hjelmgaard, K. (2016, November 8). *Only survey that correctly predicted Brexit says Clinton wins U.S. election.* USA Today. https://www.usatoday.com/story/news/world/2016/11/08/only-survey-correctly-predicted-brexit-says-clinton-wins-us-election/93415522/

7. Populus. (2016). *EU pre-referendum online poll.* Yonder. https://yonderconsulting.com/poll/eu-pre-referendum-online-poll/

8. Wells, A. (2016, June 22). *YouGov's eve-of-vote poll: Remain leads by two.* YouGov. https://yougov.co.uk/topics/politics/articles-reports/2016/06/22/final-eve-poll-poll

9. Ipsos MORI. (2016). *Ipsos MORI political monitor—EU referendum 2016—Polling day survey.* https://www.ipsos.com/sites/default/files/migrations/en-uk/files/Assets/Docs/Polls/eu-referendum-tables-final-day-update.pdf

10. ORB International. (2016). *EU poll.* https://web.archive.org/web/20160705025318/http://www.opinion.co.uk/perch/resources/poll-190616-tables-weighted-v4.pdf

11. ComRes.(2016).Dailymail/ITVnewsEUreferendumsurvey.https://web.archive.org/web/20160918003533/http://www.comresglobal.com/wp-content/uploads/2016/06/Daily-Mail-ITV-News_EU-Referendum-Survey_22nd-June-2016_61021xh5.pdf

    This source did not separate "would not vote" from "don't know" in raw data.

12. TNS. (2016). *Polling methodological note 23/06/2016.* https://www.kantarpublic.com/download/documents/190/Public_EU_Poll_TABLES_23062016.pdf

    They reported "Likely Voter" results with weights for turnout, and allowed for "Don't Know," so I kept the results as they reported.

13. Independent Digital News and Media. (2016, June 22). *Final opinium EU referendum poll shows race "too close to call."* The Independent. https://www.independent.co.uk/news/uk/politics/eu-referendum-final-opinium-poll-shows-leave-ahead-by-one-point-a7095811.html

    The Opinium Likely Voter report allowed for undecideds so I kept as they reported. Unsure if different "undecided" number in *Financial Times* reported averages is due to rounding or inclusion of "would not vote."

14. Survation. (2016, June 21). *EU referendum poll* (prepared on behalf of IG Group). https://www.independent.co.uk/news/uk/politics/eu-referendum-final-opinium-poll-shows-leave-ahead-by-one-point-a7095811.html

    The Survation results are the same in their report as reported in the *Financial Times* average.

15. YouGov/Times Survey Results. (2016). https://d25d2506sfb94s.cloudfront.net/cumulus_uploads/document/it82go26iz/TimesResults_160620_EUReferendum_W.pdf

16. Survation. (2016, June 18). *EU referendum poll part 1* (Prepared on behalf of the Mail on Sunday). https://survation.com/wp-content/uploads/2016/06/Final-18-June-Mos-EU-poll-Tables-part-1.pdf

    The Survation results are the same in their report as reported in the *Financial Times* average.

17. Huffpost Pollster. (2016). *UK European Union Referendum.* https://elections.huffingtonpost.com/pollster/uk-european-union-referendum

18. The Economist. (2016). *The Economist's "Brexit" poll-tracker.* https://www.economist.com/graphic-detail/2016/06/23/the-economists-brexit-poll-tracker

19. The Economist. (2016a). *How we calculated our extrapolated Brexit vote shares.* https://www.economist.com/graphic-detail/2016/06/23/how-we-calculated-our-extrapolated-brexit-vote-shares

20. Stone, J. (2016, June 22). *Final opinium EU referendum poll shows race "too close to call."* The Independent. https://www.independent.co.uk/news/uk/politics/eu-referendum-final-opinium-poll-shows-leave-ahead-by-one-point-a7095811.html

21. Stewart, D. (2016, June 23). *Brexit: Polls open in referendum as race too close to call.* Time. https://time.com/4379454/eu-referendum-brexit-polling-day/

22. BBC. (2016). *EU referendum results.* BBC News. https://www.bbc.co.uk/news/politics/eu_referendum/results

33

UK General Election Polls: Not Wrong Either

Important Points Checklist

- How the Margin of Error given by a poll relates to the eventual election result
- The value of recontact surveys
- What "late swing" refers to
- The amount of consideration that analysis of a poll's "wrongness" by what it "predicted" deserves
- The number of undecideds pollsters in the United Kingdom ignore
- Why the "shy Tory" effect – if it exists – cannot be considered a poll error
- What pollsters did in the 2017 general election that was somehow even worse than the Proportional Method
- The current state of poll reporting and analysis

I'll be honest with you: I did not have much interest in including UK general election analysis in this book, I intended Brexit to be my final "poll analysis" chapter. Like when I first pestered US experts with my findings in hopes they'd build on it, I hoped my work in this book regarding the Proportional Method and Brexit would motivate analysts more involved there (i.e. a citizen on that continent) to do the same. And I still hope it does. It's not that my time is particularly limited on the subject, I've done some analysis on several countries including the United States and the United Kingdom for the past several elections. Again, I'm very fun. My reason for not wanting to include all of it is this:

> If this problem is going to be addressed or fixed, it's not
> going to be me on a solo mission. I hoped explaining the problems
> with the Proportional Method would be enough to do that,
> and if not, certainly the Brexit poll analyst failure
> (not poll failure) would.

DOI: 10.1201/9781003389903-33

But in some early feedback that I've received, I'm *still* pointed to UK general elections as surefire examples of totally real, huge poll errors. To which my response is: doesn't matter (and also, probably not).

What Mintucky showed *could* happen if the Proportional Method were used, Brexit showed *did* happen. Polls and poll averages were reported as over 50% despite the fact that no polls observed that. If not for the field's tradition of lying, it would be considered scientific misconduct.

Whether or not UK general election polls – or any polls in any country – have huge errors by the current definitions is *irrelevant* to the problems of the Proportional Method. Even if the method didn't have any catastrophic failures in general elections, that's not an excuse to keep it. The goal should be to fix things *before* they collapse, not after. "So far, we have been right in every Poll" is not an excuse to keep flawed methods, even if it were true.

Unfortunately, there have been catastrophic failures.

It's unclear if experts anywhere are even aware of the methodological issues that exist with this method's "zero undecided" reporting. The numbers seem to be taken at face value with no consideration to the fact that "zero undecided" is not true.

While two-option elections like Brexit are great starting points to understand the ramifications of using this method, most UK elections are not "two-option."

In 2015, it was reported and believed that the pre-election polls failed.

Like the American Association for Public Opinion Research (AAPOR) in the United States, the British Polling Council (BPC) convened to analyze the perceived poll failure(s) of past elections.

Their characterizations are similarly flawed:

> Although the result was wrong ...

This is the same, flawed polls-as-predictions characterization, this time by Patrick Sturgis, head of the BPC review team.[1]

It doesn't appear any consideration was given to the problems of analyzing poll accuracy by how well they forecasted the result. This will continue to happen until polls and forecasts are understood as separate and distinct.

But the main problem in the view of the experts is that it's:

> ... a bit unfortunate that they (the polls) were three points over on Labour and three points under on the Conservatives.

Here, Sturgis is speaking about the 2015 UK general election, attempting to answer the question "why did the pollsters get it so wrong?"[2]

I had to work backwards to try and make some sense of this analysis.

The results[3] for the top two parties in the 2015 UK general election were

Conservative: 37.7%   Labour: 31.2%.

What the BPC reported was that compared to the eventual result, they believed the polls "predicted" Conservatives and Labour should both be around 34%.

You understand by this point why this belief is based in a misunderstanding of what polls do. But if you take the polls at face value, given the Proportional Method's requirement to lie about undecideds, it makes sense why someone would think that. Below are the results reported by the 10 polls taken closest to the election for which I could access the full report.[4]

I included a column showing how many "undecided" or "don't knows" they reported to highlight the absurdity of this method.

Conservative (%)	Labour (%)	Undecided (%)	Date[5]	Pollster
34	34	0	May 05, 2015	Populus
33	33	0	May 05, 2015	Lord Ashcroft
35	34	0	May 05, 2015	ComRes
34	34	0	May 04, 2015	YouGov
31	31	0	May 04, 2015	Survation
35	34	0	May 04, 2015	Opinium
34	34	0	May 04, 2015	YouGov
33	34	0	May 04, 2015	Survation
34	35	0	May 03, 2015	ICM
35	32	0	May 03, 2015	ComRes

It was this data that led Sturgis and others to conclude that these polls "underestimated the Conservative lead over Labour by an average of 7 percentage points."[6]

In that report, they made the erroneous observation that for all nine pollsters affiliated with the BPC:

> the notional +/–3% margin of error does not contain the true election result.

That's because the margin of error given by the sample size of the polls (+/–3%, in this case) applies to a simultaneous census. Not the eventual result.

Assuming that the margin of error given by a poll is the similar or equal to the margin of error of a forecast is an issue that continues among experts and was seen in action with the 2024 UK Mayoral Elections. One company (More In Common) provided their "voting intention" forecast and asserted a surprisingly narrow margin of error for it (approximately +/−4% for the major candidates) given the large number of undecideds in their polls. They also provided a 95% confidence interval for those forecasts. However, every election they provided a forecast for had at least one candidate outside the 95% confidence interval, and in one instance, a candidate whose 95% confidence interval ranged from approximately 2% to 4% finished near 12%.[7]

The AAPOR made the same mistake improperly citing the "margin of error" in US primary elections when they said:

> ... there is a consistent level of error that is still more than twice the "margin of error" that polls publicly report.[8]

The margin of error the BPC and the American Association for Public Opinion Research (AAPOR) are referring to is the error of the Proportional Method and Spread Method, respectively, not the one given by polls. They're conflating a poll's margin of error and a forecast's margin of error.

The margin of error given by an ideal poll can be known objectively and with certainty – the margins of error in the Spread and Proportional Methods cannot be. The fact that these pseudoscientific "poll accuracy" calculations regularly – and predictably – produce unfounded "error" results should not surprise anyone. But it's easier to just blame the polls.

This is not to say the polls *couldn't* have contained error, but to assert as they did that the polls *must* have contained a large error because that the margin of error (MOE) did not "contain the true election result" is not a statistically defensible statement because confounders were not accounted for.

If they'd like to calculate the margin of error for their methods, they can, but it is a conditional one largely based on the number of undecideds, also impacted by the number of candidates running, and has almost nothing to do with the MOE given by the sample size of the poll(s). The MOE of the poll is not impacted if there are only 2% undecided versus 20%; the MOE of any Method is.

Put another way, the population of the poll (with undecideds) was different from the population of the election (with none). You can't claim the margin of error given by the sample size applies to different populations who were asked different questions at different times.

This provides further evidence that the Simultaneous Census standard, along with understanding the differences between poll and forecasts, is needed.

Unfortunately, the numbers in the BPC report are nothing more than a measure of forecast error. The pollsters who act as forecasters in the Proportional Method, by "allocating" undecideds, are not reporting their poll's observation, nor is the analysis of how close they were to the eventual result a measure of the poll's accuracy.

This fundamental misunderstanding is best summarized with this quote from the report:

> 11 of the 12 Great Britain polls ... were some way from the true value.

The BPC believes that the "true value" that polls try to measure is the eventual election result. This is not correct.

The reality is that these pollsters took a poll, built a forecast from it using the Proportional Method, those forecasts were very bad, and due to not understanding the difference between a poll and a forecast, the BPC, media, and public were all convinced the *polls* were bad.

The Proportional Method is directly resulting in the public, and even experts, not understanding what polls tell us – and spending a lot of time on a lot of research that could be much better.

In rare instances where there are very few undecided voters, the difference between observation and reporting for the Proportional Method is smaller. In cases where there are a lot of undecided voters, the chance they make bad guesses (and blame the polls for them) increases.

So just like in the previous chapter on Brexit, I dug into the reports from those pollsters to see if the numbers of "don't knows" were substantial. They were. I recommend you compare the results the pollsters observed to the results they reported to see how deceitful the Proportional Method is.

Here are the numbers from the polls above, as they were observed[9] and should have been reported:

Conservative (%)	Labour (%)	Undecided (%)[10]	Date[11]	Pollster
29	31	10	May 05, 2015	Populus
25	26	18	May 05, 2015	Lord Ashcroft
27	27	18	May 05, 2015	ComRes
29	29	9	May 04, 2015	YouGov
28.4	29.4	11.7	May 04, 2015	Survation
31	31	10	May 04, 2015	Opinium
28	29	10	May 04, 2015	YouGov
28.6	30.5	10.3	May 04, 2015	Survation
24	27	20	May 03, 2015	ICM
27	25	20	May 03, 2015	ComRes

Decimal Points were retained in Survation's calculations because these numbers were able to be obtained directly from their report with no adjustments needed to account for "would not vote." Other pollsters put "would not vote" and "undecided" (even if the undecideds are rated as likely to vote) in the same category.

I must mention again that these recalculations are not arbitrary. Each of these pollsters identified the respondents who they believed were "likely to vote."

The numbers I posted above are nothing more than a reflection of that, allowing for people to be undecided, as a plan poll must.

Once you start trying to guess, assume, or otherwise predict how undecideds will vote, you no longer have a poll: you have a forecast. There's nothing wrong with trying to forecast – just don't confuse it with a poll.

I believe the average consumer, and hopefully experts, can understand that a poll average with a lot of undecideds, even if the poll averages in it appear close, might not end up close. And certainly, if experts relayed this reality to the public, they would be better able to do so.

In Chapter 28, I presented the following:

"It is far more likely one candidate outperforms their poll average by 10% when there are 15% undecided, than when there are only 5% undecided."

That lesson would have been useful to understand in the UK general election of 2015 and many times since – and certainly will be useful in the future too. It's a perfectly reasonable assumption to assume these analysts and experts know this, but their analysis shows otherwise.

Here is what this "simple recent average" for UK 2015 would have shown:

Conservative: 27.7%   Labour: 28.5%   Undecided: 13.7%.

Given the eventual result, one party winning by 6 (outperforming their poll average by 8%–10%, given approximately 13.7% undecided and a fair number of third-party voters) should have been considered a possible outcome. But because of the subservience to the Proportional Method, all they can say is the forecast didn't match the result; therefore, polls were wrong.

To their credit, the BPC – with the help of pollsters – did conduct a great deal of research as to the cause of the perceived poll failure. Of particular interest from their report, multiple pollsters conducted **recontact surveys** in which they contacted the same people after the election as they did before it.

This is awesome! This is the best possible data we can have to estimate Changing Attitudes and Undecided Ratio.

Unfortunately, the report refers to this "late swing" without differentiating between undecideds deciding, and who changed their mind.

> (late swing) refers most naturally to a switching from one party to another, but we also include movement from non-substantive responses ("don't knows" and refusals) to a party choice.[12]

"Don't know" should not be considered a non-substantive response. It is both substantive and valuable for estimating a range of possible outcomes. The term "non-substantive" isn't worth quibbling over, but the fact that "don't knows" deciding is a very different calculation from "switching from one party to another" is worth clarifying. Putting them into one bin is bad analysis. The probability that one candidate overperforms their poll average by 10% is much higher when there are 13.7% undecided than when there are far fewer. The analysis done on the topic betrays this fact.

Knowing how many undecideds there were close to the election is also helpful for finding the *cause* of the discrepancy between poll and eventual result. Switching from one party to another is not the same as deciding when you were previously undecided.

The analysis from the BPC report, while it incorporates the valuable recontact survey data, only considers the "Labour versus Conservative lead" in the analysis. To my knowledge, it doesn't quantify how the respondents who were undecided in the polls ultimately decided. It also appears that the "Before Election" portion was reported via the Proportional Method, and it tried to summarize what each pollster said based on its "Conservative versus Labour" spread – not the numbers each pollster reported for each option.

Using the Proportional Method is bad enough. But only considering this method's output is especially problematic as it relates to accounting for confounding variables before attempting to do an analysis.

The number of undecided voters was much larger than I expected and much larger than I believe most people would guess – which is, in a brutal source of irony – one of the main reasons we do polls in the first place.

Moving away from the BPC summary of the recontact surveys, I looked at the surveys as they were reported by individual pollsters. At least one recontact survey from the pollsters fared better in describing poll performance; Survation's report outlined each pre-election option including undecided. It showed that while uneven undecideds did not significantly impact their numbers (undecideds split approximately evenly to the major parties), changing attitudes did supply Conservatives with a net benefit of +2.2%, while costing Labour −0.4%. Perhaps surprisingly, they noted that much of the Conservative benefit came directly from Labour.[13]

Other pollsters said they found no such "late swing," which points more in the direction of an actual poll error. Whatever conclusion is true, I think it's clear that these recontact surveys provide essential data for calculating poll accuracy. Like a student whose poll with a sample size of 3 can yield an accurate conclusion, even if UK polls were wrong in 2015, the underlying methods to support that conclusion should not be given a pass just because, possibly, it got the "right answer" this time.

Past and present analysis demands that the Proportional Method pollcast (which includes incorporating assumptions about undecideds) be treated as "the poll."

To further illustrate this point, the co-founder of YouGov flatly stated that:

> All the pre-election polls in the 2015 general election were wrong, significantly overestimating Labour support and underestimating Conservative support.[14]

Overestimating or underestimating the eventual result is an error, but it isn't a poll error.

Any analysis that says a poll was wrong because it "overestimated" or "underestimated" the eventual result is wrong.

The report itself from YouGov made the same flawed assumptions that has contaminated inferential statistics discourse since political polls began:

> ... the size of the polling error—predicting a tossup when, in fact, Conservatives won by almost seven points ...[15]

Polls don't predict win probability.

Whether or not the polls were right, wrong, or "okay" is a topic that is worth debating. I think I've made my position clear and substantiated it. Believe it or not, while I'm confident in my research, I have and will continue to consider the possibility that I am wrong, and that the polls I've analyzed *were* somehow significantly wrong.

But I will not give the slightest consideration to the falsehood that failure to predict an eventual result is evidence of a poll's wrongness, nor should anyone, because it deserves none.

Until we can all agree with what is true: polls are not predictions, and polls do not try to predict the eventual "margin," nor do they attempt to estimate win probability, the debate is meaningless. A good poll plus a bad forecast do not equal a bad poll. You're free to disagree because you're free to be wrong, but that's not a standard I'm comfortable with accepting in a field that claims to be scientific.

As for the 2015 general election, I can't say for certain how much the Undecided Ratio (or Changing Attitudes) contributed to the difference between polls and results because there's not enough data. There is data to support Changing Attitudes was "net-zero" for the major parties in that both sides benefitted equally from it,[16] but most of the recontact studies seem to have put pre-election third party and undecideds in the same bin.

This supports my belief that while the existing research is useful, adopting better methods would make it even better.

Moreover, there's a big difference between trying to calculate how each party got from their observed number to final number, and trying to do so based on the "before and after" spread. My simple recent poll average said:

Conservative 27.7%, Labour 28.5%, and Undecided 13.7%.

Using this poll average as the basis, if Conservatives ended up at 30.7% and Labour at 31.5%, that would account for, at most, 6% of the undecided. Where did the rest go in this hypothetical?

Good researchers should want to know, but existing methods don't allow for it. They'd have said the polls were immaculate simply because the spread matched – despite not being able to account for all the undecideds.

And because it's a topic of some debate, I do not believe this possibility of polls underestimating Conservatives is best simplified as a "shy Tory" effect – "Tory" being a colloquial term for Conservative. There's no reason to assume that someone who is a reliable Conservative voter would not be willing to state as such any more than a Labour or Liberal Democrat voter would. It seems plenty plausible that they are genuinely undecided (or like none of the options, but always vote), **and when met with the deadline of an election, those undecideds shared some characteristic(s) that led them to favor one option.**

And *even if there is* (or has been) a shy Tory effect (i.e. polls systematically underestimating Conservative vote), that's not the poll's job to detect in the first place – it's a forecaster's.

Applying the proper Simultaneous Census standard, if your poll contains 1,000 reliably Conservative voters, but only 950 report that they plan to vote "Conservative" and the rest say they're "undecided," all that means is that **if you took a Simultaneous Census, approximately the same number of reliable Conservatives would report being undecided!**[17]

Frustratingly for people who want polls to be predictions, if some percentage of individuals in a poll lie about their voting plans, that means that if a Simultaneous Census were conducted, a similar proportion would also lie. This is a *limitation* of the polling instrument, not an error.

To use a classic example, asking a random sample of 1000 adults how often they exercise will probably overestimate adult physical activity levels.[18] Even if you were to observe the habits of those same 1000 adults and track how often they exercise, **that is not a measure of how accurate the original poll was.**

A poll that asks "how often do you exercise?" aims to measure how often that same population would say they exercise. How often they truly exercise is a separate topic. I think a similar phenomenon exists when you ask someone how likely they are to vote: many people who say they will "definitely" vote do not vote.

A good *forecaster* should try to detect a "shy voter" or possibility of "lying" and take it into account, but it has nothing to do with the accuracy of the poll. You can argue that this has to do with the *quality* of the poll – the ability to squeeze a lean or preference out of someone – but not its accuracy.

The "shy Tory effect" seems like a "lazy analyst effect" to dismiss the impact of covariance: people with similar characteristics sometimes behave similarly. Just because undecideds may have favored Conservatives in the past doesn't mean they will in the future. Panagakis didn't name the Incumbent Rule with the belief that lots of people were simply too shy to admit they were voting against the Incumbent; the best explanation was that they liked neither candidate or didn't know enough about the non-Incumbent.

It's entirely possible that some confounding variable (such as candidate quality, effectiveness of advertising, voter motivation, or many others) could explain why undecideds sometimes vote disproportionately. Regardless, undecideds eventually voting disproportionately is not a poll error.

As for whether changing attitudes and/or undecided ratio can account for a substantial proportion of the perceived error in the 2015 general election and Brexit, I can't say for certain. And if they can't, I'll happily admit I'm wrong about these polls *not* being wrong. But no one can say for certain whether that happened because they're applying flawed methods.

What I can say for certain is that the Proportional Method forbids the accurate *reporting* of data, and that "Conservatives won by 7 after polls said it was close" is not an accurate way to *measure or detect* error. It's not even good math.

Had the polls been reported and explained accurately, that there were a lot of undecided voters and the outcome could swing either way, the public and field would be in a much better place. This isn't just a critique on "what could be done better" in the past, it's something that can be immediately fixed for the future.

Analysts have a very hard job, and pollsters an even harder job, but ultimately I believe everyone, including non-experts, has the same goal: accurate data and accurate detection of errors.

My Optimistic Side

Dozens of analysts working together with pollsters is the kind of collaboration that gives me lots of hope. They had a prime opportunity in their recontact surveys to itemize not just eventual party preference but also the preferences of previously undecided. It doesn't seem that was considered for the most part, but it would provide the biggest and most valuable piece of the "were polls wrong?" puzzle.

Sometimes in research, you don't realize things you could've or should've done differently until after the fact, and that's okay. The beauty of science is that you can do it better next time, and my research is no exception. But because of the demands imposed by the "way we've always done it," whether Proportion or Spread, there can be no "better next time."

To do better research, analysts should itemize each variable that can contribute, and not remain reliant on the pseudoscientific guesswork of the existing methods.

The researchers were in a perfect position to do that, and I hope in the future – in addition to outright abandoning the Proportional Method – they will.

My Less Optimistic Side

In response to this 2015 general election "failure," in preparation for the 2017 general election, Sturgis said, "the emerging upshot is that the companies are going to have to be more imaginative and proactive in making contact with – and giving additional weight to – those sorts of respondents that they failed to reach in adequate numbers in 2015."[19]

In essence, because the polls seemed to fail in 2015 by underestimating Conservatives, the recommended solution wasn't to tell pollsters to stop trying to forecast the outcome ... it was to give more weight to Conservatives in the pollcasts.

Sturgis wasn't alone in this assessment.

While Nate Silver made a name for himself aggregating US polls and forecasting US elections, he wasn't too shy to give his analysis of UK polls. While the "Spread Method" he utilized for US elections allows for undecideds, he seemed to try to analyze the United Kingdom's "Proportional Method" polls by the same "spread" standard – piling that on top of the same misinformed belief that polls try to predict the final margin:

> On average in the U.K., the final polling average has missed the actual Conservative-Labour margin by about 4 percentage points.[20]

Silver noted that, unlike in the United States where neither party consistently "beats their polls," Conservatives in the United Kingdom had.

Matt Singh, founder of Number Cruncher Politics in the United Kingdom, had identified the same trend in his work that showed why Conservatives could outperform their poll number.[21]

So, pollsters should just give a little additional weight to Conservatives in their future pollcasts. What could go wrong?

UK General Election 2017

What has happened in the United Kingdom, and other places where the Proportional Method is used, are the ramifications of what happens when you demand pollsters be forecasters. Instead of reporting their observations, which provide valuable data, they're asked to report their pollcast. Consequently, instead of lots of independent polls from which good forecasters could forecast, and good analysts could convey the possible outcomes (and uncertainty) to the public, you have lots of forecasts based on individual polls.

Then, being unaware of the fact that the data being analyzed aren't polls but pollcasts, analysts and experts wrongly report to the public that the pollcasts being wrong was a failure of the polls. Pollsters are chided, and no consideration is given to the possibility that the methods – not the pollsters – are flawed.

Pollsters, giving in to this pressure, try to make better forecasts – which requires making lots of assumptions that have nothing to do with polls. The cycle will never end until the methods are fixed.

One such assumption was that Conservatives would beat their polls, so pollsters (with the blessing of many experts, under the disguise of "weighting") tried just that in 2017.

YouGov, among others, decided to "reallocate" don't-knows who were likely to vote "using their 2015 vote."[22,23]

For the first time in a long time, I'm without words.

Without exaggeration, this was accepted as a valid practice:

1. The pollster determines an individual is likely to vote
2. The pollster asks who that individual will vote for, and they say they "don't know"
3. The pollster asks who they voted for in the previous election
4. The pollster reports that this individual who said they "don't know" who they will vote for, said they will vote for the same party as they did previously

No wonder the public doesn't trust poll data. They've turned a challenging, inexact, but valuable science into a pseudoscience.

Reasonable people can debate whether likely voters who declare that they are "undecided" in 2017 are most likely to vote for the party they voted for in 2015. You can argue that's the most reasonable assumption, if you want! But it's still an assumption, and it's *your assumption*, not the poll's.

A good *forecaster* might take that probability into account, model a range of outcomes (e.g. an undecided voter who previously voted Conservative may be most likely to vote Conservative, but could also vote for another Conservative-leaning party, but is much less likely to vote Labour). But a poll is not a forecast.

And they didn't even do *that*!

They just allocated them to the party they voted for in 2015, then said "this is what our poll observed."

No, it wasn't.

The consensus, based on the polls, seemed to be:

"Polls point to landslide victory" for Conservatives.[24]

Landslide, the polls said. I suppose if we calibrate our tool around the assumption that Conservatives will do much better than expected, that might happen.

Similarly, I could calibrate my scale to always read 5% lower than it measures. Being wrong about my weight, in that instance, is an error – but it's not the scale's.

How did the pollcasts masquerading as polls fare in 2017?

One headline read:

How almost all the pollsters got it wrong ... again.[25]

Another said:

Why 2017 UK Election Polls underestimated Labour.[26]

Here we go again.

The trend that had been true in past elections wasn't true in this one. Maybe try the thumb on the *other* side of the scale next time. Or ... just stop trying to make forecasts out of polls, and calling the pollcast the poll.

Obligatory snark, Simultaneous Census, Proportional Method, etcetera etcetera, here's what the 10 polls closest[27] to the 2017 general election reported:

Conservative (%)	Labour (%)	Undecided (%)	Date[28]	Pollster
41	40	0	Jun 06, 2017	Survation
46	34	0	Jun 06, 2017	ICM
42	35	0	Jun 05, 2017	YouGov
43	36	0	Jun 04, 2017	Opinium
40	39	0	Jun 03, 2017	Survation
45	34	0	Jun 02, 2017	ICM
41	40	0	Jun 02, 2017	Survation
42	38	0	Jun 01, 2017	YouGov
45	34	0	May 31, 2017	ICM
42	39	0	May 30, 2017	YouGov

Again, they reported that zero voters were undecided, not true.

Compared to the final result, while the pollcasters had a large forecast error underestimating Labour (who finished at 40%), they were ironically very close on Conservatives (who finished at 42.3).

This error is not one they deserve credit for: their pollcasts *overestimated* the conservative-leaning UK Independence Party (UKIP) by nearly the amount they *underestimated* Labour. Given the ideological differences, this forecast error is not dismissible by UKIP voters switching to Labour – the pollcasters had two large misses.

This is what the analysts and media report as "the polls" being wrong.

And in case you're trying to find the best pollster from the above table, stop. The data above are not the polls – they are their pollcasts.

But pollsters are not forecasters, nor should we ask them to be.

Here is what that same poll data observed:

Conservative (%)	Labour (%)	Undecided (%)	Date[29]	Pollster
33.9	32.6	18.4	Jun 06, 2017	Survation
38	33	14	Jun 06, 2017	ICM
36	34	10	Jun 05, 2017	YouGov
37	33	11	Jun 04, 2017	Opinium
34.3	34.1	12.1	Jun 03, 2017	Survation
35	32	16	Jun 02, 2017	ICM
36.0	35.0	13.4	Jun 02, 2017	Survation
36	34	10	Jun 01, 2017	YouGov
35	33	16	May 31, 2017	ICM
38	33	9	May 30, 2017	YouGov

You'll note some aggressive "reallocations" of Conservative voters. ICM observed 38–33 in their June 6 poll but reported 46–34.

YouGov observed 36–34 in their June 5 poll but reported 42–35.

The "reallocation" logic was based on the 2015 report that said they should give more weight to Conservatives.

The average of these polls, more appropriately calculated, was:

Conservative: 35.9% Labour: 33.3% Undecided: 13.0%.

Another UK election with double-digit undecideds. How did they end up deciding?

For the first time in my UK research, there was some good data that spoke to exactly that.

Ed Fieldhouse and Chris Prosser, as part of the British Election Study (BES), followed more than 20,000 respondents for a period of several months, leading up to and after the election.[30]

They found that "more than half" of voters who hadn't made up their mind – the "don't know" and "undecided" – ended up voting Labour.

As you saw in Utah, "more than half" in an election with three or more competitive parties adds up to a lot.

Conservatives saw a particularly large "Changing Attitudes" boost from the UKIP, and some net benefit from undecideds, though far less than "proportional" or "how they voted last election."

It would be wrong to conclude this study provides proof, but the evidence points in the same direction.

The General Election polls in 2017 look very accurate.

Using the formula that I provided in Chapter 14 for estimating poll error when a simultaneous census is not available:

Conservatives 2017: $(35.9\% +/-$ Changing Attitudes + Undecided Ratio$)$ $-42.4\% =$ Total Error.

Labour 2017: $(33.3\% +/-$ Changing Attitudes + Undecided Ratio$)$ $-40.0\% =$ Total Error.

Labour saw a disproportionate benefit from an Undecided Ratio over 50%, and also a substantial Changing Attitude boost from individuals who previously expressed that they would support the Liberal Democrats. Contrary to what I would have expected (and what the existing methods demand be assumed), a lot of people changed their minds, and even a non-negligible number of Conservatives changed their minds close to the election to support Labour.[31]

No matter how beautiful or reasonable the assumption, what matters is what is true. It seems changing attitudes are not as rare as expected.

What Conservatives lost to Labour and other parties, they recouped as a large portion of previously polled UKIP supporters gradually defected to Conservative.

In total, accounting for Changing Attitudes and Undecided Ratio, I would estimate the 2017 Poll Error as follows:

Conservatives 2017: $(35.9\% + 0\% + 4\%) - 42.4\% = -2.5\%$ $(2.5\%\,\text{underestimate})$.

Labour 2017: $(33.3\% + 1\% + 7\%) - 40\% = 1.3\%$ $(1.3\%\,\text{overestimate})$.

Far from perfect, these numbers both represent very good estimates, as good polls do. Measured properly, a discussion on what caused the error(s) could be discussed. But for now, not predicting the eventual difference between the top two parties is the best that analysts can offer. Huge error.

If analysts and the media would be honest with the public about what the poll data said – including that there were a lot of undecideds and therefore the eventual winner and the size of victory is far from certain – then the shock that comes with "unexpected" results wouldn't be so unexpected. If the polls in 2015 were in fact wrong, the solution is to figure out why, so that it can be prevented from happening in the future. The solution is not to observe that polls "underestimated" Conservatives by some flawed metric, therefore Conservatives should be given additional weight in the future. The same can be said for the perceived accuracy of the 2019 General Election polls: polls that are perceived as accurate yield a lack of vigilance based on insufficient data, which is not suitable for a scientific field. Assuming there were no issues with the methodology because it produced a result close to what was expected is Literary Digest reasoning.

In 2022, Peter Kellner, the former President of YouGov, published an article regarding their final poll in 2017 – and acknowledged their fear of getting it wrong. More directly, since polls are wrongly treated as predictions, they wanted their pollcast to be accurate. Kellner reported that the process for their final poll included moving "two percentage points from Labour to Conservative."[32]

The next UK general election is scheduled for July 4, 2024. Unfortunately, that date is before this book is released but after my deadlines. While a Labour majority is expected, I'm skeptical that the popular vote margin will be as dramatic as is currently being reported in polls, with some as high as 30-points.[33]

Most polls have "undecided" as the second or third-largest group, which makes any pollcast's reported "lead" highly spurious. It appears pollsters have abandoned their "assume undecideds will vote as they did in the previous election" approach, but polls are still being reported via the Proportional Method. As mentioned earlier, 2024 Mayoral polls did not fare particularly well, not because the polls were bad, but because the data being reported as "polls" are still forecasts.

Financial Times has updated their poll aggregation methodology to exclude those who say they "don't know" from their averages, and report zero undecided.[34]

The BBC's poll tracker[35] and the Telegraph's[36] poll trackers do the same.

While the Spread Method convention in the United States has its own issues, many other countries utilize the highly problematic Proportional Method. I can't speak to countries I haven't researched, but at minimum, the Proportional Method is currently used in Canada, Australia, Spain, Portugal, France, Germany, and Italy. Both polling and forecasting have unique challenges, so those challenges must be uniquely understood by experts so they can be properly relayed to both the public and policymakers. Methods that view these distinct domains as one and the same actively misinform.

Polls remain controversial, the public largely distrusts them, and governments have even considered stricter regulations on them in the face of perceived failures.[37]

If we continue on this path of judging poll accuracy by how well they predict election results, we will continue to ride the roller coaster of "historically accurate" to "catastrophic failure" without understanding how or why, and statistical literacy will remain unimproved. Methods that "chase" past results by introducing weights that "would've" made their past polls more accurate (by the flawed polls-as-predictions standard) are not scientifically valid. And on top of these known issues, I suspect compensating error plays a larger role than many expect in how poll accuracy is measured and perceived. Analysis of past perceived poll failures should not be mistaken for purely a commentary on the past. It's analysis with applications for how things should be changed for the better going forward.

But right now, things are not getting better. The few steps I've outlined in this book could yield improvement very quickly – for analysts and the public alike.

Whether the improvement starts at the top with the experts and works its way down to the public, or in the classroom with a new generation of experts – or not at all – is not ultimately up to me. But at minimum, you have the ability to navigate the past and present analysis, flaws and all.

As for what I think the future should look like, and how educators can introduce the concepts presented in this book, the final chapter that follows offers my closing thoughts for now.

Notes

1. Clark, T., & Perraudin, F. (2016, January 19). General election opinion poll failure down to not reaching Tory voters. *The Guardian.* https://www.the-guardian.com/politics/2016/jan/19/general-election-opinion-poll-failure-down-to-not-reaching-tory-voters

2. Young, P. (2016, January 22). *GE 2015: Why did the pollsters get it so wrong?* House of Commons Library. https://commonslibrary.parliament.uk/why-did-the-pollsters-get-the-general-election-results-so-wrong/

3. BBC. (2020). *Election 2015.* BBC News. https://www.bbc.co.uk/news/election/2015/results

4. Pollsters whose data I could not access were excluded: SurveyMonkey and IPSOs.

5. Based on the earliest date in the field.

6. Sturgis, P., Kuha, J., Baker, N., Callegaro, M., Fisher, S., Green, J., Jennings, W., Lauderdale, B. E., & Smith, P. (2018). An assessment of the causes of the errors in the 2015 UK general election opinion polls. *Journal of the Royal Statistical Society. Series A: Statistics in Society, 181*(3), 757–781. https://doi.org/10.1111/rssa.12329

7. Allen, C. (2024, May 6). *UK polling aftermath.* https://realcarlallen.substack.com/p/uk-polling-aftermath

8. Kennedy, C., Blumenthal, M., Clement, S., Clinton, J., Durand, C., Franklin, C., McGeeney, K., Miringoff, L., Olson, K., Rivers, D., Saad, L., & Wlezien, C. (2018). An evaluation of the 2016 election polls in the United States. *Public Opinion Quarterly, 82*, 1–33. https://doi.org/10.1093/poq/nfx047

9. I am making the same adjustments here as I outlined in the previous chapter.

10. Includes all those the poll identified as "likely to vote" but responded "don't know" or "refused."

11. Based on the earliest date in the field.

12. Sturgis, P., Kuha, J., Baker, N., Callegaro, M., Fisher, S., Green, J., Jennings, W., Lauderdale, B. E., & Smith, P. (2018). An assessment of the causes of the errors in the 2015 UK general election opinion polls. *Journal of the Royal Statistical Society. Series A: Statistics in Society, 181*(3), 757–781. https://doi.org/10.1111/rssa.12329

13. Brione, P., & Lowe, D. L. (2015, July 14). *The general election 2015 & the polls— What happened?* Survation. https://www.survation.com/the-general-election-2015-the-polls-what-happened/

14. Shakespeare, S. (2015, December 7). *Analysis: What went wrong with our GE15 polling and what will we do to improve?* YouGov. https://yougov.co.uk/politics/articles/14102-analysis-what-went-wrong-our-ge15-polling-and-what?redirect_from=%2Fnews%2F2015%2F12%2F07%2Fanalysis-what-went-wrong-our-ge15-polling-and-what%2F

15. Rivers, D., & Wells, A. (2015). *Polling error in the 2015 UK general election: An analysis of YouGov's pre and post-election polls.* https://d3nkl3psvxxpe9.cloudfront.net/documents/YouGov__GE2015_Post_Mortem.pdf

16. Fieldhouse, E., & Prosser, C. (2017, August 1). *General election 2017: Brexit dominated voters' thoughts.* BBC News. https://www.bbc.com/news/uk-politics-40630242

17. This also applies to the misplaced belief that individuals "lying" about their voting plans constitute a poll error. If some percentage of individuals in a poll lie about their voting plans, that means that if a Simultaneous Census were conducted, a similar proportion would also lie. This is a *limitation* of the polling instrument, not an error.

    Asking 1000 people how often they exercise, then following them to observe how often they exercise, is not a measure of the original question's accuracy.

18. Brenner, P. S., & DeLamater, J. (2016). Lies, damned lies, and survey self-reports? Identity as a cause of measurement bias. *Social Psychology Quarterly, 79*(4), 333–354. https://doi.org/10.1177/0190272516628298

19. Clark, T., & Perraudin, F. (2016, January 19). General election opinion poll failure down to not reaching Tory voters. *The Guardian.* https://www.theguardian.com/politics/2016/jan/19/general-election-opinion-poll-failure-down-to-not-reaching-tory-voters

20. Silver, N. (2017, June 3). *Are the U.K. polls skewed?* FiveThirtyEight. https:// fivethirtyeight.com/features/are-the-u-k-polls-skewed/
21. Singh, M. (2015, May 6). *Is there a shy tory factor in 2015?* Number Cruncher Politics. https://www.ncpolitics.uk/2015/05/shy-tory-factor-2015/
22. YouGov. (2017). *YouGov/The Times survey results.* https://d25d2506sfb94s. cloudfront.net/cumulus_uploads/document/d8zsb99eyd/TimesResults_ FINAL%20CALL_GB_June2017_W.pdf
23. ICM Unlimited. (2017). *The Guardian—Campaign poll 10 prediction poll— PRELIMINARY.* https://www.icmunlimited.com/wp-content/uploads/2017/06/ 2017_guardian_prediction_PRELIM_1500.pdf
24. Settle, M. (2017, June 8). *Britain at crossroads as eve-of-poll survey suggests May heading for landslide victory.* The Herald.
25. Kellner, P. (2017, June 9). *How almost all the pollsters got it wrong ... again.* Evening Standard. https://www.standard.co.uk/news/politics/general-election-polls-how-the-pollsters-got-it-wrong-a3560936.html
26. Curtice, J. (2017, June 8). *Why 2017 UK election polls underestimated Labour.* The Conversation. https://theconversation.com/why-2017-uk-election-polls-underestimated-labour-79513
27. For which I had access to complete reports.
28. Based on the earliest date in the field.
29. Based on the earliest date in the field.
30. Fieldhouse, E., & Prosser, C. (2017, January 8). *The Brexit election? The 2017 general election in ten charts.* The British Election Study. https://www.britishelectionstudy. com/bes-impact/the-brexit-election-the-2017-general-election-in-ten-charts/
31. Fieldhouse, E., & Prosser, C. (2017, January 8). *The Brexit election? The 2017 general election in Ten charts.* The British Election Study. https://www.britishelectionstudy. com/bes-impact/the-brexit-election-the-2017-general-election-in-ten-charts/
32. Kellner, P. (2022, June 8). Why do polling firms like YouGov tweak polls? Because they are scared of being wrong. *The Guardian.* https://www.theguard-ian.com/commentisfree/2022/jun/08/polling-firms-yougov-tweak-polls
33. UK's Labour holds 30-point poll lead over Sunak's conservatives. Reuters. (2024, May 9). https://www.reuters.com/world/uk/uks-labour-holds-30-point-poll-lead-over-sunaks-conservatives-2024-05-09/
34. Hawkins, O., Hollowood, E., Stabe, M., & Vincent, J. (2024). *UK general election poll tracker.* https://www.ft.com/content/c7b4fa91-3601-4b82-b766-319af3c261a5
35. BBC. (2024, February 14). General election 2024 poll tracker: How do the parties compare? *BBC News.* https://www.bbc.com/news/uk-politics-68079726
36. Holl-Allen, G., & Butcher, B. (2024, February 11). *UK general election poll tracker.* The Telegraph. https://www.telegraph.co.uk/politics/2024/02/11/uk-general-election-poll-tracker-conservative-labour/
37. Sturgis, P., Kuha, J., Baker, N., Callegaro, M., Fisher, S., Green, J., Jennings, W., Lauderdale, B. E., & Smith, P. (2018). An assessment of the causes of the errors in the 2015 UK general election opinion polls. *Journal of the Royal Statistical Society. Series A: Statistics in Society, 181*(3), 757–781. https://doi.org/10.1111/rssa.12329

34

The Future, and "Try It"

Important Points Checklist

- How media-reported poll averages could improve the field's methods
- Why polling and forecasting are very different things
- What "wishcasting" refers to, and why it is a threat to the field
- How dice can also be used to think about forecasts
- "Try it" Dice and Stopwatch experiments to illustrate polls versus poll averages versus forecasts
- The modified urn problem

While the public – and apparently many experts – believe that polls and poll aggregation and forecasting are very closely related fields, they're about as related as chefs and farmers.

> When you're in Hollywood and you're a comedian, everybody wants you to do other things. All right, you're a stand-up comedian, can you write us a script? That's not fair. That's like if I worked hard to become a cook, and I'm a really good cook, they'd say, "OK, you're a cook. Can you farm?"
>
> *Mitch Hedberg*

OK, you worked hard to become a really good pollster. Can you forecast? The solution to the problems faced by these fields isn't to ask pollsters to forecast, it's to let them work in their own field.

The issues that permeate the seemingly related but very different fields of pollsters, poll aggregators, forecasters, and analysts aren't ones that will be solved by one group making a concerted effort. It will require them all working together. Whether consciously or subconsciously, these fields currently exist in a standoff; all of them point the fingers at others and have incentive to do so. Instead of working together for what would be the betterment of all involved (and inferential statistics as a whole), it seems to be a race to take

DOI: 10.1201/9781003389903-34

credit when "right" and deflect blame when "wrong." Healthy competition, even irreconcilable disagreements, can contribute to advancements in statistics and science – but only if advancement is the goal.

I haven't been shy about the direction I think this industry is going – if it hasn't already gone – but that doesn't mean it can't be fixed.

Analysts/Experts, the Future

I'm starting with analysts and experts because they're the ones who have the most power here, both over others in the field and the public. Analysts should start using an adjusted method that accounts for the fact that people sometimes change their mind and undecideds eventually decide – and none of those factors are related to poll error or poll accuracy. When I say "an adjusted method," I do not necessarily mean "one of mine" just ... something better than poll result versus eventual election result.

Moreover, they should separate the effects of polls from poll aggregators and analyze poll averages just as fervently as they do individual polls. To my knowledge, no such evaluation of poll aggregators exists; hold them to the same standards you hold pollsters. In the absence of scientific standards, misinformation will be allowed to run rampant: you can get a poll average to say almost anything you want just by averaging the right polls.

Polls, the Future

This is the easiest one to recap but the hardest to implement. The human desire to know the future won't cease to exist just because polls aren't predictions. But polls are, by far, our best tool to inform predictions. **Polling is the most time-consuming, challenging, and most important job on this list**. Having as many good, *independent* data collectors as possible is vital to the success of this field. That doesn't mean there can't be innovation or methodological differences, it means the data itself shouldn't be weighted with herding or pollcasting intent.

But because of how poll quality and accuracy are currently measured – it's better to be wrong with everyone. Pollsters have substantial incentive to herd and pollcast *because* of the bad definitions used by analysts and experts.

If the best data collectors – those who report their data independently and transparently – are squeezed out because analysts don't know how to judge poll accuracy, things will only get worse. They currently have incentive to herd and pollcast. It would be very bad if the flawed definitions used by analysts and forecasters regarding poll accuracy pushed pollsters who are more accurate (but refuse to pollcast) out of the industry.

Regardless of what analysts do, and the public wants, **pollsters should report their data as they actually observe it, undecideds and all.** This would put an end to the Proportional Method, and deservedly so.

Methodology and other weighting methods should be transparent. Conduct post-election research to analyze your own polls by a valid standard, if analysts remain reluctant to do so. Don't take them at their word about how much your poll was "off by" because they don't know. Pressure them to judge polls by the correct standard.

Poll Averages, the Future

This industry is one I believe will see the biggest growth in the coming election cycles, with the most potential for good. As the public gains understanding that poll averages are preferable to individual polls for informing predictions, considering that taking a poll average is far easier than taking a poll, data-centric organizations (both actual and perceived) will start giving them one. Like the "smoking and cancer" link was obscured from the public for decades after experts knew of it, poll aggregators have the ability to obscure poll averages for their own gain. That's not to say there is a single, best way to take a poll average, that's to say if a media outlet or "influencer's" preferred narrative doesn't fit, they can make it.

As I briefly discussed regarding RealClearPolitics' "poll aggregation" method, while it grades 100% for transparency, the most recent election featured egregious exceptions for consistency.

It's not all negative though: this power and reach, properly used, could also make the most immediate impact. Instead of reporting (or focusing on) individual polls, **media outlets can report their own poll averages – preferably using the leader reporting standard.** Report either all the numbers (including third-party and undecideds) or only the leader.

This increased competition would yield discussion, innovation, and give the public an accurate representation of what "the polls" say.

While US data analysis and reporting needs improved, the situation is far more dire outside the United States. Due to the prevalence of the Proportional Method, many poll aggregators improperly use zero-undecided polls as part of their average, with some poll aggregators increasingly reporting 0% undecided *as their poll average.*

Competition among major and even minor media outlets for the best polls and poll average and explaining to viewers that any that show 0% undecided is a lie will get rid of it faster than you can finish reading this sentence.

Forecasts, the Future

While poll averages have the most room for growth, owing to their relative ease to conduct, forecasting – properly done – is very hard. But forecasters are the only ones who actually try to predict the future, so it seems natural they will receive increased influence as well. Thankfully, given that real forecasters publish both vote share predictions and win probabilities, their error(s) can be calculated easily: Forecast versus Result.

Projected vote share versus actual vote share is straightforward: the difference is the error. No, it's not poll error. Whether poll error *contributed* to forecast error is a separate issue; if your forecast was wrong, that's a forecast error.

As for win probability, that's much harder. In the long run, forecasters who assign an outcome with a "50%" probability can be considered accurate if those things happen about 50% of the time. Likewise for 60%, 70%, 90%, and 99%.

But "in the long run" is a very long time to wait in elections. Are two election cycles enough to figure out the best forecasters? Ten? The uncomfortable answer is many more than that.

But currently, the media accepts roughly one election cycle as a way to crown the king of forecasts. If you predicted what would happen, you must be good! Who "got it right(est)" this last time is more important than better underlying methods and transparency.

Unfortunately, **thinking probabilistically is something the current standards don't allow for**; the current definitions used by experts say polls make "predictions" both about who will win and by how much.

It's unfair to ask the public, who has been wrongly convinced of polls-as-predictions, to suddenly accept a nuanced and relatively advanced understanding of the difference between polls and forecasts.

> The added layer of hypocrisy comes when forecasters brag about how many states or constituencies they "called" right (when they're "right") but want to chastise the public for not understanding "it's just a probability" when their "calls" are wrong.

With the growth of social media – where confirmation bias has the ability to run wild – I noticed another major turning point that will be dangerous for the field of inferential statistics: **Wishcasting**.

Wishcasting is the pseudoscientific offspring of forecasting and mediafication.

While pollcasting is the attempt to make a forecast out of one poll, wishcasting is simply saying what you wish to happen (or what will get you the most clicks) but with enough confidence and perceived expertise that people take you seriously. This usually comes with some sort of "guarantee," and Wishcasters will note their qualifications and expertise, but don't share anything substantive about their methodology. Yes, some influential members of the media fell victim to putting real forecasters like FiveThirtyEight and The Economist next to Wishcasters.

Like mediafication gives actual media outlets incentive to publish things to get clicks, so too in the social media age do influential individuals have incentive to do so.

This is another downstream effect that I believe is inevitable when the public isn't properly informed. A reputable forecaster says this thing is 50% to happen, but my favorite social media personality said it was a 99.99% guarantee – which one are you more likely to click on?

The desire to predict the future with perfect precision is an understandable but unrealistic one. But in the absence of accurate methods promoted by experts – methods that quite literally state being "closer to the result" means "better" – **the experts in the field have no grounds to declare wishcasters as illegitimate.**

If their guesses are better than your data, that means they were better – by your own definitions.

For readers who wish to stay informed, I hope it goes without saying, but anyone who declares with any level of certainty a specific outcome is not a legitimate forecaster. You can think of it "for entertainment purposes only" if you must. Forecasts are not as easy as dice, but the possible range of outcomes – no matter how much someone thinks they know – is just as wide, sometimes even wider.

How to Read Forecasts – With Dice

My primary goal for this book is to teach you how to read polls and poll averages to inform a prediction. I hope I've done that. But this work wouldn't be complete without a little discussion on forecasts. Building a forecast is much harder than reading them.

I have, for the most part, referred to "predictions" and "forecasts" similarly. They are similar but are not the same.

A **prediction** can be thought of as a this-or-that proposition, such as "I think undecideds with split 50/50," or "I think Red will win" or perhaps more specifically, "Red will win 52% of the vote."

Predictions can range from well supported to total guesswork.

A **forecast** is a range of outcomes that assign a probability to each possible outcome. Likewise, forecasts can range from well supported to total guesswork – but since forecasts require much more work (and the underlying methodology can be tested), people who aren't qualified to do forecasts tend to call their "predictions" their "forecasts."

Whether or not they can accurately calculate the probability, most reasonable people wouldn't "guarantee," "predict," or "call" that a roll of two dice will produce a result of 5 or more, even though that outcome is highly likely.

In general, the media and the public (and unfortunately, many experts) gravitate toward predictions. They want someone to make a **call** because calls make for better, simpler headlines. While these "calls" are easier to explain, they are almost always oversimplifications – and teach us nothing about using data to inform our predictions.

Earlier, I used two dice as an example to better understand the margin of error and what is meant by 95% confidence. Dice can also be used to understand forecasts versus predictions.

Understanding the statistics that underlie dice rolls can help people grasp the difference between a prediction and forecast – and why we should not take people who make overly confident "predictions" seriously, no matter what their "track record" is.

Here again is the breakdown of the probability of rolling each possible combination with two standard, six-sided dice.

Sum	2	3	4	5	6	7	8	9	10	11	12
Prob.	1/36	2/36	3/36	4/36	5/36	6/36	5/36	4/36	3/36	2/36	1/36

Try It: Dice

Without sharing the "true probability" breakdown, tell students/willing participants that you're having a contest to see who can forecast *your* 36 rolls of dice most accurately. Prizes optional. To assist them in their forecast, teams are given the opportunity to roll their own pair of dice 36 times.

Provide teams with two sheets: a "tally" sheet and a "forecast" sheet.

Have teams roll two dice 36 times and tally how many times each sum is made. Then, *using only their own data*, ask them to predict how many times they think *you* will achieve each outcome.

Since they're working as forecasters now, they can interpret their data however they please. It's possible that students who have a strong grasp of probability will encourage their team to "weight" numbers more closely to the true probability – that's okay.

Have each team turn in both their tallies and forecasts.

Now, before taking your 36 rolls, tally the group's (unweighted) results and create one giant chart. These numbers will be taken "as-is." How many total 2s, 3s, and so on, expressed as a percentage of the total number of rolls the whole group took.

For example, if 10 groups rolled 36 times (360 total rolls) and combined to roll "2" nine times, you would list it as "2.5%" (9/360) and so on.

This is the "roll average" for each number.

Ask each student to choose a team: they can choose to stay with their team-reported forecast, or side with the whole group results. Which do they think is better?

Take your 36 rolls and record the results. There's a very good chance your 36 rolls, like their sets of 36 rolls, don't match the known experimental value for what "should" happen. The first time I tried this, the number I rolled most was "9."

Whose forecast was best? **Remember, your 36 rolls is the "true" result they were trying to forecast.**

If the number you rolled most was "9," does that mean a team that forecasted the most 9s was best? Is that group the new king of forecasting?

How "better" is defined – whether by closeness to eventual result or underlying methodology – is an important statistical concept. This is also a level of nuance that existing definitions of "closer to result = better" do not allow for, that would benefit learners, and one I believe most people can grasp given the opportunity.

These conversations, while they seem oversimplified because I'm using dice and mints instead of elections, are at their core no different, except that there are variables that can be known.

Given a sufficient number of forecasters, it's highly likely that the best forecasters *are not* always the closest to the result. This is also true for pollsters and poll aggregators in relation to the Simultaneous Census.

In dice, the forecast involves the breakdown of how many times you roll each number.

But without an ability to rule out certain methodologies (and differentiate one or several good guesses from valid and repeatable methods), the desire to declare which forecaster is *best* will inevitably fail: whether Literary Digest in 1936 or Wang in 2016.

Now to take it even further:

What if the class tries to *predict* the outcome of *one roll*? No longer are we forecasting a range of outcomes, but we're being cornered into making a "call."

Can I get away with 7 +/− 4?

If not, the "best guess" would be seven. Seven is the likeliest of all dice rolls, occurring at a frequency of 6/36 (or 1/6). Whether or not you know the odds, you don't need to be a stats wiz to understand that, although a prediction of "seven" is the best one, there's still a very good chance you're wrong.

Ask the students about which approach – the forecast or the prediction – is a better way to think about an uncertain outcome.

Not making the "right call" says nothing about someone's ability as a forecaster. Sure, picking 2 or 12 should raise some red flags, but is "calling" 6 or 8 so bad?

While the "best bet" is seven, and all the number nerds like me "play the numbers," maybe someone realizes that by picking something *different* than everyone else, there's a chance – if that dice roll happens – they can claim they were *right* and knew more than the so-called "stats experts." Maybe 5 or 9 was "due." Maybe 6 or 8 were "hot."

If enough people apply this logic – and "predict" lots of different outcomes, it's inevitable that some of them are going to be right. What does that mean regarding their ability to predict future dice rolls? Not much.

But by the time their wrong calls miss bad enough for them to be embarrassed, the damage is already done. Someone else guessed this one right, and they're the new experts.

This applies to election forecasting more than you may think.

With one roll to determine the winner, consider there are two teams named after their respective sums: team 2–5 and team 6–12. There's a prize for being on the right team.

Without even calculating the odds, ask the participants which team they'd rather be on. There's a good chance that roughly 100% choose team 6-12.

As it happens, in this instance, we can calculate the odds of each team:

Team 2–5: 27.8% (10/36)

Team 6–12: 72.2% (26/36)

If you think of this as an election, a good forecaster would have numbers close to this. A 72% chance of winning is pretty high. No reasonable person would disagree that team 6–12 is favored. But also, no reasonable person would "call" or guarantee victory for team 6-12. Right or wrong is a matter of chance.

Saying that one side has a 72.2% chance to win is very different from saying you think they *will* win. Nonetheless, in elections, that's how results are judged. A forecaster's (and even poll's) results are interpreted as calls. It would be forgivable if the public wrongly reached these conclusions on their own, but that's not the case.

In addition to media outlets wrongly reporting probabilities as "calls", some experts include the percentage of races a pollster "calls correctly" among the factors in their accuracy.[1]

In the very real chance that the roll is 2–5, does someone who assigned "2–5" a 55% chance deserve more consideration as a great forecaster?

Existing methods would say yes because closer to result = better. I disagree.

If you choose to look at forecasts around election time, think in terms of ranges of outcomes, not calls.

While Figure 34.1 might look like a range of outcomes for dice, this is close to what a legitimate forecaster might have as vote total probabilities for a

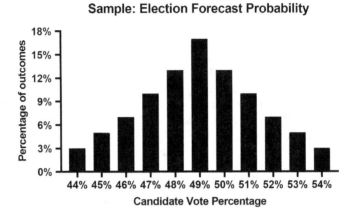

FIGURE 34.1
A chart showing a hypothetical forecast with range of outcomes. Chart by the author.

candidate. Consider what you know about dice probability before declaring which forecaster(s) are best on the basis of one or several instances of them "calling" a race correctly, or even being closest to the eventual result.

Try It: Stopwatch "Polls" and Actual Races

If the dice experiment wasn't advanced enough for your taste – or maybe you have a sport-inclined group, how about some literal races to teach statistics?

Stopwatches are excellent tools for their designed purpose but are subject to error unrelated to the tool – sound familiar?

Many individuals who wish to – or you have time for – can participate in the sprint competition. Individuals who don't wish to participate can be assigned to the "stopwatch" group, though sprinters can also be in that group when they're not running.

Participants are randomly paired to, in some near future date, race each other to a distance of 40 meters or 40 yards (m/yd), whichever is easier. The pairings can also be nonrandom to create contests that are fairer or more interesting for your group; instructors getting involved is a bonus. Nonrandom pairing doesn't have much impact on the experiment in this case.

Some number of participants, at least five, can volunteer or be assigned to the "stopwatch" group when it is not their turn to run. The stopwatch group are, in effect, pollsters for this experiment – but they can also participate in the races. It's best to keep the individuals in the stopwatch group consistent.

However, since polls are not elections, it would be somewhat unrealistic to allow each individual to be timed running 40 m/yd because going from poll to election requires forecasting. To make things interesting, I recommend a distance of 30 m/yd for the stopwatch "poll" though you're welcome to choose other distances.

Also, the pairs (even though they know who they will eventually be running against) do not run at the same time; the trial/poll round is run and timed individually. This is done to maintain similarity to political polls: you have an estimate of each individual's time but do not know with certainty – due to the potential for timing errors – who "won" the poll.

Each individual is timed, from "go" to finish line, by the various stop-watchers – for at least one trial but two or three is preferable. How quickly that individual truly ran the 30 m/yd is a subject worth discussing: it's highly unlikely any stopwatchers have exactly the correct time, or the same time. Using the average of all of them – or an average that first removes the best and worst times – are fine approaches; discuss the pros and cons of each.

After all individuals have completed their trials, the results are reported and provided to each student. They're asked to forecast both the winner of the race, and the eventual times (in seconds and tenths) for each participant.

Since all of the trials were over 30 m/yd, but the actual race is 40 m/yd, there's some forecasting to be done. There are a few pretty big steps between 30 and 40 m/yd. Will the person reported as having the better 30 m/yd time always win? If they don't, does that mean the stopwatchers were wrong?

Moreover, how can you project the spread by which they will win? Can we assume their spread will remain the same in the race as it was in the trial? Proportional? Are any discrepancies between these assumptions and eventual result attributable solely to stopwatch error?

In simplest terms: what does the 30 m/yd time predict about the 40 m/yd time? It's a valuable piece of data, but it is not itself a prediction. I'd much rather have a 39 m/yd time than 30 m/yd because it's easier to build a prediction with! But like in elections, we don't get to choose how many undecideds there are: and pretending the 30 m/yd time is a 40 m/yd time doesn't make it so.

> Making a piece of data *appear* more valuable than it is doesn't help inform a prediction – it actively misinforms it.

All of these lessons can be taught and understood to any person who wants to learn it, and many of those lessons are in this book. They're not particularly complicated at their base level, but the existing methods do not effectively understand or explain them.

This "race" example, which is easy to implement but hard to forecast, provides another way to teach statistics in a hands-on way.

Like polls in real elections, we can use the stopwatch time to form a "base of support" from which we build a forecast: it's highly unlikely the person runs 40 m/yd faster than they ran 30 m/yd, for example. That likelihood can change if the stopwatch race is to 35 or 39 m/yd. There's a "finish line" analogy in there, and while it's another silly analogy, it is extremely appropriate. I would estimate the probability of someone winning a race as much higher if I had data to support that they ran 39 m/yd a little faster, compared to just 30 m/yd. And if my prediction is wrong that doesn't mean the stopwatchers were wrong, though it's possible it contributed, the tools needed to measure each are not the same. The only thing that can be said with any level of certainty in this case is that my forecast was wrong.

To know how accurate the stopwatchers' reported trial times were – a separate calculation – if you don't have access to laser timing, you would most feasibly require a recording of the trials, slowing the video to slow motion to "start" and "stop" the timer at a more precise spot. You could then compare the stopwatchers' times to the video-recorded one. A fair amount of work, but a valid metric.

It is not accurate to make uniform assumptions about what a 30 m/yd time "should" produce in a 40 m/yd one, any more than it is accurate to do with Changing Attitudes and Undecided Ratio. While a qualified practitioner could certainly derive a beautiful estimate, assigning "error" calculations in every specific case based on the average or median case is spurious. For example, a runner who slips or trips in the actual "race" is an obvious confounder that would impact that assumption – and like poll-to-election confounders, needs to be controlled for. It's also possible that running a competitive race versus a non-competitive one has an effect, which would also need to be controlled for.

My proposal – analogous to the simultaneous census standard – is that to more closely estimate how accurate a past metric was, since it is impossible to *know*, we must account for the variables that changed.

The actual 40 m/yd races – being the "most important" piece of data – should be recorded on video, and times calculated as certainly as possible to mimic elections. With the "final result" known, comparing 30 m/yd times to "projected" 40 m/yd times is still insufficient.

Now, each stopwatcher has the task to find the "final 10m/yd" times for each participant. This should be done at full speed from video, with the same number of trials and stopwatchers as before the race. This reflects the imprecision inherent in recontact surveys.

That final 10 m/yd time – subject to some error – can then be used to "piece together" the difference between the actual race time, which is known with a high level of precision, and the previously taken 30 m/yd times. The closeness of those "recontact" numbers is not a perfect measure of stopwatch accuracy, nor can it be, because the only tool that can achieve that is a simultaneous census. In this experiment, the simultaneous census is analogous to each participant's true 30m/yd time. Like in elections, in this sprint experiment, that value literally cannot be known, but it exists, and is indisputably the time stopwatchers were trying to measure. That approach is intended to be a *more accurate approximation*, and one that is statistically valid – than assuming Changing Attitudes (the time it takes the participant to complete the same distance at a later date) are zero and Undecided Ratio (the time it takes them to complete the final 10m/yd) must be equal for all participants, or proportional to their 30m/yd performance.

Try It: Modified Urn Problem(s)

As you're aware, Mintucky and the modified urn problem are very similar. But explaining that to people who haven't read this book, you'll get a look, "What's Mintucky?"

I have found this is a valuable mental exercise to introduce the concepts of ideal polls, simultaneous census, present polls versus plan polls, and Spread versus Proportional Method, in addition to the lessons already present in the urn problem.

The urn problem has two options: black or white. Hundreds of years later, it remains a perfect way to explain many concepts in statistics and probability.

The fact that results can be and are compared to a simultaneous census in this problem leaves a gap for learners to understand what the "true value" means in inferential statistics.

It's not that the urn problem takes this fact "for granted" it just isn't covered: if someone took a random sample from the urn with a margin of error of +/−4%, and after the fact someone dumped out a few hundred marbles, then replaced them with a few hundred marbles from a different population, it's possible the data from the original sample is fine. That sample is certainly better data than no sample – but that margin of error can't be said to apply to this new, changed population with the same level of confidence because the population of the census would be different from the one in the poll, which leads to the modified urn problem.

I've presented it a few different ways (including with mythical pebbles that change color) and while I believe that hypothetical works fine, it feels a bit *too* unrealistic. So I tend to favor something closer to Mintucky. You are more than welcome to modify this problem as you see fit.

Your instructor presents to you an enormous bin, which he says contains "a very large number" of marbles. The instructor knows the exact contents of marbles in the bin, but no one else does. The marbles within the container are all red or blue, but some number of marbles in the bin are covered with a thin layer of foil to obscure their "true color." The foil cannot be removed except by the instructor, thus remains unknown to anyone who takes a sample.

He says you may take one sample that totals 600 marbles, record your answer, and place them back in the bin.

Assume that your sample is truly random, and that each color marble (red, blue, and silver) has a chance of being drawn equal to its proportion within the bin.

You (Student 1) have two main assignments:

1. Estimate the proportions of the colors currently in the bin.
2. Estimate the proportions of the colors that will be in the bin after the silvers are unwrapped.

Your sample produces the following:

Blue: 43%

Red: 43%

Silver: 14%

From this, you can calculate the margin of error for any confidence level you'd like, though I'd recommend the 95% confidence level.

Given this data and only this data, what can you say about the composition of the bin?

I believe the answer should be something like: "39%–47% blue, 39%–47% red, and 11%–17%[2] silver, with 95% confidence."

No problem, right?

Now consider how the Spread and Proportional Methods would characterize it:

Spread Calculation:

Blue: 43% (+0)

Red: 43%

Proportional Calculation:

Blue: 50%

Red: 50%

You can use these methods, or your own, to inform your prediction about the *eventual composition of the bin* (after the silvers are unwrapped). A few discussion topics:

1. What does your sample predict about the eventual composition of the bin, after the silvers are unwrapped? In other words, your sample gives you a margin of error for "known" reds and blues, how can your poll account for the future preferences of the ~14% silver, and should it?
   (I believe this illustrates the major difference between a poll and a forecast, and why they should remain discrete).

2. If asked to make a prediction, how would your prediction, or forecast, relate to the accuracy of your poll? Certainly, a good poll gives you a better chance of having a good forecast, but does it guarantee it? Would an inaccurate poll guarantee that your prediction is bad?

3. What would be the best way to measure the accuracy of the respective poll and forecast?

A Second Sample

A second student is permitted to take a sample, with all the same rules as above. To remain independent, you do not share any of your data. Their sample produces:

Blue: 50%

Red: 44%

Silver: 6%

They attempt to answer the same problems as above, starting with "Given this data and only this data, what can you say about the composition of the bin …?"

Spread Calculation:

Blue: 50% (+6)

Red: 44%

Proportional Calculation:

Blue: 53.2%

Red: 46.8%

4. Assuming you know both students' samples, how would you calculate whose is more accurate?

Your instructor knows both the true red versus blue contents of the bin *and* the current percentage of red, blue, and silver currently in the bin.

If your goal is to know whose *poll* was more accurate, and you can only ask your instructor about one of those, what question do you ask: the true contents, or the current percentage?

The fact that we have conducted a "plan poll" in which we want to know some future preference, doesn't mean that the accuracy of our sample is determined by how closely it matches that future result.

The Current Composition of the Bin

You ask, and your instructor reports that the current composition of the bin is:

Blue: 45%

Red: 40%

Silver: 15%

The fact that you can know each student's accuracy (via whatever method you deem appropriate) *without knowing the result* reinforces or demonstrates the simultaneous census standard, and the "true value" a poll attempts to measure.

Unwrapping the silvers *and then* trying to figure out the accuracy, while it may be the closest we can get to a simultaneous census "in real life," is not a "true value," and in many cases, it is not even a good approximation. The reasons why it may not be a good approximation is a worthwhile discussion too.

One response I received when presenting this problem was something like:

> I have no idea what reasoning, if any, the person who wrapped the silvers used, so I can't really assume the possibilities fit neatly under a bell curve.

I found refusal to make assumptions – especially in the face of being presented with a probability question – refreshing. I think you could say something similar about undecided voters.

When I put my "forecaster" hat on, I want to find the best data I can regarding undecided voters. If someone who says they are currently undecided voted Conservative in the last election, that's a pretty good place to start. But it is a deeply misplaced and untrue analysis to say that person "should" or "will probably" end up voting Conservative – or else the poll was wrong.

It's extremely presumptuous to assume that undecided voters and "unknowns" *must* fit similarly with decided or "known" ones – as both the Spread and Proportional Methods do. Forecasters can argue what they think the "best assumption" would be – you cannot argue that the assumption is the poll's.

Simply put, the first student's sample was much more accurate than that of second student. The second student's sample is on or outside the margin of error for all three options, and way outside of it for Silver.

I hope that we can all agree that, regardless of the eventual result, Student 1's sample was more accurate.

In terms of forecasting the eventual result, we now know the "current composition" of the bin. This may only narrow down the ranges of eventual results a little compared to what we may have forecasted from the polls, but it greatly increases the confidence we can assign to that range. You can discuss the differences in forecasting from a "known" simultaneous census value, compared to only the poll value which comes with a margin of error.

Also worthwhile to discuss is the difficulty of forecasting even if you know the population's current composition with certainty, due to the uncertainty of undecideds.

The Red Versus Blue Result

After making your forecasts/predictions, the instructor shares the red-to-blue composition of the bin:

Blue: 53%

Red: 47%

Because you knew the composition of red-blue-silver, you can work backwards and discover the silvers were 8% blue and 7% red, unremarkable.

But consider what you would have thought if you only knew the eventual result (and not the bin's composition including silvers).

It is a very insightful experiment to present some classes with the "red-blue-silver" Simultaneous Census numbers, and have a discussion, while presenting other classes with only the final "red-blue" numbers, to see how the discussion differs.

In my experience, interested learners seem much more capable of understanding the disconnect between poll and eventual result when presented with the "current composition" *before* the eventual result. Presenting the "current composition" first, learners seem baffled by the possibility that anyone would try to use the eventual result – largely dependent on how silvers turn out – to calculate poll accuracy.

But when presenting only the eventual result, never mentioning the idea of a "Simultaneous Census" or that the composition of the bin at the time the poll was taken is or can be known, which is what we get in real life, reception is much more mixed. People seem happy and eager to calculate each poll's accuracy from the eventual result. Rarely is the question "do we know what proportion of silvers ended up being blue versus red?" asked. Even when *offering* the composition of the bin at the time the poll was taken, the eventual result remains as something of an "anchor" in evaluating the poll accuracy, despite being given the correct "true value." I think a study that presents this data the two different ways described above, to two different classes, would be a very interesting topic to inform best practices for teaching the "simultaneous census" concept.

More often than not, based only on the results, most people (and current methods) would passionately assert that Student 2 was much more accurate, despite objectively insufficient data to prove that this was the case.

A wonderful debate could be had as to whether, based only on the result, the *most reasonable hypothesis* is that Student 2's sample was more accurate. There is no debate to be had, however, around the fact that reaching this conclusion based only on the given data – as the Spread and Proportional Methods do – is a demonstration of how the methods can fail.

Proponents of the Proportional Method would characterize Student 2's numbers as nearly perfect, while Student 1 was three points over on red and three points under on blue.

Spread proponents would reach the same conclusions about whose sample was more accurate.

To be perfectly honest, the modified urn problem is less of a brain teaser, and more of a reverse brain teaser: a math problem in which people who *don't* do advanced math for a living tend to do better. Several experts kind enough to respond to this thought experiment said that they had to overcome the urge to "model it out" accounting for red and blue probabilities, and assigning a range of outcomes for silvers – and others *did* respond only with a model to forecast this, even when only asked for the bin's current composition. This is a well-meaning temptation that is independent of the sample's observation.

What colors are in the bin – and what does this data tell us about it? That's the poll. End of observation.

What colors will be in the bin after silvers are unwrapped? Forecast.

These are separate calculations whose quality is independent. Most year 1 stats students can calculate "1" with great precision. The second question is much harder, one which even experts would justifiably have different answers for.

Being bad at (or disinterested in) forecasting does not mean you can't reliably analyze and understand poll data! Having a forecast that is far from the eventual result, likewise. The modified urn problem illustrates that.

Another way to present the problem, as Bernoulli did, is to *start* from the "known bin proportion" in this case, including silvers. Given that information, what would you expect a sample to produce?

Moreover, what does that sample tell us about the eventual result?

Especially eager calculators could introduce many different colors beyond red and blue to represent elections with many parties. Even more eager calculators could also introduce random (or nonrandom) population changes – such as replacing some red marbles with blue ones – to represent Changing Attitudes.

Regardless of the setup, you'll find that the current methods for calculating accuracy are only concerned with how predictive a tool not intended to be predictive is. These methods only exist due to a misunderstanding or misconception of the "true value" a poll tries to measure. That inability to account for undecideds and mind-changing, or outright apathy toward their potential impact, directly leads to public distrust in understanding the range of possible outcomes. Just because the poll is "close" doesn't mean the eventual result must be – and just because the poll isn't close, doesn't mean the eventual result can't be.

A simpler example

In my experience, I have found few examples better encompass the importance of avoiding unjustified assumptions than the hypothetical below.

Poll 1	Poll 2	Simultaneous Census
Candidate A: 46%	Candidate A: 44%	Candidate A: 45%
Candidate B: 44%	Candidate B: 46%	Candidate B: 45%
Undecided: 10%	Undecided: 10%	Undecided: 10%

By any metric, Polls 1 and 2 are equally accurate. Given that they are each very close to the population's true value, they are very accurate. I think everyone would agree that, given this data, this is all true.

But what happens when the simultaneous census is not known or knowable is that the election result becomes a deceptive placeholder.

Poll 1	Poll 2	Simultaneous Census	Result
Candidate A: 46%	Candidate A: 44%	Candidate A: ??%	Candidate A: 51%
Candidate B: 44%	Candidate B: 46%	Candidate B: ??%	Candidate B: 49%
Undecided: 10%	Undecided: 10%	Undecided: ??%	

Now, two polls that had very small errors – and were equally accurate – are viewed incompatibly. Poll 1 would be rated as "perfect" (+2 in poll and +2 in result), while Poll 2 would be rated as "off by 4" (−2 in poll and +2 in result). As you now understand, election math is never truly as simple as zero changing attitudes and zero frame error, but seeing how undecideds favoring Candidate A by an imperceptible 60/40 proportion could result in such drastic discrepancies is a powerful bit of knowledge.

It seems reasonable to assert that the Poll 1 was more likely to be the most accurate, given the data available. Maybe it is the most reasonable assumption. But the fact is that the result alone is absolutely insufficient evidence to reach that conclusion with any sort of certainty. It is a "not enough information" problem.

This example also demonstrates the current, misplaced obsession with polls being "directionally" correct, that is, reporting the eventual winner was polling higher. Most polling is done in elections that are expected to be close. Given a population where the two candidates' supporters are equal in number, all else equal, producing a directionally correct poll number is literally chance.

Many perceived "poll failures" are not poll failures at all – but analyst failures, a few of which I outlined in the book, and there are many more. How often polls have been wrong, and how wrong they were, is a very important area of research – poll errors do happen – but the terms must be properly defined before they can be reliably studied. It seems likely that by sheer quantity and chance, that there have been major elections in which the polls appeared accurate but were inaccurate. The field would benefit from increased vigilance regardless of what outcome is assumed.

It would be wonderful if we could ask a crystal ball instead of surveying people or marbles, but that tool doesn't yet exist, as far as I know. The survey tools we do have are pretty good, but they must be used and understood properly, limitations and all. In real life, like in the modified urn problem, the best answer we can possibly get from a poll is an estimate of the population's *current* composition. How accurately we can detect the current composition gives us better ability to make a forecast or prediction about its *eventual* composition. A perfectly accurate poll is not a guarantee of a good forecast, nor is an inaccurate poll a guarantee of a bad one, they are distinct entities. How we measure the accuracy of polls versus forecasts – the "true value" those separate tools attempt to measure – is something well within the grasp of everyone who has stomached my analogies and puns, and most who haven't or couldn't. I believe it would mark a huge advancement of the field to teach and incorporate these differences into the current and future work.

Whether you opt for beans, pebbles, marbles, mints, real people, or none of the above ... or if you're still somehow undecided ... these experiments all demonstrate something you now understand: polls are observations. Your weight on a scale is an observation. The amount of time it takes for someone to run a certain distance is an observation. Those observations aren't good or bad based on how closely they predict your future weight, who wins a race, or who wins an election and by what amount – those observations are good or bad based on how close they were to the correct observation. Observations do not try to predict eventual results, and polls do not try to predict elections.

If you do these experiments, or find inspiration from "What's for Lunch?", or "Mintucky" or anything else, I'd love to hear about your results.

All elections are unique, and some are more unique than others; treat "historical data" accordingly.

Vote.

Notes

1. Rakich, N. (2023, March 10). *The polls were historically accurate in 2022.* FiveThirtyEight. https://fivethirtyeight.com/features/2022-election-polling-accuracy/

   "On each pollster's *individual* pollster-ratings page, you'll find the percentage of races it called correctly, the most recent cycle in which it polled, a list of all its qualifying polls and how accurate they were as well as whether it participates in the American Association for Public Opinion Research's Transparency Initiative or shares its data with the Roper Center for Public Opinion Research."

2. Slightly smaller margin of error for silver since it is farther from 50%. Most people will answer 10%–18%, which is fine, and the expectations of precision can be adjusted for the level of the lesson.

Index

Note: Locators in *italics* represent figures and page numbers followed by "n" refer to notes.